Ergonomics and Safety
of Intelligent
Driver Interfaces

HUMAN FACTORS IN TRANSPORTATION

A Series of Volumes Edited by
Barry H. Kantowitz

Ergonomics and Safety of Intelligent Driver Interfaces

Edited by

Y. Ian Noy
Transport Canada, Ottawa

LEA LAWRENCE ERLBAUM ASSOCIATES, PUBLISHERS
1997 Mahwah, New Jersey

Lawrence Erlbaum Associates, Inc., Publishers
10 Industrial Avenue
Mahwah, New Jersey 07430

Library of Congress Cataloging-in-Publication Data

Ergonomics and safety of intelligent driver interfaces / edited by Y.
Ian Noy.
      p.   cm.
   Includes bibliographical references and indexes.
   ISBN 0-8058-1955-X (c : alk. paper). — ISBN 0-8058-1956-8 (p :
alk. paper).
      1. Automobiles—Automatic control. 2. Intelligent Vehicle Highway
Systems. 3. Automation—Human factors. I. Noy, Y. Ian.
TL152.8E74   1996
629.26—dc20   96-32074
                                                                   CIP

Books published by Lawrence Erlbaum Associates are printed on acid-free paper,
and their bindings are chosen for strength and durability.

Printed in the United States of America
10 9 8 7 6 5 4 3 2 1

To my wife, Bella,
and to my children, Evie, Brian, and Shawna

Contents

Series Foreword

Barry H. Kantowitz
Battelle Human Factors Transportation Center
Seattle, Washington

The doman of transportation is important for both practical and theoretical reasons. All of us are users of transportation systems as operators, passengers, and consumers. From a scientific viewpoint, the transportation domain offers an opportunity to create and test sophisticated models of human behavior and cognition. This series covers both practical and theoretical aspects of human factors in transportation, with an emphasis on their interaction.

The series is intended as a forum for researchers and engineers interested in how people function within transportation systems. All modes of transportation are relevant, and all human factors and ergonomic efforts that have explicit implications for transportation systems fall within the series purview. Analytic efforts are important to link theory and data. The level of analysis can be as small as one person, or international in scope. Empirical data can be from a broad range of methodologies, including laboratory research, simulator studies, test tracks, operational tests, fieldwork, design reviews, or surveys. This broad scope is intended to maximize the utility of the series for readers with diverse backgrounds.

I expect the series to be useful for professionals in the disciplines of human factors, ergonomics, transportation engineering, experimental psychology, cognitive science, sociology, and safety engineering. It is intended to appeal to the transportation specialist in industry, government, or academia, as well as the researcher in need of a testbed for new ideas about the interface between people and complex systems.

The present book is devoted to the leading edge human factors technologies for ground transportation. In recent years there has been intense interest in Intelligent Transportation Systems around the world. Human factors technology as applied to the driver interface has been advancing rapidly in Europe, North America, Japan, and Australia. A major goal of this series is to offer a forum for transportation researchers to exchange ideas and this book accomplishes that goal by bringing together new concepts from several nations. Although the emphasis is upon navigation systems and route guidance, there is discussion of both applied issues, such as crash avoidance, as well as theoretical issues, such as attention and fatigue. This accomplishes yet another key goal in this book series: linking theory and application. The book editor has served the field of transportation human factors well by compiling such an interesting and provocative assembly of international papers on advanced driver interfaces.

Preface

The advent of intelligent transport systems (ITS) is about to transform the driving experience. Already the application of advanced technologies to automative engineering has made available on-board digital maps, extensive information databases, vehicle positioning and tracking, and integrated visual and auditory displays. Once perfected and in widespread use, ITS is expected to improve safety, alleviate traffic congestion, decrease transportation costs, increase economic productivity, and reduce the environmental damage caused by the use of motor vehicles. Whereas intelligent transport systems are safety neutral in that they are not inherently beneficial or detrimental, there is the very real danger that, unless the driver–ITS interface is well designed, ITS will be a safety hazard.

The aim of this book is to present knowledge that will contribute to the design of ergonomically sound driver–ITS interfaces. It is based, in part, on papers presented at the Symposium on the Ergonomics of Intelligent Vehicle-Highway Systems (IVHS) held in Toronto during the 12th Congress of the International Ergonomics Association in August 1994. Although intelligent transport systems comprise many different elements, the Symposium focused on the driver–ITS interface in order to ensure that it supports the driving task, rather than distracts from it, and presents information to the driver in a reliable, accurate, and easy-to-understand way.

A fundamental obstacle to ensuring that the full benefits of ITS are realized is that technological advances are increasing at a much faster pace than knowledge in the behavioral sciences. For instance, the driver's role and the interaction of the driver with the other elements of the road transport system are not well understood. In addition, our current knowledge of ergonomics and traffic psychology, although considerable, does not permit us to determine how well drivers will handle ITS-related functions or how they will adapt to new human–machine interfaces. More important, the current state of ergonomics theory and the available empirical data permit only a preliminary formulation of principles and guidelines upon which to base the design of driver–ITS interfaces. Whether ITS will become an adjunct or an impediment to safety will depend largely on the specific technologies that are invoked, the manner in which they are incorporated into the vehicle, and how they are presented to the driver.

The individual chapters of this book, which have been subjected to peer review, present material that is related to a variety of ITS applications, such as route guidance and collision avoidance. In many instances, the theoretical constructs, discussions, and research findings have been updated and are described more fully than in the papers presented at the Symposium. In selecting the contributions, an effort was made to provide an international sample of research; no attempt was made to treat specific topics in a systematic or comprehensive manner.

It is with deep gratitude that I wish to thank the International Ergonomics Association for providing the forum for the Symposium; the U.S. National Highway Traffic Safety Administration for its sponsorship; the authors for the time and energy they devoted to preparing their manuscripts; the reviewers for their thoughtful comments and suggestions; and the publisher, Lawrence Erlbaum Associates, for undertaking to produce this book. All those who have contributed to this volume have done so in the hope that their work will help to improve the usability of driver–ITS interfaces and thereby enhance traffic safety.

—*Y. Ian Noy*

Contributors*

Motoyuki Akamatsu, National Institute of Biosciences and Human Technology, Agency of Industrial Science and Technology, MITI-1-1, Higashi, Tsukuba, Ibaraki, 305, Japan

Håkan Alm, Department of Mechanical Engineering, Linköping Institute of Technology, S-581 83, Linköping, Sweden

Linda L. Bossi, Defence and Civil Institute of Environmental Medicine, P. O. Box 2000, 1133 Sheppard Avenue West, Downsview, Ontario, M3M 3B9, Canada

James R. Buck, Industrial Engineering, 4131 Engineering Building, University of Iowa, Iowa City, Iowa 52242-1527

Gary Burnett, HUSAT, Loughborough University, The Elms, Elms Grove, Loughborough, Leicestershire, LE11 1RG, UK

Tatsuru Daimon, Department of Administrative Engineering, Keio University, 14-1 Hiyoshi, 3-chome, Kohoku-ku, Yokohama, 223, Japan

Stephen H. Fairclough, HUSAT, Loughborough University, The Elms, Elms Grove, Loughborough, Leicestershire, LE11 1RG, UK

Robert Graham, HUSAT, Loughborough University, The Elms, Elms Grove, Loughborough, Leicestershire, LE11 1RG, UK

Stephen Hirst, HUSAT, Loughborough University, The Elms, Elms Grove, Loughborough, Leicestershire, LE11 1RG, UK

Wiel Janssen, TNO Human Factors Research Institute, Kampweg 5, 3769 ZG, Soesterberg, The Netherlands

Barry H. Kantowitz, Battelle Human Factors Transportation Center, 4000 N. E. 41st Street, Seattle, WA 98105

Kenji Kimura, Toyota Motor Corporation, Component and System Development Center, Ergonomics Vehicle Engineering Division, 1 Toyota-cho Toyota-chi, Aichi, 471, Japan

Guy Labiale, Département de Psychologie Cognitive, Université Paul Valéry, Route de Mende, BP 5043, 34032 Montpellier, France

John D. Lee, Battelle Human Factors Transportation Center, 4500 Sand Point Way N. E., Seattle, WA 98105

Michael A. Mollenhauer, Center for Computer-Aided Design, University of Iowa, Iowa City, Iowa 52242-1527

Simon A. Moss, Department of Psychology, Monash University, Wellington Road, Clayton, VIC 3168, Australia

Andrew Parkes, Institute for Transport Studies, University of Leeds, Leeds, LS2 9JT, UK

John Richardson, HUSAT, Loughborough University, The Elms, Elms Grove, Loughborough, Leicestershire, LE11 1RG, UK

Tracy Ross, HUSAT, Loughborough University, The Elms, Elms Grove, Loughborough, Leicestershire, LE11 1RG, UK

Raghavan Srinivasan, School of Aviation and Transportation, Dowling College, Oakdale, NY 11769

Leonard Stapleton, HUSAT, Loughborough University, The Elms, Elms Grove, Loughborough, Leicestershire, LE11 1RG, UK

Kathryn Wochinger, Science Applications International Corporation, 1710 Goodridge Drive, MS 1-6-6, McLean, VA 22102

David M. Zaidel, Transportation Research Institute, Technion-Israel Institute of Technology, Technion City, Haifa, 32 000, Israel

---

*The names and addresses listed here are only those of the lead contributor for each chapter in this volume.

Chapter 1

Driver Reliability Requirements for Traffic Advisory Information

Barry H. Kantowitz
Richard J. Hanowski
Susan C. Kantowitz
Battelle Human Factors Transportation Center, Seattle

Transportation engineers would like to provide reliable traffic information to motorists, but there are many perturbations in the highway system that make it difficult to achieve this goal. Congestion, delays, and accidents combine so that the information provided to motorists may no longer be entirely accurate when it is received. Although it seems reasonable to suspect that unreliable information causes drivers to discount or even to ignore posted traffic messages (e.g., on road signs), there are little empirical data concerning how reliable the information should be. In some domains, even a single incident is sufficient to extinguish behavior (as when a vending machine fails to operate). How many people continue putting coins into defective parking meters? Other systems may be more robust. Will drivers tolerate some errors in route guidance systems without losing trust (Lee & Moray, 1991) and ignoring the information provided?

PREVIOUS ATIS AND ROUTE GUIDANCE RESEARCH

An important goal of Advanced Traveller Information Systems (ATIS) is to allow travellers to drive safely and efficiently by providing real-time route guidance and traffic advisory information. Because the impact of ATIS on driver behavior is not yet fully understood, human factors research is required to fill this gap. Before describing the results of two ATIS-related experiments conducted at the Battelle Human Factors Transportation Center, it is worthwhile to review related research. Each of the following three experiments used a low fidelity, part-task simulator to investigate driving performance and the effects that in-vehicle traffic information systems might have on driver route-choice behavior.

Adler and Kalsher (1994) used an interactive travel choice simulator, FASTCARS, to investigate the influence that ATIS technologies have on driver behavior. Their objective was to determine whether or not in-vehicle information systems provide travel performance benefits to drivers. Two ATIS information types were investigated: route guidance information and traffic advisory information. Drivers were assigned to one of four groups and presented with the following ATIS information: no information (control), only route guidance information, only traffic advisory information, and both route guidance and traffic advisory information. The primary dependent variable was the origin-to-destination travel time.

Using FASTCARS, drivers "drove" through a hypothetical traffic network and received information from their assigned ATIS. Following a practice session to become familiar with FASTCARS, drivers received instruction about the ATIS they would be using and then completed 10 trips from origin to destination. Five more trips followed where no ATIS information was provided to any of the groups. One reason for including the last 5 trips was to test for possible learning effects within the control group.

The results indicated that drivers in the no ATIS control group had longer mean travel times than any of the ATIS groups. The mean travel times for the three ATIS groups were equivalent. Performance over the last five trials, where no ATIS were available, improved for all groups. Two possible reasons for this finding were suggested: First, drivers received the most benefit from ATIS in unfamiliar traffic networks. Second, the traffic network used for this experiment was not complex enough, which allowed drivers to learn the optimal routes.

Allen, Ziedman, Rosenthal, Stein, Torres, and Halati (1991) conducted a laboratory assessment of driver route diversion in response to in-vehicle navigation and motorist information systems. The objective of this research was to investigate the effects that four route navigation systems had on driver route diversion behavior in congested traffic situations. A part-task simulator was used to measure driver performance and behavior. Four types of navigation systems were examined: static map system—map display with vehicle position but no congestion information; dynamic map system—static map with congestion information; advanced experimental system—dynamic map with alternate route, text information, and auditory instructions; route guidance system—non map system using directional symbols, and showing exit distance, estimated arrival time, and destination distance. Traffic network familiarity was also included in the experimental design; drivers were either familiar or unfamiliar with the real traffic network. Also, each driver experienced four driving scenario traffic-congestion conditions. These consisted of a far destination (23 miles) with an 11-min delay, a far destination with an 18-min delay, a far destination with a 30-min delay, and a near destination (9 miles) with a 30-min delay. Age and commercial driving experience were also considered. Drivers were categorized as either young (18–29 years), middle (30–54 years), or old (> 55 years), and as either having or not having a commercial driver's license.

The driver's task was to travel to a destination as quickly as possible. A cash reward–penalty structure was implemented to motivate drivers to travel quickly. Larger rewards were earned for faster travel times. Drivers were assigned one of the four navigation systems or, alternately, no system (control group). Prior to beginning the first of four simulated journeys, drivers made two practice trips, one without use of the assigned navigation system. In order to optimally traverse a route and avoid traffic congestion, drivers were required to divert from their present route.

The results indicated that driver route diversion patterns varied for the different navigation systems. In general, greater compliance to route diversion suggestions were found for the more sophisticated systems. The advanced system was found to be best, followed by the route guidance system, and then the dynamic map. The static map and control conditions were comparable. The extra information presented by the advanced system allowed drivers to better anticipate and avoid congestion. No statistically significant differences were found with traffic network familiarity. A significant difference was found with the driving scenario traffic-congestion independent variable: Whether the destination was near or far, drivers were more likely to divert from their route when the delay was 30 min, as compared to the 11-min and 18-min delays. Finally, a significant age effect was found indicating that older noncommercial drivers were the least likely to divert.

Using a part-task simulator, interactive guidance on routes (IGOR), Bonsall and Parry (1991) investigated drivers' compliance with route guidance advice. The objective of their research was to examine the effect that the quality of route guidance advice had on drivers' compliance with that advice. The independent variables in their experiment included the objective quality of the route guidance advice, the quality of previously received advice, and driver's familiarity with the traffic network. Objective quality of advice was defined as the ratio of the minimum time to reach a destination by means of the advised route to the minimum time by any route. The quality of the previously received advice was defined as, on average, being very good or very poor. The traffic networks were hypothetical and developed to represent a typical small town. Being novel to all drivers, traffic network familiarity was based on practice and successive journeys to destinations within a network.

The results of this experiment indicated that the quality of the route guidance advice impacted driver acceptance; as the quality of advice decreased, so did acceptance. Driver acceptance was also influenced by the quality of previously received advice. Drivers were more apt to accept advice—even if it was poor—if previously received advice had been very good. Previously received advice that was very bad had a negative impact on current advice, especially if it too was very bad. Acceptance of advice was also influenced by traffic network familiarity. Drivers unfamiliar with a network were more apt to accept advice as compared to drivers familiar with a network. In other words, advice acceptance generally decreased as network familiarity increased.

To summarize, each of the three experiments outlined used a low fidelity, part-task simulator to investigate driver behavior and performance while using an ATIS. Each study found that drivers can benefit, through reduced travel times, from traffic-related information provided by an ATIS.

Three substantive issues described in these studies are particularly germane to the present experiments. First, these studies have shown that low fidelity, part-task driving simulators can be used as effective and efficient tools for investigating ATIS-related issues. As outlined by Bonsall (1994), these types of simulators can effectively examine driver route choice behavior because they can test new or novel systems not yet implemented; as a laboratory-type experiment, the researcher can control the experimental design; and a variety of issues can be investigated that could not be done so efficiently or safely in the field. Following this insight, the Battelle Route Guidance Simulator—a low fidelity, part-task simulator that can be used to investigate driver behavior, route choice, and other ATIS-related issues—was developed and used for both Experiments 1 and 2.

The second relevant consideration is that each of the three studies describe results related to driver's familiarity with the traffic network. The experiments by Adler and Kalsher, and Bonsall and Parry, both found network familiarity to be important; as compared to the familiar network, drivers in the unfamiliar network either received greater benefit from the ATIS or were more likely to accept the advice provided. In contrast, Allen et al. found no differences between drivers who were familiar or unfamiliar with the network. However, they suggested that these results may be attributed to a similar knowledge of the traffic network between the familiar and unfamiliar driver groups.

To further investigate this issue, we made two improvements on the three studies described. First, network familiarity was investigated in Experiment 2 using a familiar, actual network, and a matched unfamiliar, hypothetical network. As an improvement over the three studies outlined, where matched networks were not used, Experiment 2 used familiar and unfamiliar networks with identical topographies. The primary difference between the familiar and unfamiliar networks in Experiment 2 was that the travel direction was from east to west in the familiar network, and from north to south in the unfamiliar network. The same topography and roadway structure was maintained in the unfamiliar network by simply rotating the familiar network 90 degrees.

Second, only drivers who were very familiar with driving in Seattle (the familiar network) were used as subjects for the present experiments. Using subjects who were frequent drivers in the familiar network allowed network familiarity to be examined as a within-subject variable (unlike Allen et al.) and without confounding practice (like Adler and Kalsher, and Bonsall and Parry).

The third issue concerns the quality, or reliability, of the ATIS information. Recall that Bonsall and Parry investigated system acceptance as a function of information quality. The notion of system use and acceptance and system reliability is an important one, particularly for ATIS designers. Will drivers use systems that provide unreliable information? And, how much unreliable information is too much? Will drivers be tolerant of a system that occasionally makes an error, or will they switch off such a system? Bonsall and Parry considered these ideas by examining "very good" and "very poor" ATIS advice. Experiment 1 and Experiment 2 take this idea one step further by specifying different degrees of reliability. Across both experiments, overall system reliability of 100%, 77%, 71%, and 43% were examined. It might be expected that drivers will tolerate some unreliability in a system, but not extreme unreliability.

TRUST

One method of measuring ATIS acceptance is based on driver trust in the system. In a formal sense, the driver's decision about accepting the advice of an automated route guidance system is very similar to the dynamic allocation of function decision made by an operator controlling some industrial process (B. H. Kantowitz & Sorkin, 1987). In both cases, the system operator either lets the automation make the decision or manually makes the decision. The operator's subjective feelings about trust in the automated system play an important role in the dynamic allocation of function decision (Lee & Moray, 1991); indeed, a quantitative model of operator trust has been developed for process control. Additional research using this model (Lee & Moray, 1994) has shown that better predictions of operator behavior are made when subjective self-confidence is also taken into account. In general, when trust exceeds self-confidence, operators accept automated control. Conversely, when self-confidence exceeds trust, operators use manual control.

THE BATTELLE ROUTE GUIDANCE SIMULATOR

Part-task simulators are an effective tool for studying how operators interact with large systems in general (B. H. Kantowitz, 1988), and route guidance systems in particular (Bonsall, 1994). Simulators combine many of the benefits of both laboratory and field research while avoiding some of the disadvantages associated with applying each method individually.

Driver behavior was investigated using the Battelle Route Guidance Simulator (RGS; S. C. Kantowitz et al., 1995), which consisted of two linked Intel 486 computer systems and two video displays (see Fig. 1.1). One monitor provided drivers with a real-time windshield view of the traffic scene from the driver's perspective. Additional information displayed on this monitor below the windshield included speed of the vehicle, queried information (both written on the video screen and "spoken" by a DecTalk system), other traffic information (always extraneous to the chosen route), current time, goal time, bonus amount, and current location (street location written on the screen). The second computer, equipped with a touch screen input device, displayed a computer-generated map of the traffic network. A moving dot indicated the current location of the simulated vehicle. Drivers used the touch screen to select route options from the displayed map and to query the system about the traffic congestion on any route segment ("link"). System queries for traffic congestion informa-

FIG 1.1. Battelle Route Guidance Simulator.

tion could be done at any time and querying a link was required prior to traversing it. When a link was chosen, the appropriate video was displayed in real time on the first monitor.

Although most of the links had "light" traffic, several had "heavy" traffic as defined in terms of *level-of-service* (Transportation Research Board, 1992). Light traffic, *level-of- service A*, represents a free flow of traffic where individual drivers are unaffected by others present in the traffic stream. Heavy traffic was defined as *level-of-service E* or *level-of-service F*. *Level-of-service E* occurs when operating conditions are at or near capacity level and all speeds are reduced to a low, relatively uniform value, and *level-of-service F* occurs when operations within a traffic queue are characterized by unstable stop-and-go traffic.

Previous research on route guidance (Adler & Kalsher, 1994; Bonsall & Parry, 1991) used an artificial traffic network and found that acceptance of system advice generally decreased as familiarity with the network increased. The present study used a real Seattle traffic network. Other research (Allen et al., 1991) used slides to display real intersections. The present research goes beyond this by displaying real-time video under computer control.

Experiment 1

Method

Subjects. A total of 48 licensed drivers, 24 males and 24 females ranging from 18 to 75+ years of age, participated in this experiment (Table 1.1). They were recruited from the University of Washington, and community and senior citizens' centers in the surrounding neighborhoods. Each driver was paid $10.00 per hour plus a cash bonus that depended on performance (maximum of $20.00). Subjects were initially screened by telephone using a questionnaire to establish a minimum Seattle driving frequency of two times per week.

Design. Traffic scenes from the origin (Westlake Center in downtown Seattle) to the destination (Bellevue Square Mall) had been videotaped, digitized, and stored on hard disks as part of the Battelle Route Guidance

TABLE 1.1
Age Group and Number of Subjects

Age	Number	Mean Years Living in Seattle	Mean Years Driving in Seattle	Mean Miles Driven Annually
18–54	24	16.6	10.5	11,042
54–75+	24	42.1	39.4	8,854

Simulator (RGS). Twenty-six traffic links were displayed, with each link varying in length from one to several streets. Twenty-nine different routes were possible on a variety of roads, including city streets, four-lane state roads, and interstate freeways in an urban setting. Due to the topography of Seattle, the driver must cross Lake Washington to reach the destination. Because there are only two bridges across the lake, the experimenters retained some control over traffic congestion encountered regardless of route selected. Simulated trips took about 22 min in light traffic and 37 min in heavy traffic.

The major independent variable was *accuracy* of traffic information during each simulated trip. Information was either *reliable* (accurate on 100% of the links) or *unreliable* (accurate on 77% of the links). Unreliable information could be either *harmful* (e.g., traffic reported to be light when actually heavy) or *harmless* (e.g., traffic reported heavy when actually light). A repeated-measures design was used consisting of two 100% reliable trips followed by two 77% reliable trips.

Procedure. The driver's task was to travel from origin to destination by choosing the links with the least amount of traffic and shortest travel time. This combination of links comprised the optimal route. Drivers were encouraged to use the ATIS by querying the system to obtain information on traffic conditions on upcoming links at a nominal cost of $.10 per query. A monetary penalty reducing the bonus was used to motivate drivers to select the fastest routes and avoid heavy traffic. Frustration due to traffic delays was simulated by a bonus decrement whenever a nonoptimal route was selected. Drivers who had the misfortune of encountering heavy traffic had their bonus immediately cut in half. Drivers were paid the largest bonus achieved during the four trips.

To become familiar with the RGS, each driver practiced using the simulator prior to the start of the four simulated trips. The driver was instructed that all trips took place at 4:30 p.m. on a Friday, a time of high congestion in Seattle. Before each scenario commenced, the driver indicated a preferred route on a paper map. Outlining their preferred route served both to orient the driver to the traffic network and to define the baseline for measuring convergence: the extent to which drivers followed their baseline route. The driver used the RGS to select each route link and to obtain traffic information about that link. To help prevent unintentional link selections, drivers were required to confirm each link choice by pressing the confirm button located on the touch screen. After each link was traversed, drivers rated on a 100-point scale their trust in the system, and their self-confidence in their own ability to accurately anticipate traffic conditions.

Results

Figure 1.2 shows mean penalty costs as a function of reliability. On the second repetition for 100% accurate information, drivers were able to reduce their penalty costs. But, when unreliable information was presented, penalty costs increased. This interaction is significant, $F(1, 46) = 6.56$, $p < .005$. Thus, the drivers were able to benefit from system information when it was accurate but not when it was unreliable. Because the costs incurred for purchasing information were the same for both reliable ($.76) and unreliable ($.79) conditions, this result cannot be explained by drivers' unwillingness to buy unreliable information.

A convergence score of 100% indicates that drivers perfectly followed their preferred route marked on the paper map; a score of 0% indicates no common links between the preferred path and the route selected. When information was reliable, drivers were influenced by system information and departed from their preferred routes (51% convergence). But when information was unreliable, drivers departed even more with lower convergence scores (32% convergence), $F(1, 46) = 4.72$, $p < .05$.

Figure 1.3 shows rated trust. It increased for the first two trials, then decreased for the last two when inaccurate information was presented for the middle links; trust then increased following accurate information on the last two links. Figure 1.4 shows that only harmful inaccurate information decreased trust, $F(3, 624) = 36.6$, $p < .001$.

Figure 1.5 shows information cost as a function of driver age for older drivers. On the first repetition, all drivers incurred approximately the same

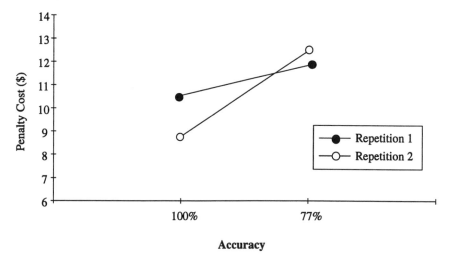

FIG. 1.2. Mean penalty costs as a function of accuracy and repetition.

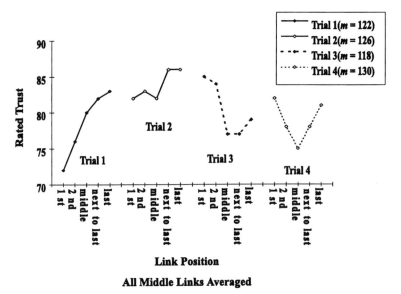

FIG. 1.3. Mean rated trust as a function of information accuracy and link position.

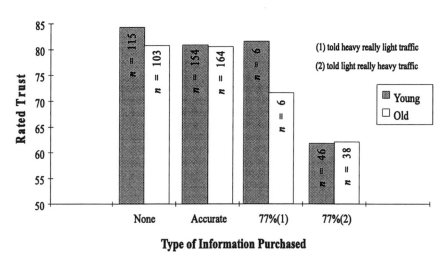

FIG. 1.4. Mean rated trust as a function of age given the type of information purchased for Trials 3 and 4.

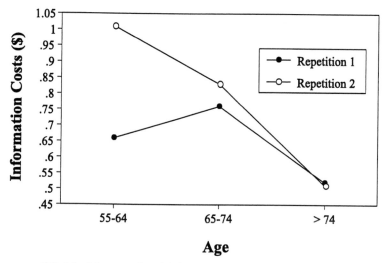

FIG. 1.5. Mean purchased information costs as a function of age.

costs to purchase information. However, the oldest drivers did not use the ATIS on the second repetition, whereas the 55- to 64-year-old age group did improve; this interaction between age and repetition was significant, $F(2, 20) = 3.43$, $p = .05$. This is consistent with the general finding that acceptance and use of new technology diminishes with age (B. H. Kantowitz, Becker, & Barlow, 1993).

Figure 1.6 is a bar graph outlining mean trust ratings for younger and older drivers based on purchased link information. When information was

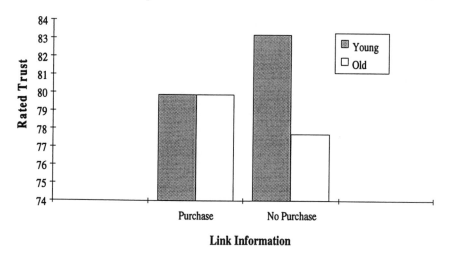

FIG. 1.6. Mean rated trust as a function of age and purchased link information.

purchased, mean trust ratings were identical (79.9) for younger and older drivers. When no information was purchased, mean trust ratings were 83.2 for younger drivers and 77.7 for older drivers. An analysis of variance (ANOVA) indicated a significant age effect, $F(1, 1260) = 5.13$, $p < .03$, and a significant Age × Link Information interaction, $F(1, 1260) = 9.01$, $p < .005$.

Self-Confidence. Figure 1.7 displays mean self-confidence ratings for younger and older drivers on purchased link information. When information was purchased, mean self-confidence ratings were 66.2 for younger drivers and 73.8 for older drivers. When no information was purchased, mean self-confidence ratings were 73.2 for younger drivers and 69.6 for older drivers. An ANOVA indicated a significant age effect, $F(1, 1260) = 11.3$, $p < .001$, and a significant Age × Link Information interaction, $F(1, 1260) = 25.8$, $p < .001$.

Trust Minus Self-Confidence. Figure 1.8 displays a post-hoc dependent variable created by subtracting rated self-confidence from rated trust. This new variable is shown for younger and older drivers as a function of link information. When information was purchased, mean rated trust minus self-confidence was 13.7 for younger drivers and 6.0 for older drivers. When no information was purchased, mean rated trust minus self-confidence was 10.0 for younger drivers and 8.1 for older drivers. An ANOVA indicated a significant age effect, $F(1, 1260) = 27.3$, $p < .001$, and a significant Age × Link Information interaction, $F(1, 1260) = 6.48$, $p < .001$. A *t* test was conducted on the Age × Link Information interaction means. Significant

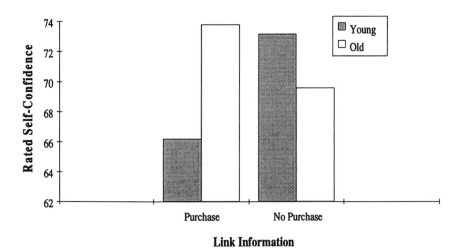

FIG. 1.7. Mean rated self-confidence as a function of age and purchased link information.

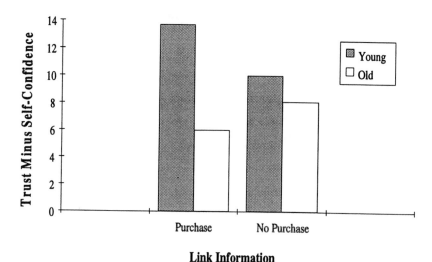

FIG. 1.8. Mean rated trust minus self-confidence as a function of age and purchased link information.

differences were found between purchase and no purchase conditions for both older drivers, $t(1260) = 2.80$, $p < .01$, and younger drivers, $t(1260) = 4.93$, $p < .001$.

Mean rated trust minus self-confidence is also shown for younger and older drivers given the type of information purchased during Trials 3 and 4 (Fig. 1.9). When no information was purchased, rated trust minus self-confidence was 9.9 for younger drivers and 6.6 for older drivers. When the information purchased was accurate, rated trust minus self-confidence was 14.3 for younger drivers and 6.5 for older drivers. When drivers were told that the traffic was heavy when it was actually light, rated trust minus self-confidence was 11.7 for younger drivers and 5.0 for older drivers. And when drivers were told that traffic was light when it was actually heavy, rated trust minus self-confidence was −2.4 for younger drivers and −6.4 for older drivers. An ANOVA indicated main effects of age, $F(1, 624) = 12.7$, $p < .001$, and type of information purchased, $F(3, 624) = 12.2$, $p < .001$. Student–Newman–Keuls multiple comparison tests found that the harmful inaccurate information purchased condition differed from all other conditions ($p < .05$).

Summary of Experiment 1 Results. Table 1.2 summarizes the most important results of this experiment. Both objective (penalty cost and information cost) and subjective (trust and self-confidence) measures are reliably influenced by independent variables that have important implications for transportation engineers providing traffic advisory information to drivers.

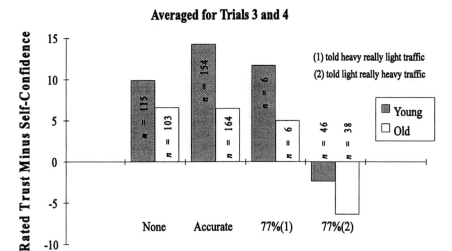

FIG. 1.9. Mean rated trust minus self-confidence as a function of age given the type of information purchased for Trials 3 and 4.

TABLE 1.2
Summary of Results (Experiment 1)

Dependent Variable	Figure	Key Finding
Penalty cost	2	Drivers learn to benefit from accurate information
Trust	3, 4	Inaccurate information decreases trust
Trust	6	Purchasing information is associated with increased trust for older drivers but decreased trust for younger drivers
Information cost	5	Oldest drivers do not benefit from repetition
Self-confidence	7	Purchasing information is associated with increased self-confidence for older drivers but decreased self-confidence for younger drivers
Trust minus self-confidence	9	Trust minus self-confidence is negative only for harmful inaccurate information

Experiment 2

The goal of Experiment 2 was to replicate and expand the preceding results. This was accomplished by adding another level of reliability and also by manipulating the familiarity of the environs. Bonsall and Parry (1991) used a simulated artificial traffic network to investigate the quality of advice defined as the ratio of the minimum time to reach a destination by means of the advised route to the minimum time by any route. They found that user acceptance declined with decreasing quality of advice in an unfamiliar network. As familiarity with the network increased, drivers were less likely to accept advice from the system. However, Allen et al. (1991) found that familiarity did not affect route choice behavior. These researchers used a real traffic network, as opposed to the artificial network created by Bonsall and Parry (1991). Allen et al. (1991) explained their results by speculating that perhaps both familiar and unfamiliar driver populations may have been more similar than intended. Thus, these two experiments yielded conflicting results about the effects of familiarity on driver choice. Because a comparison of these two experiments confounds familiarity with real versus artificial traffic networks, additional research is required. The present experiment compares a familiar real traffic network with an unfamiliar artificial network that has been carefully matched to the topography of the real network.

Method

Subjects. Subjects were 24 males and 24 females, ranging from 18 to 35 years of age, recruited from the University of Washington and the surrounding community. Each was paid $5.00 per hour, plus a cash bonus (maximum of $13.00). Prospective participants were administered a screening questionnaire by telephone to ensure that all had a driver's license, were familiar with driving in the Seattle area, and drove at least twice per week. To determine subjects' driving experience, data were collected on the following: years lived in Seattle ($M = 10.8$), years driven in Seattle ($M = 4.8$), and total miles driven annually ($M = 9219$).

Experimental Design. Videotaped traffic scenes of the various routes from Westlake Center in downtown Seattle to Bellevue Square Mall were used for Seattle, the familiar network. The unfamiliar network, "New City," was created by combining video clips of streets and highways throughout northwest Washington. All video clips were digitized and stored on hard disk. The traffic network structures for both familiar and unfamiliar cities were identical (i.e., 31 links were used for each scenario). Thus, the topography was identical for both familiar and unfamiliar networks (see Fig. 1.10); note that Fig. 1.10b is a 90-degree rotation of Fig. 1.10a.

Thirty-three different routes were possible on a variety of roads, including city streets, four-lane state roads, and interstate freeways in an urban setting. Completing a route required traversing seven links, regardless of the path chosen. The present methodology is an improvement over the previous experiment where inaccurate information was path dependent on a trial-by-trial basis and hence not directly controlled by the experimenter. In this experiment, information accuracy was controlled independent of the path selected.

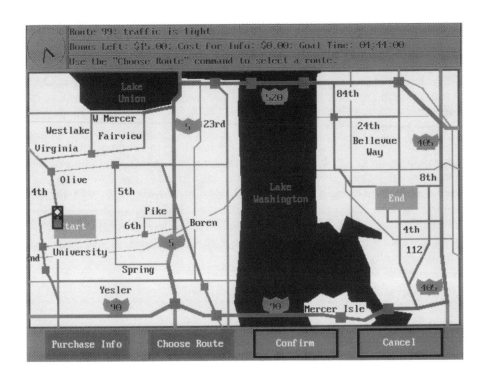

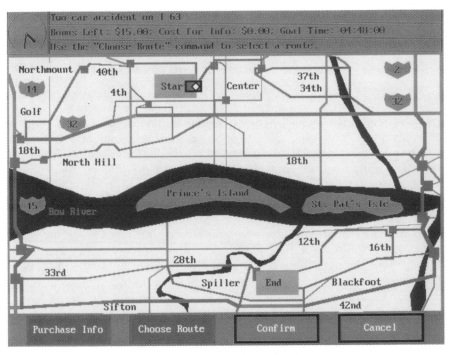

FIG. 1.10b. Topography of an unfamiliar city.

The two independent variables of major interest were traffic *network familiarity* and *information accuracy*. Network familiarity, a within-subject variable, had two levels: familiar (Seattle) and unfamiliar (New City). Information accuracy, a between-subjects variable, also had two levels: 71% accurate and 43% accurate. In addition, all subjects experienced 100% accurate information for the first two simulated trips. Inaccurate information was either harmful (e.g., traffic on link reported to be light, but was actually heavy) or harmless (e.g., traffic was reported to be heavy, but was actually light). All subjects experienced four trials with information always being 100% accurate for the first two trials and inaccurate (either both trials 71% or both 43%) for Trials 3 and 4.

Procedure. The procedure for Experiment 2 was similar to Experiment 1 except that traffic information was provided without any monetary cost. The system would not allow drivers to traverse a link until congestion information pertaining to that link had been obtained.

Results

Objective Variables. Two objective dependent variables are reported: penalty cost and system query frequency. Figure 1.11 shows mean penalty costs were lower when traffic information was accurate, $F(1, 44) = 108$, $p < .001$. This result is consistent with the previous experiment and shows the

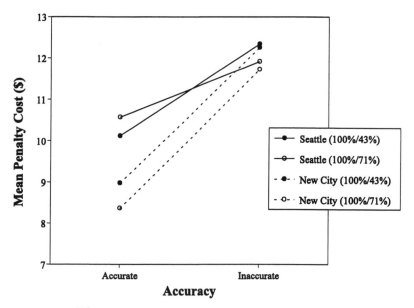

FIG. 1.11. Mean penalty costs as a function of accuracy.

greater benefit of using the simulated ATIS when it was accurate. Penalty costs were higher in the familiar Seattle network, $F(1, 44) = 15.5$, $p < .001$, suggesting that drivers were more likely to follow ATIS advice in an unfamiliar setting. When traffic information was 100% accurate, penalty costs were higher in the familiar setting ($10.34) than in the unfamiliar setting ($8.67), $t(44) = 4.02$, $p < .01$. The interaction shown in Fig. 1.11 between accuracy and familiarity while significant, $F(1, 44) = 6.81$, $p < .02$, is unimportant; the less rapid rise in cost for the familiar setting is probably a ceiling effect (B. H. Kantowitz, Roediger, & Elmes, 1994, p. 335) due to a maximum penalty of $13.00 on any one trial. Penalty costs were reduced on Trial 2 ($9.20) versus Trial 1 ($9.80), $t(44) = 2.34$, $p < .02$, showing that repetition allowed drivers to use the simulated ATIS more effectively.

The higher penalty costs for the familiar network cannot be explained by a lower frequency of queries. Query frequency did not differ between Seattle (32.3) and New City (29.8), $F(1, 44) = 3.95$, $p > .05$. Thus, although the same amount of traffic information was received in the familiar setting, it was not used effectively. However, more queries were made when information was inaccurate (34.8) than when accurate (27.3), $F(1, 44) = 34.7$, $p < .001$.

Subjective Measures. Three subjective dependent variables were analyzed: trust, self-confidence, and trust minus self-confidence. All F ratios involving the link position independent variable used the Greenhouse–Geiser correction for repeated measures.

Figure 1.12 shows mean rated trust as a function of link position and information accuracy. Trust was higher when information was accurate, $F(1, 44) = 31.6$, $p < .001$. When information was inaccurate, trust was higher for the 71% condition than for the 43% condition, $F(1, 44) = 5.02$, $p < .03$. Whereas there was a significant effect of link position, $F(6, 264) = 11.9$, $p < .001$, this effect was due to the inaccurate information on Trials 3 and 4 rather than the accurate information on the first two trials, $F(6, 264) = 6.17$, $p < .001$. When inaccurate information was presented, trust recovered on subsequent links when accurate information was presented. There was no main effect of familiarity, $F(1, 44) < 1.0$, on trust.

Trust did not differ according to type of inaccurate information: harmless (71.3) versus harmful (68.8), $t(1338) = 1.89$, $p > .05$. Although this differs from the Experiment 1 results, the present results are more definitive. In the previous experiment, type of inaccurate information depended on the path taken and was based on a small number of observations. This experiment controlled the type of inaccurate information.

Rated self-confidence was higher in familiar (76.6) versus unfamiliar (71.7) settings, $F(1, 44) = 6.92$, $p < .02$. Trust minus self-confidence was negative only for the familiar Seattle setting with 43% inaccurate informa-

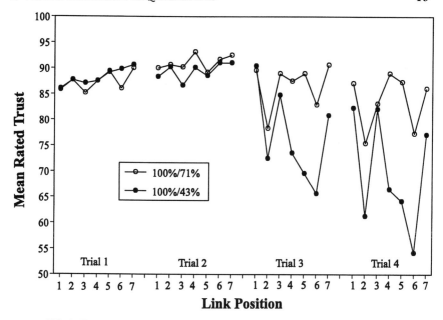

FIG. 1.12. Mean rated interlink trust as a function of information accuracy and link position.

tion. Trust minus self-confidence was higher for the unfamiliar city (12.3) versus the familiar city (7.6), $F(1, 44) = 6.10$, $p < .02$.

Summary of Experiment 2 Results. Table 1.3 summarizes the most important results of this experiment. Key findings replicate those of Experiment 1. A more detailed set of results can be found in B. H. Kantowitz, Hanowski, and S. C. Kantowitz (in press).

TABLE 1.3
Summary of Results (Experiment 2)

Dependent Variable	Figure	Key Finding
Penalty cost	11	Inaccurate information leads to greater penalties. Penalties were higher in familiar Seattle setting.
Trust	12	Inaccurate information decreases trust. Very inaccurate information decreases trust even more.

CONCLUSIONS

The goal of these experiments was to provide an initial answer to the following two questions asked by ATIS designers:

1. How reliable must traffic information be for motorists to trust and use it?

2. How does the familiarity of the setting influence trust and use of unreliable traffic information?

For several dependent variables results showed that whereas 100% accurate information yields best driver performance and subjective opinion, information that is 71% accurate remains acceptable and useful. Drivers are willing to tolerate some error in a simulated ATIS. However, when information accuracy drops to 43%, driver performance and opinion suffer. Thus, information accuracy below 71% is not recommended to system designers. Future research is needed to evaluate information accuracy between 44% and 70%.

Drivers did not use simulated ATIS accurate information as effectively in the familiar setting as in the unfamiliar setting. Inaccurate traffic information was more harmful in the familiar setting. These results may imply that commercial success for in-vehicle ATIS will be easier to accomplish in unfamiliar settings (e.g., for use in rental vehicles for visitors), than in one's home city. Because drivers have greater self-confidence in familiar settings, they are more critical of ATIS advice and hold to a higher standard of user acceptability when they know the area geography. Thus, to achieve user acceptance, in-vehicle systems intended for purchase by a driver in a private passenger vehicle will likely have to meet higher standards than systems intended for commercial use.

Driver trust was decreased by inaccurate traffic information but recovered when accurate information was received. However, the more likely that information was inaccurate, the less the recovery. For the familiar Seattle 43% accuracy condition, trust minus self-confidence became negative, implying that drivers would prefer their own solutions to those offered by the simulated ATIS device. So, although drivers do not demand perfect traffic information, high degrees of inaccuracy will cause drivers to ignore system advice, especially in familiar settings.

Age Effects. A very interesting difference in trust patterns emerged for the two cohorts. Younger and older drivers exhibited equal trust for links about which information was purchased (Fig. 1.6). But, trust decreased for older drivers for links where no purchase was made, whereas younger drivers showed increased trust for those links. It appears that younger drivers did not need to purchase information if rated trust was high. However, when older drivers did not buy information, their rated trust was lower.

For older drivers, self-confidence was higher for links where information was purchased, whereas the opposite result was obtained for younger drivers (Fig. 1.7). Younger drivers had greater trust minus self-confidence differences, which is consistent with a preference for automated technology in younger cohorts (Fig. 1.8). Younger drivers had higher difference scores when they purchased link information, whereas older drivers had higher scores when they did not purchase link information.

The subjective data considered together suggest that the purchase of information is either motivated differently or produces different feelings for the two cohorts. Perhaps younger drivers use the ATIS depending first on their subjective feelings, whereas older drivers use the ATIS system to alter their subjective feelings. This speculation implies that younger drivers use internal states to control their use of automated systems, whereas older drivers use the system to modify their own internal states. If true, this hypothesis has important design implications.

Methodological Comparisons. It is interesting to compare these results with those of Bonsall and Parry (1991) and Allen et al. (1991). When familiarity is defined as knowledge of a local geography, we found that familiarity was harmful because penalty costs were higher; Allen et al. (1991) found no effect of familiarity. When familiarity is defined as learning over repeated trials, we found no decrement in driver trust over the first two repeated trials. However, penalty cost did decrease on the second trial. This conflicts with results of Bonsall and Parry, who found drivers to be less likely to accept advice over repeated trials. Note that experiments using artificial networks to some extent necessarily confound familiarity with the traffic network and familiarity with the simulated ATIS device, because the network is learned by using the new device. Although artificial networks can be useful microworlds for the ergonomics researcher, we believe that they are most useful when topographically matched to a real network as in Experiment 2.

Finally, note that this experiment manipulated information reliability in only one of several ways that might be meaningful to drivers. In this experiment, accuracy was based on level of service. Drivers judged whether the actual level of service on a particular traffic link matched their expectations based on information provided by the simulated ATIS. Other traffic dimensions may be equally salient for drivers. For example, Janssen and Van Der Horst (1993) presented drivers in a simulator with length of congestion in kilometers, delays relative to normal travel times in minutes, and travel times in minutes. Reliability was manipulated by altering the variability of these estimates. They found that travel time information was most resistant to degradations in reliability. So the present results may not generalize to all of the possible formats and types of information traffic engineers can offer to drivers.

ACKNOWLEDGMENTS

The research reported here was funded by the Federal Highway Administration under contract DTFH61-92-C-00102; M. J. Moyer was the FHWA technical representative.

REFERENCES

Adler, J. L., & Kalsher, M. J. (1994). Human factors studies to investigate driver behavior under in-vehicle information systems: An interactive microcomputer simulation approach. In *Proceedings of the Human Factors and Ergonomics Society 38th Annual Meeting* (pp. 461–465). Santa Monica, CA: Human Factors and Ergonomics Society.

Allen, R. W., Ziedman, D., Rosenthal, T. J., Stein, A. C., Torres, J. F., & Halati, A. (1991). Laboratory assessment of driver route diversion in response to in-vehicle navigation and motorist information systems. *Transportation Research Record, 1306,* 82–91.

Bonsall, P. (1994). Simulator experiments and modelling to determine the influence of route guidance on drivers' route choice. In *Proceedings of the First World Congress on Applications of Transport Telematics and Intelligent Vehicle-Highway Systems* (pp. 964–972). Paris: Artech House.

Bonsall, P., & Parry, T. (1991). Using an interactive route-choice simulator to investigate driver's compliance with route guidance advice. *Transportation Research Record, 1306,* 59–68.

Janssen, W., & Van Der Horst, R. (1993). Presenting descriptive information in variable message signing. *Transportation Research Record, 1403,* 83–87.

Kantowitz, B. H. (1988). Laboratory simulation of maintenance activity. In *Proceedings of the 1988 IEEE Fourth Conference on Human Factors and Nuclear Power Plants* (pp. 403–409). Monterey, CA: Institute of Electrical and Electronic Engineers.

Kantowitz, B. H., Becker, C. A., & Barlow, S. T. (1993). Assessing driver acceptance of IVHS components. In *Proceedings of the Human Factors and Ergonomics Society 37th Annual Meeting* (pp. 1062–1066). Seattle, WA: Human Factors and Ergonomics Society.

Kantowitz, B. H., Hanowski, R. J., & Kantowitz, S. C. (in press). Driver acceptance of unreliable traffic information in familiar and unfamiliar settings. *Human Factors.*

Kantowitz, B. H., Roediger, H. L., & Elmes, D. (1994). *Experimental psychology* (5th ed.). St. Paul, MN: West Publishing.

Kantowitz, B. H., & Sorkin, R. D. (1987). Allocation of functions. In G. Salvendy (Ed.), *Handbook of human factors* (pp. 355–369). New York: Wiley.

Kantowitz, S. C., Kantowitz, B. H., & Hanowski, R. J. (1995). The Battelle route guidance simulator: A low-cost tool for studying driver response to advanced navigation systems. In *Proceedings of the Sixth International Conference on Vehicle Navigation and Information Systems (VNIS 95)* (pp. 104–109). Piscataway, NJ: Institute of Electronic and Electrical Engineers.

Lee, J. D., & Moray, N. (1991). Trust and allocation of function in human–machine systems. *Ergonomics, 35*(10), 1243–1270.

Lee, J. D., & Moray, N. (1994). Trust, self-confidence, and operators' adaptation to automation. *International Journal of Human Computer Studies, 40,* 153–184.

Sorkin, R. D. (1988). Why are people turning off our alarms? *Journal of the Acoustical Society of America, 1,* 1107–1108.

Transportation Research Board. (1992). *Highway capacity manual* (Special Rep. 209; 2nd rev. ed.). Washington, DC: National Research Council.

Chapter 2

Design Decision Aids and Human Factors Guidelines for ATIS Displays

Michael A. Mollenhauer
Melissa C. Hulse
Thomas A. Dingus
Steven K. Jahns
Cher Carney
University of Iowa

Advanced traveler information systems (ATIS) will incorporate functions that enable drivers to access information that was previously unavailable or was retrieved from domains outside of vehicle. More specifically, the proposed ATIS functions will encompass in-vehicle routing and navigation systems (IRANS), in-vehicle motorist services and information systems (IMSIS), in-vehicle signing and information systems (ISIS), in-vehicle safety and warning systems (IVSAWS), and commercial vehicle operation (CVO) information systems. When giving drivers access to such systems inside the vehicle, designers must consider not only usability, but also safety (i.e., overloading or distracting the driver). One of the primary concerns is that ATIS will inadvertently reduce safety instead of improving it. The overall impact of ATIS on driving will depend largely on the driver's ability and desire to successfully utilize the information provided. If ATIS is not developed to be driver friendly and easily utilized by the majority of drivers, its overall effectiveness will suffer.

This chapter explores the decisions designers must make when developing ATIS displays. It also describes a design support process that has been developed to help formulate answers that reflect current human factors research and accepted design principles. Examples of decision tools that make up this process will be provided along with a description of how these tools can be used together to aid in the design process. In addition, specific results (generated using these design tools to analyze a large set of proposed ATIS information items) are also presented and discussed.

Designers must answer the following when developing displays for ATIS, which will affect or have an impact on both the safety and usability of the system:

1. What information should be included in the ATIS that is being developed? Current and future advances in technology will provide the capability to present an immense amount of information to drivers. Developing a cohesive and usable system will require designers to make decisions about which individual pieces of information will best help the driver accomplish the tasks related to the functions that ATIS is trying to support.

2. What functions of the ATIS should the driver be allowed to use while driving? It is clear, in the case of fully functional ATIS systems, that driver access to all functions in-transit may pose a potential safety risk, and is often unnecessary or undesirable for effective system operation. A driving trip can be divided into three parts: predrive (the vehicle is in Park), zero-speed (the vehicle is not in motion, but is in a gear other than Park), and in-transit (the vehicle is in motion). This issue is further discussed later. The allocation of information to different parts of the trip (predrive, zero-speed, and in-transit) is referred to as *trip status* for the remainder of this chapter.

3. To which sensory modality (e.g., auditory, visual, tactile) should information items be allocated? Sensory modality allocation can greatly affect both the safety and usability of ATIS. For example, excessive visual presentation can overload the visual sensory system, which already receives roughly 90% of the information used by drivers. Excessive auditory information can also result in a system that is unusable, frustrating, and annoying.

4. What format (e.g., text, map, tone, voice) should be used to present the information? Display format can also affect a system's safety and usability. For example, a moving map for navigation may be distracting, and may become too complex to be legible. Alternatively, a short list of textual directions may not provide all information that a driver needs to effectively use the system (e.g., an effective description of a maneuver at a complex intersection).

There are other questions associated with displays that can be included in this list, such as display location and the timing associated with information presentation. However, the process discussed here concentrates solely on the content of the display and does not address these additional issues. Wherever possible, similar systematic methods of applying human factors design principles and research should be used to answer the questions excluded from this process.

ATIS DISPLAY DESIGN PROCESS

This section defines the design tools that can be used to effectively answer the questions already described for a particular system concept. The examples provided and the results discussed later are based on a concept of general ATIS functionality including IRANS, ISIS, IMSIS, IVSAWS, and CVO-specific capabilities.

Figure 2.1 depicts the design process suggested for use in aiding the development of ATIS information displays. The individual tools that make up the overall process include the information items for a given conceptual system, the functional information grouping, the sensory modality allocation, the information type categorization, the information criticality assessment while driving, the trip status allocation, and the display format trade study analysis.

The procedure for using the process in Fig. 2.1 involves assessing individual pieces of information, or "information items," that will be provided to the driver by the ATIS system. Once established, each item is systematically assessed through the use of each of the tools provided. An example of an information item for an ATIS system is the distance to the next turn along a guided route. Such an information item is important in an IRANS application so that the driver can plan the next maneuver in advance while navigating to an unknown destination. Another important information item for this example is the direction of the next turn. Other information items that could be included in such a case are name of the turn street, name of the current street, and distance to the destination. However, these items may not be required by the user to navigate, depending on such factors as the accuracy of the system. Assessing the items that are "required" or "desired" and the best method of presentation is the purpose of this process. Upon completion of the process, a list of information items important for inclusion in a system design will have been identified, along with the suggested trip status allocation, sensory mode of presentation, and display format for each information item.

The ultimate goal of the process shown in Fig. 2.1 is to develop usable design tools that generate guidelines for ATIS designers. To achieve this goal, the traditional human factors design process has been modified to provide decision criteria in a form that can potentially be used by system designers naive to human factors. Design criteria are provided in a series of decision aids. All the criteria developed for the tools in this analysis relate only to one aspect of the design—namely, the human factors issues of safety, usability, and preference/acceptance. In practice, designers might want to modify the design decision aids to include other design criteria or system constraints such as cost. To allow for such modifications, the tools are specifically designed to provide multiple selections (where prac-

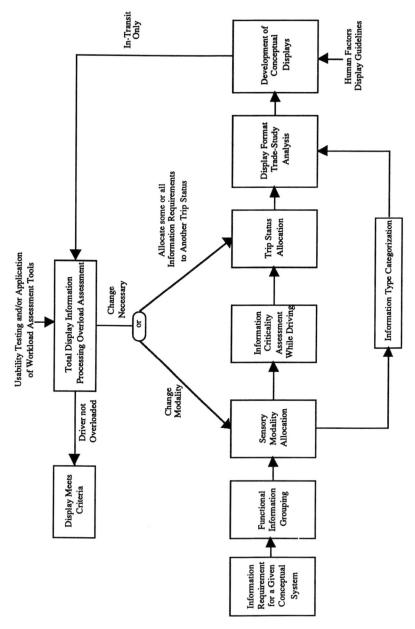

FIG. 2.1. Design process for an ATIS display.

tical) without greatly compromising the human factors criteria. If other criteria are also important, due to particular design goals or system constraints, then these criteria must also be properly weighed and considered as part of the overall design process.

The design aids shown in Fig. 2.1 are made up of several different types of traditional and nontraditional human factors design tools. It is necessary to include multiple tools because of the diverse questions that a designer must answer while developing ATIS systems. Advances in display technologies have increased the number of feasible options that are available to designers and have made the task of creating a system that meets multiple design criteria difficult. The ATIS designer who is naive to human factors design principles will benefit from having comprehensive tools that systematically reduce the number of feasible options based on user-defined design goals.

TRADE STUDY ANALYSIS AND DECISION TREES

The tools developed as part of this process are presented in two major forms. A "decision tree" approach is used to enhance the usability of the design criteria used by the system designer. In instances where the decision is potentially difficult to assess for a naive user, examples are provided to guide the designer in the correct direction. The decision tree is simply a tool used to help make decisions about the characteristics of an ATIS information item and how it might be displayed. To analyze the information format options, a more standard "trade study" analysis is used to aid in design decisions. These analyses serve as systematic aids for complex decision making. These tools allow designers to choose between a variety of design alternatives based on relative importance of system-critical performance criteria, which are determined prior to the establishment of the alternative designs under consideration.

This section presents an overview of the trade study process and discusses the steps to follow for the best design configuration with a given set of criteria. It also introduces a decision-making protocol called the Kepner–Tregoe analysis (KTA). Though the KTA method is easy to use, it has some inherent limitations (discussed at the end of the section). It is important to keep these limitations in mind when performing and reporting trade study analyses.

Due to constraints imposed by cost, techniques, and so on, designers cannot create designs that are optimal in every way. To meet one objective, they often must compromise another. Thus, throughout product development, designers must determine the combination of criteria that best satisfies the design objectives. The trade study analysis is a systematic approach developed to identify and prioritize criteria to optimize design.

Using trade studies, designers attempt to consider all the design factors and objectives, then decide which objectives to compromise in favor of others. The overall goal is to maximize the design advantages. By using trade studies to evaluate alternative solutions, designers can avoid the tendency to go directly to a design based on their past experiences. Instead, they can implement the product design that best satisfies the users' requirements. To ensure that evaluations are as rational and unbiased as possible, trade studies use a structured procedure as a framework.

Researchers have developed several methods that apply decision theory and multi-attribute utility theory to trade study analyses. The following are steps for completing a trade study:

Define objectives. A trade study's objectives must be defined in clear terms. This will provide a basis for selecting criteria.

Establish alternatives. A trade study should consider all feasible approaches to achieving the design objectives.

Establish decision criteria. The criteria provide a logical basis for making selections between alternative solutions. For many systems, especially those that are relatively complex, there are many possible criteria. An optimal design for one criterion is not necessarily optimal for another. Typical criteria used in trade studies might include: efficiency, cost, user acceptance, safety, and error potential.

Assign weights to criteria. Criteria weights range from 1 to 10, with larger numbers indicating the more important criteria as they relate to the system objectives. These weights must be assigned as objectively as possible, even though their assignment is a subjective process. To facilitate this process, it is often valuable to have multiple system experts independently weight the criteria and negotiate a final weighting.

Score alternatives. Two methods are commonly used to score alternatives. The rating method simply has one or more experts score the design on a scale of 1 to 10 for each of the criteria. The ranking method has the experts rank the alternatives from highest (best) to lowest (worst). Particularly for complex analyses with large numbers of criteria and alternatives, ranking, performed either outright or by paired comparisons, appears to be an equally reliable and easier method. This is the most difficult and subjective aspect to the trade study process. Using multiple, expert raters appears to provide the most reliable outcome.

Generate a trade table. It is possible to create a trade study table that helps calculate the weighted scoring. To start, all possible options are listed in the left-hand column. The criteria, developed in a previous step, are then listed at the top of the subsequent columns with their assigned weightings. The rankings (or ratings) for each feasible option are then filled into the table by completing one criterion column at a time. Once all criterion columns have been completed, the rankings for each within each criteria column are then filled into the table.

The individual rankings for each option are then multiplied by the appropriate criterion weights for the given criterion and totaled for the row to generate a final score for the option. An example of a trade study table may be seen in Table 2.1. This table evaluates display format options. The individual rankings within one criterion are the nonbold numbers under the column labeled R for ranking. The product of that ranking multiplied by the weighting for that particular criterion can be found in the $R \times W$ column in boldface type. A total score for a given option is shown on the right in bold, with the final rank ordering next to it in non-boldface type. The final ranking suggests which is the preferred display format given the information items within the functional grouping.

Analyze sensitivity. In this step, the decision's sensitivity to changes in the values of attributes, weights, costs, and subjective estimates is determined. This analysis verifies that changes in the weights or scoring will reverse the decision, and assesses sensitivity to changes in system requirements. As a rule of thumb, for a distinction in candidate scores to be meaningful, there should be at least a 10% difference in the total score.

Predict adverse consequences. The adverse consequences of selecting the candidate are examined as the final step to ensure that its selection will not adversely affect the overall design.

Prepare documentation. This report should also describe changes in weights, scoring, or requirements that would revise the selection. From these steps, designers can obtain the configuration that best meets their criteria.

The trade study analysis is a systematic and widely used tool for aiding designers in making a complex design decision. The KTA method is easy to use, and has some advantages: It provides an explicit decision model, it accounts for mandatory criteria to eliminate alternatives, it accounts for varying importance of criteria, and it provides a single score for each alternative. However, this method also has some limitations: It relies on subjective data, the decision quality varies with experience and biases of the expert raters, and it requires that the designers of the system rescale objective data.

This method enforces two primary objectives of good design practice. First, the designer considers more design options. Because an entire step in the trade study process involves identifying feasible design alternatives,

TABLE 2.1
Trade Study Matrix Example

		Assigned Weights														
		9		10		7		5		6		5			
		Criterion 1		Criterion 2		Criterion 3		Criterion 4		Criterion 5		Criterion 6			
Option	General Option Description	R	R×W	R	R×W	R	R×W	R	R×W	R	R×W	R	R×W	Total	Rank
A		5.0	45.0	2.0	20.0	5.0	35.0	1.0	5.0	3.0	18.0	2.0	10.0	133.0	5.0
B		4.0	36.0	5.0	50.0	5.0	35.0	3.0	15.0	5.0	30.0	6.0	30.0	196.0	2.0
C		3.0	27.0	6.0	60.0	5.0	35.0	4.0	20.0	6.0	36.0	5.0	25.0	203.0	1.0
D		2.0	18.0	3.0	30.0	3.0	21.0	6.0	30.0	4.0	24.0	4.0	20.0	143.0	4.0
E		6.0	54.0	4.0	40.0	2.0	14.0	5.0	25.0	1.0	6.0	1.0	5.0	144.0	3.0
F		1.0	9.0	1.0	10.0	1.0	7.0	2.0	10.0	2.0	12.0	3.0	15.0	63.0	6.0

Note. R = Ranking; R × W = Ranking times Weight.

a greater number of options are considered. And second, the designer avoids bias generated from preconceptions about the design. In many instances, a designer starts to conceive the design as soon as the design problem is given. In such cases, the conceptual design is often formulated with little serious consideration given to feasible alternatives. In addition, once a designer has established a conceptual design, there is often a strong bias against deviation from the original concept.

Despite the advantages of the trade study process, it is not a panacea that guarantees a good design. Selecting criteria, defining criteria, weighting criteria, and ranking alternatives are all subjective processes and can be subjects of considerable debate. In addition, considerable expertise is required to effectively carry out these subjective processes.

DESIGN TOOLS USED FOR THIS ANALYSIS

Each of the tools included in this analysis, as shown in Fig. 2.1, are discussed in the following paragraphs. The tools are defined by including their inputs and outputs, the assumptions associated with each, and by an example of each that was used to determine some of the results discussed in the next section. The reader should bear in mind that these tools were created for a human factors analysis. The use of these tools to create a design based on goals other than safety, usability, and acceptance would require modifications to the criteria chosen and the weights assigned to those criteria.

Information Items for a Given Conceptual System

In assessing the information display options, it is necessary to understand the tasks that the user must perform and, more specifically, what information the driver requires to perform them. The first step in developing ATIS displays is to determine which individual pieces of information, or *information items*, will need to be displayed. Information items can be determined from the design goals for a system, a task analysis of driver activities, an exhaustive brainstorming exercise identifying all possible items, or any combination of these and other methods. Examples of information items include a distance to destination (IRANS), the current speed limit (ISIS), a low tire pressure indication (IVSAWS), and the services available at the next exit (IMSIS).

For the results listed in the next section, approximately 400 information items were generated and reviewed. Some of the individual information items are listed in Table 2.2. When identifying information items to be

TABLE 2.2
Example of Information Items for a Given Conceptual System

IRANS Trip Planning
Information Items
Current criteria for automated trip planning
Time to get to each destination from previous destination
Cost of each toll along the route
Total toll charges along the trip
Total time for trip
Estimates of mileage
Location of attractions and points of interest
Forecast weather information
Historical traffic information
Street or roadway names on the route
States, regions, communities and districts that the route will traverse
Landmarks or topographical features along route
Number of turns or roadway changes required
Types of roads used on route (interstate highway, two lane street)
Distance to each destination from previous destination
Distance to specific attractions

used in this process, it is important to define them to be as simple as possible. Each information item is also listed under a heading of the ATIS subfunction to which it belongs (i.e., IRANS, as seen in this example). The output of this tool is a complete list of all information items that may be displayed by the ATIS. This output serves as the input for the next step in the process which is the functional information grouping task.

Functional Information Grouping

The functional information grouping categorizes each information item into one of six functional groups. Table 2.3 is an example of this process. Functional groupings were devised by combining similar ATIS subfunctions. The purpose of this grouping was to create broad categories of information items that should be assessed using like decision criteria. For example, the decision tree for a hazard warning signal would most likely be quite different from that of a communication regarding motorist services information due to differences in criticality and required driver response. This level of grouping was used as an input for the subsequent sensory modality allocation analysis to provide generalizable results across similar ATIS information items.

TABLE 2.3
Example of Functional Information Groupings

Route Planning and Coordination
  Trip planning
  Multimode travel coordination and planning
  Predrive route and destination selection
  CVO-specific (route scheduling)
  Destination coordination
Route Following
  Dynamic route selection
  Route guidance
  Route navigation
Warning and Condition Monitoring
  Immediate hazard warning
  Road condition information
  Vehicle condition monitoring
  CVO-specific (cargo and vehicle monitoring)
Signing
  Roadway guidance sign information
  Roadway notification sign information
  Roadway regulatory sign information
  CVO-specific (road restriction information)
Communication and Aid Request
  Message transfer
  Manual aid request
  Automatic aid request
Motorist Services
  Automated toll collection
  Broadcast services/attractions
  Services/attractions directory

Sensory Modality Allocation

The sensory modality allocation tools are individual decision trees developed for each function information grouping shown in Table 2.3. Figure 2.2 shows a sensory modality allocation decision aid for the route planning and coordination functional grouping. Multiple decision aids allow for the use of specific criteria that are appropriate for a given type of information. The goal of each of these decision aids is to help designers select among the feasible sensory modality alternatives.

To determine the correct sensory modality for presenting each information item, the designer first selects the correct decision tree (of the trees that exist for each functional grouping). Next, the designer answers each of the questions about the information item from left to right on the tree until a final box has been reached at the right. The branches of the decision trees represent the categorization of information that is most

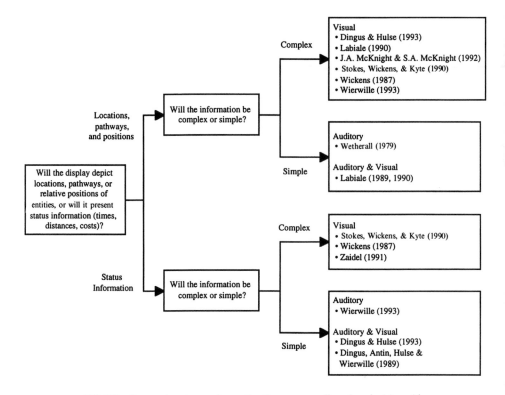

FIG. 2.2. Route planning and coordination sensory allocation decision aid.

often observed in consensus in ATIS research literature. The literature reviewed to determine these categorizations and subsequent branches are listed in the rightmost boxes of the tree.

The final boxes on the right give the suggested sensory mode for presentation and several sources of research that support that decision. Specific citations of the research and how it supports design decisions are given in the results section. Some of the criteria used for decisions may not be clear to a designer naive to human factors. Therefore, examples of correct decisions associated with the trees were also created. An example for the route planning and coordination decision aid has been provided in Fig. 2.3.

The sensory modality allocation results provide input for two additional tools. The first has been designated information type categorization. The second is the display format trade study analysis. The information type categorization will determine the appropriate trade study table to use in later analyses. The sensory modalities that were suggested by each of the

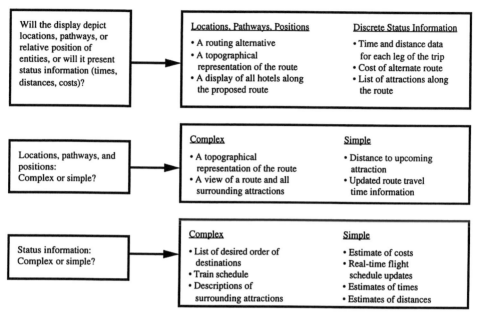

FIG. 2.3. Route planning and coordination sensory allocation decision aid examples.

end boxes will also determine which display formats will be included in each of the display format trade study analysis tables.

Information-Type Categorization

The information-type categorization is nothing more than the process of recording the labels from the branches of the sensory modality allocation process that were chosen for each information item. For example, noting that an information item provides "simple status" information, as depicted in Fig. 2.3. This type categorization is important because it will be used later in this process when selecting the appropriate display format trade study analysis table to use.

As highlighted in Figs. 2.2 and 2.3, sensory modality decisions are made using the assessment of the features inherent to the information items. Features such as information type (e.g., spatial versus verbal) and information complexity are determined and should be recorded as a result of using this tool. This determination proves invaluable in the assessment of the display format trade study analysis. Segregating the information items by specific features allows for the creation of trade study criteria, which are specific enough to be useful without requiring the very laborious task of creating criteria on an information item-by-item requirement basis.

Information Criticality Assessment While Driving Analysis

The information criticality assessment while driving analysis determines whether each information item is required, desired, or not required/desired while the vehicle is in motion. The entries in the right-hand column of Table 2.4 show some examples of this classification. If the information is required while driving, it is automatically allocated to an *in-transit* trip status. In-transit is operationally defined as any time when the vehicle is in motion. If the information is neither required nor desired while driving, the information item is allocated to a *predrive* status. Predrive refers to the time prior to starting the trip when the vehicle is still in Park or Neutral. Therefore, the output from the information criticality assessment is a determination of whether the information should be made available to drivers in-transit or only during predrive circumstances for information items.

Some of the information items are not required, but they might be desired by the user of the ATIS system while driving the vehicle. The most appropriate trip status for those information items is analyzed on a case-by-case basis in the next step of the process, which is the trip status allocation analysis.

TABLE 2.4
Example of Information Item Criticality Assessment While Driving

IRANS Trip Planning	
Information Items	*Information Item Criticality*
Current criteria for automated trip planning	neither required nor desired
Time to get to each destination from previous destination	desired
Cost of each toll along the route	desired
Total toll charges along the trip	desired
Total time for trip	desired
Estimates of mileage	desired
Location of attractions and points of interest	desired
Forecast weather information	desired
Historical traffic information	desired
Street or roadway names on the route	required
States, regions, communities and districts that the route will traverse	neither required nor desired
Landmarks or topographical features along route	desired
Number of turns or roadway changes required	neither required nor desired
Types of roads used on route (interstate highway, two lane street)	desired
Distance to each destination from previous destination	desired
Distance to specific attractions	desired

Trip Status Allocation

The information items designated as desired in the information criticality assessment while driving are the inputs for this process. The trip status allocation decision aid is shown in Fig. 2.4. The goal of this decision aid is to provide designers with guidelines, based on information-processing and manual control requirements, for choosing the most appropriate trip status (predrive, zero speed, or in-transit) for displaying these remaining information items.

To use this tool, the designer would answer a series of questions about each information item separately, starting from the upper leftmost box until there are no more decisions to be made and a final box has been reached. The final boxes contain the suggested trip status allocation for the information item being analyzed. The decisions that must be made while performing this process require the designer to make some assumptions about the complexity of the information and the types of control inputs that might be required to interact with the system. To provide some frame of reference for these assumptions, existing data has been compiled (Table 2.5), showing the average glance length and number of glances for a number of general tasks that drivers perform while driving (Dingus, 1987). Using this type of information as a guide minimizes the dangers of making poor assumptions during this process. Another safeguard inherent to the process (Fig. 2.1) is an iterative feedback loop where designers must critically evaluate the resulting designs based on assumptions made while utilizing each tool.

Display Format Allocation Analysis

At this point in the design process, the designer will know the trip status mode, functional grouping, and sensory modality in which the information item should be presented. However, within one sensory mode there may be several different categories of display format that can present the information. For example, if the mode of presentation is visual, there are two global ways to present information: in a graphical or pictorial fashion and in a text format. Within these global classifications there are categories that apply specifically to the information items. For example, a graphical presentation of location information could be a full map with a lot of detail, a partial area map, a three-dimensional representation, or even a simple iconic arrow. The presentation medium can also change; it can be printed on paper or computer generated on a video screen.

Ideally, the best guidance provided to a designer would go one step further than format and find the best design features within a category. Considerations of different line weights, screen brightness, color, legibility,

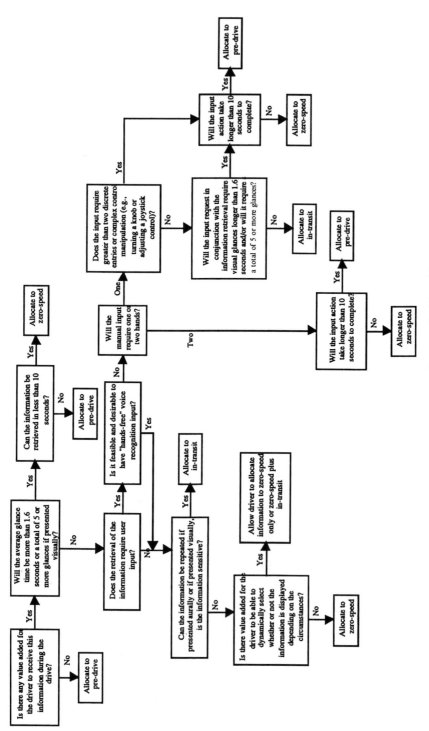

FIG. 2.4. Trip status allocation decision aid.

38

TABLE 2.5
Average Display Glance Times and Number
of Glances for Some General Automotive Tasks

Task	Average Display Glance Time	Average Number of Glances
Signal vehicle is turning	0.30	0.63
Read speedometer	0.62	1.26
Check for following traffic	0.75	1.31
Adjust vent	0.62	1.83
Remaining fuel level	1.04	1.52
Turn on fan	1.10	1.78
Turn on defrost	1.14	2.51
Orient and insert a cassette tape	0.80	2.06
Adjust cruise control	0.82	5.88
Adjust power mirror	0.86	6.64
Tune the radio	1.10	6.91

user preference, and so on, all need to be compared to find the best method of information presentation. However, decision aid development at this level of detail is beyond the scope of this chapter. The user of this tool should bear in mind that it was developed to work at a generic, categorical level and therefore sacrifices some decision-making resolution concerning the presentation of specific information items.

An example of the trade study tool used in this analysis is shown in Table 2.6. This trade study suggests a format category that allows for a high degree of standardization across the functional groupings based on specific characteristics of the information. However, results obtained from this trade study may not match the standardization strategy that has been adopted by a designer. Therefore, deviation from the suggested format may be necessary. Engineering judgment should be used in the application of the recommendations when the information item is specific and tends to be unique in comparison to other items in its functional grouping.

At this point in the process, the trip status, sensory modality, and functional grouping decisions have already been made, so some of the available possibilities for display format are not applicable. For this reason, not every display format method was rated for every functional group.

The criteria used in the display format trade studies fit into three separate categories: safety, usability, and preference. These categories are broken down into specific elements that are operationally defined as being mutually exclusive from one another. The user of this trade study should keep in mind that the criteria were created to fit broad topic areas and not one particular display configuration. Therefore, the criteria only consider display advantages and disadvantages in the context of cognitive information-processing aspects. It is not the intent of this trade study to

TABLE 2.6
Display Format Trade Study Matrix Example: Warning Modality—High Priority, External Environment, Context Not Required

		Assigned Weights														
		8		10		8		4		6		4		6			
Display Mode	General Display Description	Distraction Potential		Attention Demand		Postretrieval Workload		Efficiency		Error Potential		Driver Acceptance		Annoyance Potential		Total	Rank
		R	R × W	R	R × W	R	R × W	R	R × W	R	R × W	R	R × W	R	R × W		
Tactile presentation (tactile)	Automatic control manipulation or change in tactile feedback	6.0	48.0	5.0	50.0	2.0	16.0	4.0	16.0	2.0	12.0	4.0	16.0	2.0	12.0	170.0	3
	Vibration of controls alerting driver	5.0	40.0	6.0	60.0	1.0	8.0	2.0	8.0	1.0	6.0	3.0	12.0	3.0	18.0	152.0	4
Speech combined with graphical presentation (auditory, visual)	Iconic or graph representation plus voice	3.0	24.0	2.0	20.0	6.0	48.0	6.0	24.0	5.5	33.0	6.0	24.0	1.5	9.0	182.0	2
Speech combined with text presentation (auditory, visual)	Description on video screen with voice	1.0	8.0	1.0	10.0	5.0	40.0	5.0	20.0	5.5	33.0	1.0	4.0	1.5	9.0	124.0	6
Tones combined with graphical presentation (auditory, visual)	Iconic or graphical representation with simple tones, chimes, etc.	4.0	32.0	4.0	40.0	4.0	32.0	3.0	12.0	4.0	24.0	5.0	20.0	5.5	33.0	193.0	1
Tones combined with text presentation (auditory, visual)	Description on video screen with tones	2.0	16.0	3.0	30.0	3.0	24.0	1.0	4.0	3.0	18.0	2.0	8.0	5.5	33.0	133.0	5

Note. R = Ranking of Display; R × W = Ranking times Assigned Weight.

distinguish the single best method of information presentation down to the levels of detail such as font size, amount of information per screen, sound level, and so on, all of which are sensory related issues. To make comparisons at this level of detail requires direct evaluations of detailed design specification information. Instead, criteria have been developed to guide the designer toward one or two format types that best fit the requirements of the given situation. Consistent with the trade study process described in a previous section, the criteria were weighted based on their perceived importance to the human factors objectives of ATIS systems. Operational definitions of the format trade study criteria and their associated weights are as follows:

Distraction potential (10). The potential that a display of information will divert the driver's attention away from the primary task of driving.

Attention demand (10). The amount of information-processing resources required to retrieve the displayed information in a safe manner within the confines of the driving environment. Note that this is retrieval only, not the processing time needed to act on the information (which is covered under postretrieval workload).

Postretrieval workload (8). After the sensory aspects of retrieval, the amount of information-processing and decision-making resources required to respond to the displayed information.

Efficiency (4). The amount of time required to utilize the displayed information in order to perform the required task. This includes only the task completion time when circumstances are well below the threshold of unsafe conditions.

Error potential (6). The potential of the displayed information to cause confusion and thereby increase the likelihood of errors. This includes the ease with which the information can be standardized and the expectancy of the naive user.

Driver acceptance (4). The relative partiality of the driver to a given format. This is operationally defined as including any perceived positive aspects of a display format and as excluding any aspects of annoyance.

Annoyance potential (6). The potential for a negative reaction toward the display due to the frequency, reliability, or sensory characteristics of the information. Annoyance differs from distraction in that it describes the level of irritation or frustration that a display might cause a driver.

The criteria that fell into the category of safety tended to have the highest weighting implications of increasing accident potential in circumstances of distraction or overload. In cases where the display presentation occurs during a predrive trip status, safety no longer becomes an issue because both the time critical and visual overload elements of attention demand are removed with the absence of the driving task.

Attention demand was given the highest weighting of 10 because this factor can interrupt visual scanning, which in turn creates the greatest accident potential.

Distraction potential and postretrieval workload were weighted as 8 because they still can create accident potential. If a display were to cause a high degree of peripheral view distraction (e.g., while using a moving map), it could divert driver attention away from the primary task of vehicle control and guidance. Distracting auditory displays may cause a startle response and have the same detrimental effect on driving performance. If the information presented to the driver is difficult to understand or requires a high degree of cognitive processing, the driver will have fewer and possibly insufficient resources to allocate to the driving task. This type of distraction is less critical than visual distraction, but it still has the potential to affect the driving task due to the limited amount of information a driver can process. For example, if the driver is trying to choose between several suggested alternatives, there is potential for the driver to become overloaded and make a poor judgment about following distance or gap acceptance.

Efficiency was weighted as a four because increasing the speed that the driver can retrieve, process, and respond to the displayed information makes the system more usable. Being able to navigate throughout the system quickly and efficiently will also increase the total number of tasks that can be performed in a given time period. This is different from postretrieval workload and attention demand in that the time to process the information is not a safety issue. Take, for example, displays presented in predrive; attention demand as a safety issue is no longer relevant because the car is not moving. The most efficient display format would win out in this trade study based on usability issues. The relatively low weighting of a 4 is assigned because usability at this level does not have the higher costs associated with safety issues or other criteria that lead to lack of use of the system.

Error potential was weighted as a 6 because a system that evokes a high error rate adversely affects the system's usability. Differences in the types or numbers of errors are likely to occur using one method of display presentation over another. If the error rate is high, the user becomes frustrated and might avoid using some functions in the future. Users tend to have preconceived notions about the way a device should work, so it is important that the presentation of information match these notions. Closely related to this is the ability to standardize the method of presentation across several functions of the system. The ability to provide uniformity will aid in the reduction of errors.

Driver acceptance was weighted as a 4 because it deals only with positive aspects of driver preference and is not really a performance issue. It is given a low weighting because of its lack of safety or nonutilization effects on driver performance, but it is still an important factor in that increased acceptance will lead to an increase in the use of the system.

Annoyance potential, in contrast to driver acceptance, received a higher weighting of 6. This weighting was due to the potential for the driver to ignore or disable the entire system if the presentation of the information was annoying. A system that annoys drivers to the point that they ignore it is similar to having no system at all. Annoyance caused by only one element in a subsystem could cause the entire system to be deactivated. Care must be taken to account for user preference issues that can cause these severe negative reactions. Based on a review of pertinent literature (much of which is cited and discussed in the results section), the different display formats were rank ordered from most desirable to least desirable within each functional grouping. Most research performed to date discusses the comparison of a subset of all the possible format options within a functional category. None of the research provides a complete, comprehensive review of all possible format options that are available, given the state of today's display technology. Because of this, ratings were developed based on general advantages and disadvantages that were evident across all the published literature that was reviewed.

As might be expected, in attempting to establish appropriate criteria that would lead designers to reasonable solutions, it was difficult or impossible to make a number of decisions. This limitation is due to the current state-of-the-art in human factors design guidelines and principles and/or ATIS research. These instances are discussed later in order to highlight the need for additional research. Note that although many decisions are difficult, particularly in the ranking of format alternatives, this trade study technique presents the authors' best collective judgment as to the appropriate decision. The alternative of providing criteria and weightings while leaving the matrices blank in order to demonstrate the process was deemed far less constructive.

Development of Conceptual Designs

Once the display formats have been selected, designers have received considerable guidance about the information items for a given conceptual system. They can then use this guidance, along with additional human factors display guidelines and any additional system-related constraints, to develop the integrated conceptual designs. Because decisions about modality, trip status allocation, and format have been made, the conceptual display design should proceed much more effectively.

Total Display Information-Processing Overload Assessment

Designers have to assess the in-transit display's potential for driver overload. It is anticipated that this assessment will consist of usability testing, workload, and attention demand measurement in the laboratory. If an information-processing overload assessment reveals that an overload condition exists, designers can reduce the amount of information allocated to the in-transit trip status, or they can change the modality of the information from a chosen option to some other option.

RESULTS

This section discusses the results of performing the previously defined trade study process on a list of over 400 general ATIS information items. Specific results and design implications are provided, along with comments about the applicability and usability of the tools themselves.

Display Modality Selection Results

The previous analyses highlight the complexity of the modality selection decision for ATIS information. Given the complexity of this issue, it is important to treat individual items of information on a case-by-case basis while considering the aforementioned data to make an appropriate modality selection. However, in conducting a detailed analysis of over 400 potential ATIS information items, general trends of modality selection were found, depending on the type of information provided. Although the specific recommendations of this analysis are beyond the scope of this chapter, the general findings are provided here.

Route Planning. Route planning information enables the driver to select or review a route and to determine the relative positions of roadways, destinations, topographical features, and attractions. In general, route planning information includes the presentation of spatial locations for different trip entities relative to each other, and larger, more complex quantities of data. The nature of this type of information suggests that visual displays would be the most efficient method of presentation, having the highest probability of conforming to the user's expectations. ATIS should, however, facilitate route planning during the predrive segment of a trip due to the complexity of this task. Some system designs (e.g., TravTek) have even "locked out" most route planning features while the vehicle is moving to ensure safe system use.

Route Following. The types of information presented by the route following functions can be separated into two distinct groups. The first group includes information specific to the navigation task, such as spatial location or directional instructions. Specific examples include the direction of the next turn and the vehicle's current position. The second group includes discrete status information such as times and distances. It is necessary to distinguish between these two types of information because the navigation displays will generally convey spatial information and the status displays will generally convey verbal information.

Guidance information should be presented in the form of instructions to help avoid driver confusion or navigational errors. Spatial information, such as a direction to turn or current position, is often best presented through a visual display. However, routing instructions will typically be presented to the driver before every required turn. Because this information will be presented while driving, an auditory signal or redundant cue should be included to notify the driver that the information has changed. Also, displays of information presented while the vehicle is moving should be made as simple and efficient as possible to avoid driver overload and confusion. Several authors (Labiale, 1990; Parks, Ashby, & Fairclough, 1991; Streeter, Vitello, & Wonsiewicz, 1985) have stressed the importance of limiting the amount of information presented to drivers as they are driving.

Warnings and Signals. The warning portion of ATIS is intended to improve driving safety. The system will provide information about obstacles in the roadway, advanced warnings, and weather and road conditions. The designer of a warning information display must provide a design that will alert the driver but not startle, confuse, or annoy him or her.

The warning causal factor could influence which sensory mode is chosen for presentation. Because designers know the source of a potentially dangerous situation, they can design the display to direct the driver's attention in a way that is most advantageous for resolving the situation. The auditory sensory mode offers an advantage for warning displays because it is omnidirectional; that is, it gains the driver's attention regardless of where the driver's initial attention is directed.

Now that more sophisticated information systems will be available in vehicles, designers can take advantage of this capability to present more useful information. Warning messages can now describe in more detail the problem severity, as well as actions that drivers should take to resolve it. The addition of contextual information to warning messages may have an impact on the choice of sensory mode because the amount and complexity of the information has been increased. Using traditional methods of warning display, such as a dash light or a red zone on an analog display, will be inadequate for displaying additional contextual information. The details of problem context might require displays using a visual text or a speech description. In most cases, warning displays can be best presented through some combination of an attention-gaining auditory signal and a corresponding visual message that will provide additional content. Warning message sensory mode decisions will also be heavily influenced by the message priority considerations discussed earlier.

Signing. Drivers are presented with road sign information that falls into two different categories. The first includes the information a driver needs to operate the vehicle safely and legally, or regulatory information. Items that do not fall into this category make up the second category, informational messages. The reason for segregating these two groups is to distinguish the relative importance of the information being presented.

In a study to assess the effects of visual or auditory sensory mode presentation for in-vehicle signing displays, Mollenhauer, Lee, Cho, Hulse, and Dingus (1994) conducted a simulator study using a prototype in-vehicle signing system to present road sign information to drivers. Drivers were presented with road sign information through either an in-dash visual or auditory display. Drivers' recall of information and driver performance was evaluated to determine the effects of the different sensory mode displays.

The results indicated that the auditory display provided the most effective means of maximizing a driver's ability to recall relevant road sign information. The recall test results indicate that drivers were able to remember more information that had been presented with auditory displays. Driving performance measures were found to be significantly better with the use of visual displays. Because the primary task of driving can be visually demanding, it was expected that any additional visual tasks (i.e., reading the sign information from the display) would cause a greater degradation in driving performance than the addition of an auditory task. However, this was not the case. A further review of the study shows that the auditory information was being displayed rather frequently during the drive. The researchers hypothesize that it was the intrusiveness and attention-demanding characteristic of the auditory display that caused the degradation in driving performance. This study provides an example of the potential for continuous automatic auditory displays to cause a degradation in driving performance.

The most critical regulatory information (e.g., a height restriction for a commercial vehicle) should be presented to the driver in a salient manner because the information can affect the safe operation of the vehicle. A salient display in this case should include some auditory component to gain the driver's attention, although auditory messages would not be ideal for information where the expected presentation frequency is high. Informational messages could be displayed in a less salient visual format or filtered out and not displayed at all so the driver could attend to only those that are desired.

It should be noted that in selecting a sensory modality for the signing function, it is important to determine the effects of interactions with other ATIS components. This could become important when some other ATIS component is displaying auditory messages; it may no longer be optimal or even feasible to display additional auditory messages. This would also be true for the visual channel; it could become overloaded with messages from too many system components.

Messaging. Message content displays must be able to present longer, detailed items of information. In a study that measured truck drivers' preferences in messaging system display formats currently used in Europe, Huiberts (1989) found that the users (both dispatchers and drivers) preferred a visual text-based system over a speech-based system. The higher preference rating was based on four factors: the text-based system's ability to set up standardized messages, its ability to integrate with other communications systems, its advantages for setting up an electronic mail system, and the higher comprehensibility of messages.

One possible function of a messaging system would be to notify the driver that a message has been received and is ready to be displayed. This function requires gaining the driver's attention and could be accomplished by an auditory display (e.g., tone or chime). The message could then be displayed through a visual presentation when the driver feels it is safe to do so. The priority of the message might also have an effect on the sensory mode chosen for display. If the message is urgent, it could be presented through an auditory display that would require little or no visual attention and would not distract the driver from the driving task.

Very little research has addressed specific topics related to communications and aid request systems. Much of the literature reviewed for developing the criteria discussed here was intended to define general human factors guidelines for display design.

Note that when developing a tool such as the sensory allocation decision aid for the communication and aid request functions, no attempts were made to determine the effects of interactions with other ATIS components. This could become important when some other ATIS component is displaying auditory messages. It may no longer be optimal or even feasible to display additional auditory messages. This would also be true for the visual channel. It could become overloaded with messages from too many system components.

Motorist Services. The motorist services portion of ATIS is designed to provide the driver with information about services and attractions. Using current technology, the driver would need to consult the yellow pages of a phone book; see information on a road sign; or use a telephone to get information about hotels, attractions, restaurants, and available services. Although this information is useful to the driver, it is neither as urgent nor as important as safety information from other ATIS subsystems. There is also a greater potential to display a greater amount of complex information given the number of diverse services available. Thus, the displays for this function will need to be designed so that they are less intrusive and can display information efficiently to ensure availability for display of safety information. The characteristics and requirements of this type of information indicate that a visual display, available while the car is stopped, might be the best option.

Trip Status Analysis Results

The components of ATIS can provide a wealth of useful information to drivers. However, care must be taken to limit the amount of information displayed to drivers while they are driving. The secondary task of receiving complex, untimely, or distracting information could endanger drivers while they are operating the vehicle. It is apparent that some drivers are very good at self-limiting their secondary tasks, such as looking for a cassette or disciplining children in the back seat, whereas others are not. One only has to look around during a morning commute to see drivers swerving while combing hair or cleaning up spilled coffee. Given that driving is an overlearned task that is highly automatic, it is common for drivers to overestimate the secondary task workload they can handle and still safely operate the vehicle. In fact, the leading accident causal factor is driver inattention (National Highway Traffic Safety Administration, 1991). For this reason, it will be necessary to limit the availability of some system functions to times when the risks associated with divided attention are not present.

The times when a driver must access information from the system have been divided into three separate categories. As mentioned previously, these are predrive, zero speed, and in-transit. Predrive information is allocated for presentation only before the drive has started. At this time, the vehicle is stopped and in Park, which allows the driver to fully allocate information-processing and manual control capability to the operation of the system, without concern for the driving environment.

The zero-speed category is similar to predrive in that the vehicle is stopped. However, the driver is in an active traffic situation and must devote some attention to the driving environment (e.g., while waiting at a stop light, some attention is required to monitor its status) (Fleischman, Carpenter, Szczublewski, Dingus, Krage, & Means, 1991). In this situation, drivers are still able to devote nearly full attention to the system, but the time available for work with the system is limited by the duration of the traffic control device or any other cause of the zero-speed condition. Therefore, operations available during a zero-speed situation must typically take less time than those available during a predrive situation.

The in-transit condition occurs when the vehicle is in motion and the driver is required to perform the driving task. All efforts must be made to limit the functionality of the in-transit mode to those tasks that do not significantly interfere with the driving task, have convenience benefits that outweigh the cost (in terms of required driver resources), and will be used relatively frequently.

The following paragraphs discuss a series of decision criteria that were used to create a decision aid for the allocation of information items to the trip status categories of predrive, zero-speed, and in-transit. The decision aid is depicted in Fig. 2.4. As shown, decisions are based on information value, information retrieval difficulty, manual control requirements, and time sensitivity of the information.

All the information that a well-designed ATIS can present will be considered useful, but the system does not need to present all of it to drivers while they are operating the vehicle. The value and cost (in terms of information-processing and control requirements) of providing the information during any given trip status must be determined. Some types of information, such as guidance instructions, should be presented in-transit because they have a high value while driving. Other types of information, such as full trip planning functions, are complex, require more attention, and are therefore high cost. Such information should be displayed only when the vehicle is stationary.

The value of some types of information may vary with context. To achieve the system goals, it may be necessary to display ATIS information that is not currently important. An example is messages coming from a CVO message transfer system. Lower priority messages can usually be held and presented when the vehicle is stopped or parked. If the message has high priority (e.g., if it will change the path of the driver or a sequence of deliveries), it might be necessary to display the message immediately. Such a variety of situations dictates that the mode of information display be based on trip circumstances.

The amount of effort required to retrieve information from a display should influence which trip status mode is chosen for that display. For a visual display, valuable measures of effort are the amount of glance time and the combined number of glances required to retrieve the information. Some research has been performed to determine the amount of time that drivers can safely direct their attention away from the roadway. A visual display that requires frequent and lengthy glances might prevent the driver from adequately monitoring the driving environment. In fact, research has shown that deviation from the roadway lane center increases with longer eye-off-the-road time (Zwahlen & DeBald, 1986). French (1990) determined that glances away from the roadway average around 1.28 sec for normal drivers, and recommended that glances of more than 2.0 sec be avoided. French's recommendations agree with a study performed by Zwahlen, Adams, and DeBald (1987) in which driver deviations from the centerline were measured while the drivers performed operations on a touch screen display. Based on this study, the authors stated that it should not take more than four glances to completely retrieve information from a display. These values also agree with the average length and number of glances required for drivers to perform most standard in-car tasks, such as adjusting the fan (Dingus, Antin, Hulse, & Wierwille, 1989). Therefore, any display that requires more than four glances, or requires glances longer than 2 sec, would require more visual attention than a driver could safely allocate, in at least some circumstances.

Another variable that should influence when the information is displayed is the impact of any required manual input. Dingus, Antin, Hulse, and Wierwille (1988) found that the total required display glance time more than doubled when one or more button presses were required to access information from a moving map navigation system. In one circumstance, subjects had to perform one or more button presses to change the map zoom level for accessing the name of the next roadway along a route. Results showed that drivers' average total display glance time was 12.1 sec when button presses were required, compared to 4.6 sec when the information was immediately available. Zwahlen et al. (1987) studied the effects of increased workload while making control inputs. They measured how much centerline deviation resulted when drivers performed an input operation on a CRT touch screen display. The results showed that the control inputs increased the amount of time that a driver's eyes were fixed inside the vehicle, and they increased the chance of lane deviations large enough to cause an accident.

Recall that a zero-speed situation exists when the vehicle is stopped during the normal drive. It is impossible to evaluate all of the zero-speed conditions that exist in normal driving maneuvers. The vehicle may stop because of traffic congestion, traffic lights, or simply because the driver has elected to pull over and stop. The most common cause of a zero-speed event is a traffic signal. The average length of a red stoplight is approximately 20 sec. However, because vehicles may arrive when the light has been red for a while, one cannot assume that all drivers will have the full 20 sec. An estimate of the average stop duration at a red light is about 10 sec. This number is used as a criterion for allocating information to the zero-speed category, because in many circumstances it will result in successful retrieval. In addition, required retrieval times that are significantly less than 10 sec can be allocated to the in-transit case (i.e., four glance maximum × 2 sec per glance maximum = 8 sec).

As discussed earlier, some of the information presented to a driver is time sensitive. In order to accomplish the system's functional goal, some components of ATIS must present information to a driver while the vehicle is in motion. Navigation instructions, such as distance to the next turn and direction of the next turn, should be presented serially, giving the driver enough time to react and carry out each part of the instruction. It would not be feasible to wait until the vehicle is stopped to give the driver an instruction about an upcoming turn, because a series of turns might need to be accomplished before the vehicle encounters a stopping situation.

Like navigation instructions, most warning messages should also be presented immediately. Messages presented through the auditory channel are transient and give the driver only a single opportunity to grasp the information. A driver may be confused by a message or may miss all or part of the message because of noise. Therefore, if information is presented through the auditory channel while the vehicle is in motion, the driver needs the ability to repeat the message.

A major difficulty in developing the criteria for trip status allocations was the definition of an acceptable zero-speed task duration. This category is designed to allow for the display of more complex information without compromising the guidelines for visual attention demands. Some of the ATIS functions that drivers will desire or require during the drive will necessitate complex or detailed displays. Rather than reserving these functions for predrive situations, designers could make them available to drivers during zero-speed situations when visual attention is not required for the driving task. The 10-sec duration chosen for this decision aid was determined using what is currently known about the driving environment. The amount of useful time that drivers will normally have available during zero-speed situations is a possible topic for further research.

Another issue not addressed in the trip status allocation decision aid is individual driver differences. Numerous studies have shown age-related differences in driver performance and ATIS performance. In addition, performance differences will undoubtedly occur between commercial and private drivers because of differences in training, age range, and experience. In both of these cases, the data is insufficient to establish criteria differences based on population differences. Even though some quantifiable performance differences do exist, several issues of how to logistically allocate information features based on individual differences while maintaining safety still remain. These issues constitute a research gap that may need to be addressed in order to optimize the benefits of ATIS across a wide range of users.

Sensory Modality and Trip Status Trade Study Analyses
Results

The sensory allocation and trip status decision aids are used to analyze each individual information item and determine acceptable modes of presentation. To perform the analysis, each question on the decision aid tool must be answered and the proper branch of logic must be followed. While performing this analysis, it is possible that some of the decisions could take either available branch of logic at a decision point based on preconceived ideas the designer had about a display configuration. In cases such as these, it is important for a designer to take a close look at the decisions and ensure that when two different branches are feasible, they are both investigated. Delving into the decision aid logic at a deeper level may reveal which path is correct for a given information item. The end result may be a suggestion for a single sensory mode or a combination of two modes. If more than one sensory mode choice is suggested by the decision aid results, designers must weigh the positives and negatives of each choice and make a decision based on their engineering judgment.

Several of the information items reviewed could successfully take another branch of a given decision. For example, in cases where a decision is based on complexity, the information item could feasibly be displayed with high and low levels of complexity. An upcoming turn could be displayed as a simple verbal instruction such as "turn left on Colson Road," or as a complex verbal instruction such as "turn left 500 ft ahead at the stoplight from Giles Street to Colson Road." In this case, the designer must choose the branch that would most effectively support the goals of the overall system. A good general recommendation is to keep the presentation as concise and meaningful as possible.

As stated earlier, there may also be points in the decision aids where the designer will want to follow both branches of logic and make a note of any differences that are identified and under what conditions they are valid. For example, there is a question on the signing function decision aid that asks whether the information has been requested by the driver or if it will be displayed automatically. This decision could be answered either way, depending on the functional goals of the particular system. The designer should note the differences in the sensory mode that would be chosen and the conditions that make them valid. Later, when the display is actually being developed, the designer could make the proper sensory choice based on the result that is most applicable for the case being examined.

The results of analyzing the information items with the sensory allocation tools of this chapter show that using the auditory channel is desirable when the information is simple or there is urgency associated with the message. The results also show that information items containing routing instructions should be displayed using the auditory channel or a combination of the auditory and visual channels together. The motivation behind the use of the auditory channel is that the driver can be presented with useful information without compromising the visual attention required for driving.

The visual channel is the best choice when information is complex or spatial. Combinations of the auditory and visual channels can also be effective means of reducing postretrieval processing time, reducing errors, and providing a more salient display that can command the driver's attention.

After performing the analysis of the information items, there was some concern that the sensory allocation choices included at the end of each branch did not always contain all the feasible possibilities. For example, a branch may end with a box that suggests auditory presentation only, when in fact most designers could probably develop successful systems that do not use auditory presentation at all. The choices included were determined by examining the available research. These trade study tools are not designed to evaluate every display design possibility that could exist, but rather to suggest one or a combination that have been determined to be effective.

Display Format Trade Study Analysis Results

The output of the display format allocation tool is a recommendation for a display format or formats, based on the functional grouping and type of information identified in the sensory modality allocation analysis. A separate trade study matrix was developed for each of the information groupings. The recommended display format(s) are then provided on an information item-by-information item basis by simply selecting the appropriate matrix and assessing the scores provided.

The format trade study matrices were developed using the trade study techniques already discussed. The display format options were then ranked for each of the criteria. The individual scores were then multiplied by the appropriate weightings and totaled in order to generate a final ranking of the feasible display formats. Table 2.1 shows an example of this process. The final ranking suggests which is the preferred display format, given the information items within the functional grouping. For specific display modality recommendations, it would be best to consider the top two rated displays, possibly even the top three, if the total scores are within about 10% of one another. The rankings, as well as the weightings, are subjective and the measurement tool has a relatively large degree of variance associated with it. Therefore, close scores should not necessarily be seen as definitively different.

The following paragraphs discuss the general results of the trade study, as well as considerations and exceptions that were identified during construction of some matrices.

Route Planning and Coordination. The information types that are included within the route planning and coordination functions are intended to be performed in the predrive trip status. Because the driver will not be engaged in the driving task while using the system, the safety criteria were not included in making a choice of optimal display format.

In developing ratings for the format alternatives for the routing and coordination functions, it became clear that there were several areas requiring further research in order to provide definitive answers to format selection questions. Three-dimensional map displays have been mentioned in papers exploring alternative methods of displaying navigation information. However, there is little empirical research available to support judgments about the effectiveness of this type of display in meeting each criterion. Due to this, the rankings were based on the researchers' human factors judgment, given their knowledge of similar types of displays.

The sensory allocation decision aid suggested that the auditory channel could be used to display information in several routing and coordination situations. The use of simple tones for the presentation of information was not considered for routing and coordination functions because it does not provide context or transmit complex message information. Even simple information in this functional grouping cannot be effectively conveyed through the use of tones or other nonspeech auditory options.

For location, pathway, or position type information that was either in a complex or simple format, full or partial route video maps are the most desirable. Adding text is beneficial in complex situations where the information might not be fully understood with just a picture. The redundant or supplemental use of text will help to provide context to ease information transfer. In simple information situations, the use of speech improves the speed at which the supplemental information can be presented. Iconic representations of complex information were ranked low because of their inability to fully and efficiently convey the needed information.

For complex information that is status related, however, text and icons with text were rated highest. Because spatial elements are no longer included, the information can often be presented best with iconic information that includes a text description. Simple status information had three formats that were closely ranked: messages presented as speech, iconic information with speech, and text with speech.

Route Following. Route following functions are almost entirely performed during the in-transit trip status, so all the criteria were used when ranking the available display format options. The navigation information that is typically displayed by the route following function can be displayed on maps or graphic displays, through verbal instructions, or a combination of these two. In the case where a hard copy has been printed, it is assumed that the map or list was created during some predrive planning operation and would generally contain the same level of detail as the video display.

The route following function contains a large number of feasible options to choose from, making the task of ranking the displays very difficult. Choices were made for the different rank values by applying relevant results from previous research findings. No single study contains empirical results that cover the full range of display options. Therefore, extrapolations had to be made across differing experimental methods, thus providing a potential source of error in the rankings.

For complex navigation information requiring position information, the use of a partial-route video map was generally the format of choice based on the criteria used. As the information complexity decreased, the trade study results pointed to simpler display formats that used iconic representations of the information supplemented with speech or tones. For simple routing instruction information, the use of voice was ranked high (with respect to safety criteria) as an aid to keep the driver's attention focused on the driving task. The top two ranked formats for simple routing instructions included voice, but also required a visual element that could be useful for referencing the information in more detail or for preventing the voice message from being constantly repeated. Specifically, the presentation of simple routing instructions should include an iconic presentation with supplemental voice instructions. Visual text descriptions can also be used if the information does not lend itself well to iconic representation.

The rating of formats for discrete status information highlights the elimination of the need for maps. Complex information of this type should use iconic or graphical representations that include text descriptions. As the information becomes less complex, however, only icons or text are useful to convey the information and a combination of the two is not required.

There was no clear choice for continuously displayed status information. Text and iconic combinations, iconic or graphical representations, or text alone could all be used successfully if properly implemented. It is assumed that continuously displayed information would be very simple (i.e., a count of accumulated travel time). Such information could be handled by a text display that only contained a few digits.

Warning and Condition Monitoring. There are gradient levels of information priority that fall into the warning and condition monitoring functional group. High, medium, and low levels of priority will often change the basic format of the display based on the need to capture the driver's attention. This was taken into account for the ranking of the different display formats. Tactile displays were deemed appropriate as an option for this functional grouping. Tactile displays were grouped into vibration displays, such as shaking the gas pedal for a proximal incident warning display and control-change displays. Control-change displays were operationally defined as tactile or proprioceptive displays that changed the resistance or feel of a vehicle control. An example of a control-change display would be increasing the resistance of the gas pedal in a reduced-speed construction zone.

There is very little research that tests vehicle systems with control-change displays. This is simply because there are not many systems available to test. Where context is not required, the results of the trade study show that control-change is rated fairly close to audible warnings combined with an icon-type presentation as a desirable format option. However, a research study comparing the three options is required to validate this result.

Another area of unclear distinction is high priority displays requiring context information to be understood or acted on. An auditory modality presentation was recommended, which has only one feasible option (speech). However, the rankings between the three suggested that options for this case were very close. For a given design situation, therefore, the differences between the displays may not be all that significant. Further empirical research is needed to distinguish between the different auditory methods of high priority presentations where context is required.

Signing. The display of signing information will likely be a dynamic process that has the potential to provide the driver with a very active in-vehicle display. Therefore, it is anticipated that filtering of selected sign information will be required. Many of the more complex map displays were not considered as a presentation option for this information due to the amount and complexity of the sign information, especially when integrated with a map display. Partial maps could perhaps be used in some instances to provide integrated information.

When complex sign information is necessary for safe vehicle operation, it was clear that an iconic or graphical representation with supplemental voice was the preferred method of presentation, based on the criteria selected. For presentation of complex, requested informational messages, iconic representations with a voice description was the clear choice over other methods of presentation.

In the case of less complex requested informational messages, which are not related to safety, the use of an iconic presentation was rated high and approximately equal to speech alone. Therefore, using icons/tones for more frequently occurring information and voice for more infrequent and important events should be considered the standard in the case of complex sign messages. As the sign information becomes simple, the use of auditory information alone is recommended by the sensory allocation tool. A voice message with an alerting tone was the selected option. Note, however, that this recommendation must be limited to relatively high priority and low frequency information or the driver annoyance level will become high. Therefore, one design solution is to provide an auditory option that can be switched off if desired by the driver.

Communication and Aid Request. There is very little research available addressing message transfer in the vehicle environment. The formats selected as options for this trade study were deemed to be feasible based on comparable systems and the goals of this type of system. According to the results of the trade study, actual messages that contain critical information are best presented by text on a video screen with an alerting tone to let the driver know that a message is present. For lower priority messages, some standard information items could be relayed in the form of an icon with a text description provided on request.

The trade study found that for critical message events (e.g., "a message has been received"), text descriptions on a video screen should be considered, but with the inclusion of voice or tone aids. This will alert the driver to specific events and allow him or her to read about the message event.

Motorist Services. There was a wide variety of display format options to consider in the motorist services functional group. Many different types of maps were considered in addition to other visual display types. Combinations with audible displays were also considered. The format option that should be used can be highly situation specific. Therefore, due to the wide variety of differing information items within this functional grouping, the designer should take special care in applying these recommendations. Specific trade studies comparing systems at a finer sensory level may be necessary for motorist services information.

For complex, position-oriented motorist services information, three display methods should be considered: partial route maps with a text description, partial maps alone, or some type of iconic presentation with a text description. The choice between the three may be determined by the available screen size. This constraint was not considered by this trade study and may be a good topic for a comparison research study.

As the information becomes more simplified, the addition of voice or tones to an icon or partial route map may reduce the need to concentrate on the screen to receive the presentation. Once again, the choice of a display format may depend on the scale of the information being presented. If the information is related to the position of the vehicle in relation to streets, cities, and so on, then maps may be more useful than other options. In such cases, the information may be too complex for a display using icons.

Continuously displayed information involving status is best devoted to text presentations on a video screen and often as an alphanumeric display of low textual density. Intermittent information that is complex should be displayed graphically (when practical) with a textual explanation. As the intermittent information becomes simplified, the use of tones with icons or even tones alone should be considered.

CONCLUSIONS

These recommendations are general guidelines that do not apply to every information item present in an ATIS system. Thus, the guidelines must be applied with care, particularly when an individual requirement differs from the general trend of the rest of the requirements within a functional grouping. For example, in isolated instances, a route planning information item specifies an in-transit drive mode. The normal trip status for route planning is predrive. As a result, the decision aid was developed without the consideration of the safety criteria associated with motion of the vehicle, so adjustment in the decision aid (i.e., adding safety criteria) and recommended display format may be necessary.

Note that the guidelines often include highly functional ATIS systems for which both visual overload (due to safety) and auditory annoyance (due to frequency of potential messages) are of primary concern. In a lower functionality system, both of these concerns, as well as some of the trip status allocation issues, are less critical.

During predrive trip status, the sensory modality that the allocation tool selects is split between visual-only presentations and visual and auditory combinations. The complexity of information being presented is the key to this difference. Because the hazards of visual distraction and workload have been removed, there is little reason to limit the use of visual-only information. In route planning situations, some spatial relations between current location and desired destination are simply easier to show on a map than to explain in a verbal message. It is reasonable to plan a route before departure anyway, so the need to program the vehicle while stationary is not really a disadvantage. The visual component used in route planning is not always a map. On the contrary, text was frequently recommended to describe locations. There are a number of information items that are too specific to be handled by a graphic alone. In addition, a written explanation often helps users who are not spatially adept to plan and execute routes.

Once the destination has been determined and the driver begins the journey, the use of visual displays with no supplemental auditory information decreases in frequency. Because the driver must now devote a large amount of visual attention to driving, as well as vehicle navigation and guidance, the use of visual-only displays must be curtailed. Note, however, that auditory displays alone are not necessarily what the driver requires. The results show that having a visual component to which the driver can safely refer will often prevent the audible message from requiring repetition. The use of iconic representations of information with voice instructions for navigation will reduce the added visual attention required by maps. Another benefit of iconic representation is that it reduces the distraction potential generated by peripheral motion from a moving-route map.

There are a few areas where auditory-only displays may be considered, namely, message transfer situations and in-vehicle signing requested by the driver. Message transfer could potentially be very visually distracting to a driver. However, the length of a message will play a key role in the feasibility of visual presentation. A major problem with auditory presentation is the overload and chatter that integrated ATIS will likely generate. This will be of particular concern for higher frequency information, such as ISIS. If ISIS makes extensive use of the auditory channel with no option for selection, it will be annoying. Given the amount of sign information currently in the environment, the potential for overload and annoyance will be great if these subsystems are not properly integrated.

The motorist services and information systems grouping covered the most diverse range of trip modes. Unlike the other groupings, there was no one clear trip status within this category. Much information provided by this function will be valuable in-transit. However, it must be emphasized that the inclusion of desired features requires a global system perspective and significant attention paid to the possibility of overloading the driver. There are many visual-only sensory presentations that are recommended within IMSIS because the information is divided into small units. This information can be safely displayed to the driver through the use of graphic icons and simple maps or text. The danger comes not from the small informational units, but from the overload caused by having too many small units. There are many services that could be displayed, and drivers could easily request more information than they really should receive while driving. Some method of automatic filtering should be employed to prevent this potential hazard.

Warnings is a functional grouping that presents information to drivers while they are in-transit. The display formats tend to be auditory and include more tones than other functional areas. This is due to the simple nature of the information being presented. Often the driver needs only to react to some discrete event that does not require context information to be understood. Because of this, the use of tactile presentations is seen as a method of information transfer. However, this is an area that requires research and advancements in technology before it can be viewed as a serious alternative to simple auditory warnings.

REFERENCES

Dingus, T. A. (1987). *Attention demand requirements of automotive, moving-map navigation systems.* Unpublished doctoral dissertation, Virginia Polytechnic Institute and State University, Blacksburg, VA.

Dingus, T. A., Antin, J. F., Hulse, M. C., & Wierwille, W. W. (1988). Human factors issues associated with in-car navigation system usage: An overview of two in-car experimental studies. *Proceedings of the Human Factors Society 32nd Annual Meeting* (pp. 1448–1452). Santa Monica, CA: Human Factors Society.

Dingus, T. A., Antin, J. F., Hulse, M. C., & Wierwille, W. W. (1989). Attentional demand requirements of an automobile moving-map navigation system. *Transportation Research, 23A*(4), 301–315.

Dingus, T. A., & Hulse, M. C. (1993). Some human factors design issues and recommendations for automobile navigation information systems. *Transportation Research, 1C*(2), 119–131.

Fleischman, R., Carpenter, J., Szczublewski, F., Dingus, T., Krage, M., & Means, L. (1991). Getting information to the driver: Human factors in the TravTek Intelligent Vehicle Highway System (IVHS) demonstration. *Proceedings of the Human Factors Society 35th Annual Meeting* (pp. 1115–1119). San Francisco, CA: Human Factors Society.

French, R. L. (1990). In-vehicle navigation—status and safety impacts. *Technical papers from ITE's 1990, 1989, and 1988 conferences* (pp. 226–235). Institute of Transportation Engineers.

Huiberts, S. J. C. (1989). How important is mobile communication for a truck company? *Institute of Electrical and Electronics Engineers* (CH2789-6/89/0000-0361, pp. 361–364).

Labiale, G. (1989). Influence of in-car navigation map displays on driver performances. *SAE technical paper series* (No. 891683, pp. 11–18). Warrendale, PA: Society of Automotive Engineers.

Labiale, G. (1990). In-car road information: Comparisons of auditory and visual presentation. *Proceedings of the Human Factors Society 34th Annual Meeting* (pp. 623–627). Santa Monica, CA: Human Factors Society.

McKnight, J. A., & McKnight, S. A. (1992). *The effect of in-vehicle navigation information systems upon driver attention.* National Public Services Research Institute. Washington, DC: AAA Foundation for Traffic Safety.

Mollenhauer, M., Lee, J., Cho, K., Hulse, M., & Dingus, T. (1994). The effects of sensory modality and information priority on in-vehicle signing and information systems. *Proceedings of the Human Factors Society 38th Annual Meeting* (pp. 1072–1076). Santa Monica, CA: Human Factors Society.

National Highway Traffic Safety Administration (1991). *General estimates system 1991: A review of information on police reported traffic crashes in the United States.* U.S. Department of Transportation.

Parks, A. M., Ashby, M. C., & Fairclough, S. H. (1991). The effect of different in-vehicle route information displays on driver behavior. *Proceedings of the Vehicle Navigation and Information Systems Conference* (pp. 61–70). Warrendale, PA: Society of Automotive Engineers.

Stokes, A., Wickens, C., & Kyte, K. (1990). *Display technology: Human factors concepts.* Warrendale, PA: Society of Automotive Engineers.

Streeter, L. A., Vitello, D., & Wonsiewicz, S. A. (1985). How to tell people where to go: Comparing navigational aids. *International Journal of Man–Machine Studies, 22,* 549–562.

Wetherall, A. (1979). Short term memory for verbal graphic route information. *Proceedings of the Human Factors Society's 23rd Annual Meeting* (pp. 464–468). Santa Monica, CA: Human Factors Society.

Wickens, C. D. (1987). Information processing, decision-making, and cognition. In G. Salvendy (Ed.), *Handbook of human factors* (pp. 549–574). New York: Wiley.

Wierwille, W. W. (1993). Visual and manual demands of in-car controls and displays. In B. Peacock & W. Karwowski (Eds.), *Automotive ergonomics* (pp. 229–320). London: Taylor & Francis.

Zaidel, D. M. (1991). *Specification of a methodology for investigating the human factors of advanced driver information systems.* Ontario, Canada: Transport Canada.

Zwahlen, H. T., & DeBald, D. P. (1986). Safety aspects of sophisticated in-vehicle information displays and controls. *Proceedings of the Human Factors Society 30th Annual Meeting* (pp. 256–260). Santa Monica, CA: Human Factors Society.

Zwahlen, H. T., Adams, C. C., & DeBald, D. P. (1987, September). *Safety aspects of CRT touch panel controls in automobiles.* Paper presented at the Second International Conference on Vision in Vehicles, Nottingham, UK.

Chapter 3

A Functional Description of ATIS/CVO Systems to Accommodate Driver Needs and Limits

John D. Lee
Battelle Human Factors Transportation Center, Seattle

Advanced traveler information systems/commercial vehicle operations (ATIS/CVO) systems have the potential to dramatically change the task of driving and navigating vehicles. One reason for this dramatic change is the diversity of systems that can be classed as ATIS/CVO. These systems range from navigation aids to trip planning and hazard notification systems. Although technology exists to develop many of these systems, it is unclear whether the systems alone or in combination will enhance or degrade driver safety and efficiency. Understanding how intelligent transport systems (ITS) technology might be deployed to enhance driver performance requires both analytical and empirical research. This chapter takes an analytic approach to examining the functionality of ATIS/CVO systems and their implications for driver performance and acceptance.

Successfully integrating a diverse set of features so that they enhance rather than hinder drivers' abilities requires a functional description of the overall system. A functional description can be defined as a representation that addresses generic capabilities, requirements, and processes of a system; however, this representation does not specify physical configurations or mechanisms. Thus, a functional description explains how different system configurations may satisfy system requirements, but it does not identify the specific, physical mechanisms used to satisfy system requirements. Therefore, any number of combinations of physical mechanisms might combine to create a function. For example, a function such as route guidance can be delivered by mechanisms including voice, icons, or a highlighted route on an electronic map.

This chapter contains two important applications of a functional description. It describes human factors issues associated with ATIS/CVO functions. This chapter also uses the information flows derived from the functional description as input to a network analysis of the interactions among functions. In this way a functional description facilitates a driver-centered design philosophy that identifies system capabilities needed to make the system useful to the driver. This stands in contrast to a technology-centered design, which incorporates features that may be easy for the designer to develop, but may not make the system useful to the driver. Considering the functional requirements needed to achieve driver goals will help ensure a useful and accepted system. This chapter presents a functional description of ATIS/CVO that examines the range of ATIS/CVO functionality, potential human factors concerns, and the interactions among the functions.

The information used to develop the functional description of ATIS/CVO systems was collected from several sources, including a survey of current literature; a review of current technology presented at trade shows and conferences; interviews with industry and government representatives; and a series of three focus groups involving private and commercial drivers, dispatchers, industry representatives, and representatives of regulatory and enforcement agencies. The functional description included in this chapter summarizes a large amount of information drawn from a variety of sources, using a variety of methods. However, this functional description represents only a snapshot of the current and potential of ITS technology. The rapidly advancing nature of ITS ensures that any functional description will require continual assessment and revision to reflect the increasing range of functions ITS technology can support.

A DEFINITION AND FRAMEWORK
FOR THE FUNCTIONAL DESCRIPTION

One approach to defining the functions of an ATIS/CVO system is to describe the system at different levels of abstraction and aggregation (Frey, Rouse, & Garris, 1992; Rasmussen, 1986; Rasmussen, Pejtersen, & Goodstein, 1994). The dimension of aggregation specifies the level of detail or scope included in a system description. The dimension of abstraction specifies the relation between the purpose of the system and its physical implementation.

The dimension of abstraction forms an abstract–concrete continuum, with descriptions of system objectives being most abstract and specific physical configurations being most concrete. A very abstract description of an ATIS/CVO system would focus on why the system was constructed and would describe the system in terms of high-level performance objectives. A more concrete description would indicate how the system operates and describe the system in terms of the physical characteristics of specific devices. A description that includes several levels of abstraction helps to document the relations between the high-level objectives, the functions used to achieve those objectives, and the physical implementation that supports those functions.

A functional description that is linked to multiple levels of abstraction has several practical benefits. For example, a description of the system at higher levels of abstraction identifies the information flows and generic functions that comprise the system. Using these descriptions, a choice of specific physical components can be made with the knowledge of how the components will interact in the broader system context. Specifically, a component may be selected from several alternatives to support route guidance based on interactions with other functions.

The dimension of aggregation forms a part–whole continuum, with a description of the entire system considered as a single entity at one extreme, and a fine decomposition of component parts at the other. A description based on a fine decomposition of the overall system might include symbols used on an electronic map, whereas a description at a greater level of aggregation might address the characteristics of collections of many components, as in the comparisons of in-vehicle routing and navigation system (IRANS), in-vehicle motorist services information system (IMSIS), in-vehicle sign information system (ISIS), in-vehicle safety and warning system (IVSAWS), and CVO-specific subsystems of ATIS/CVO.

Describing a system using multiple levels of detail has several benefits. A description confined to a low level of aggregation, one that focuses on minute system components, may fail to address the overall system operation (fail to see the forest for the trees). Similarly, descriptions based solely on the higher levels of aggregation may fail to identify human limits concerned with details of the system (e.g., map symbols and location and size of individual buttons and knobs). Thus, describing the ATIS/CVO system at different levels of aggregation shows how details of specific components (e.g., illegible graphics on an electronic display) affect overall system performance, and how interactions between aggregates of system components affect system performance (e.g., IRANS interfering with IVSAWS, and IMSIS facilitating IRANS).

Figure 3.1 summarizes the distinctions between different levels of aggregation and abstraction by showing several different pictorial descriptions of a potential ATIS system. The representation at a high level of abstraction shows information flows that represent the relation between the driver and the ATIS, whereas low levels of abstraction show physical location and characteristics of the system. Viewed at a high level of aggregation, system components are combined and a broader representation of the system is shown. At a low level of aggregation, the picture shows details of individual components.

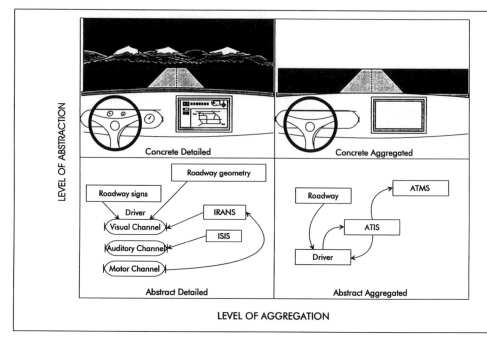

FIG. 3.1. A graphical depiction of a navigation system at different levels of aggregation and abstraction.

DEVELOPMENT AND APPLICATION OF THE ATIS/CVO FUNCTIONAL DESCRIPTION

The functional description presented here describes ATIS/CVO functions and links them to different levels of aggregation and abstraction. Each of these descriptions provides different information and insight into the functional coupling that defines the ATIS/CVO system. Figure 3.2 summarizes the different elements of ATIS/CVO at different levels of abstraction and aggregation. The focus here is on describing the functional characteristics. It begins with a context that relates ATIS/CVO features to objectives and the physical features of the system. This is followed by a summary of potential human factors issues and a network analysis of the information flows that link the ATIS/CVO functions.

Objectives

This level of system description addresses the purpose of the system (system objectives) in terms of the relative priorities of objectives and their interactions. The description of the system at this level identifies the relation

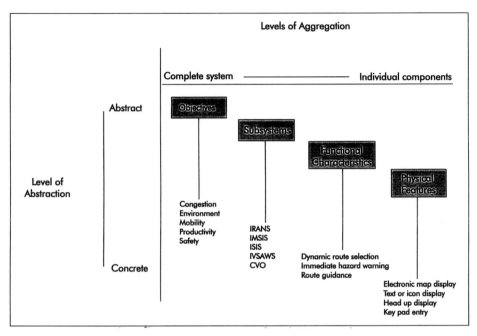

FIG. 3.2. A summary of system descriptions based on different levels of aggregation and abstraction.

between specific performance measures and the reasons for system design. As such, this description addresses properties of the environment with which the system interacts. In the case of the ATIS/CVO system, this includes the ability of the system to influence congestion, safety, mobility, environmental quality, and economic productivity.

Subsystems

Description of ATIS/CVO at this level differs from that based on the objectives in terms of the level of abstraction and aggregation; it describes the system in a more concrete and detailed manner. The description involves broad collections of functional capabilities required to meet system objectives. These collections of functional capabilities include routing and navigation, motorist information, in-vehicle signs, warning and safety information, and commercial vehicle-specific functions. These subsystems have been identified previously (IRANS, IMSIS, ISIS, IVSAWS, and CVO-specific functions). These categories of system functions describe the overall system through their interrelation, their relation with the objectives of the system, and their relation to a more detailed decomposition of system functionality.

Functional Characteristics

Description of ATIS/CVO at this level differs from the subsystems description along the dimension of aggregation. That is, functional characteristics are the elements that comprise each of the ATIS/CVO subsystems. The description at the level of subsystems provides little detail concerning specific functional capabilities, but a description of functional characteristics provides a detailed account of the various functions that may be available in ATIS/CVO systems. At this level, system description involves general system functions independent of any specific implementation; functions can be implemented in a variety of ways. For example, the function *route guidance* might be implemented using a detailed electronic map to display the intended route, voice commands, icons indicating turns, or simple depictions of the route direction relative to the surrounding streets; however, the description of the function is independent of these alternatives. A description of ATIS/CVO at the level of functional characteristics describes route guidance in terms of information flows, general decision-making processes, cognitive characteristics that may limit a driver's ability to work with the system, and interactions with other functions. Examples of subsystem functions include *route guidance, route planning,* and *vehicle condition monitoring.*

A total of 26 ATIS/CVO functions were identified. Each of these functions fall into one of five subsystems: in-vehicle routing and navigation system (IRANS), in-vehicle motorist information system (IMSIS), in-vehicle signing information system (ISIS), in-vehicle safety and warning systems (IVSAWS), and CVO-specific functions (see Table 3.1). A set of attributes that includes information flows, decision-making elements, human constraints, and interactions with other functions describe each of the 26 ATIS/CVO functions. See Lee, Morgan, Wheeler, Hulse, and Dingus (1993) for complete details.

Physical Features

Description of the ATIS/CVO at this level differs from the other descriptions in both the level of aggregation and abstraction. This level of description is the most concrete and the most detailed, because it identifies specific physical features of individual system components. Thus, system description at this level specifies the physical configurations of the system needed to support the functions. The description of physical features identifies the data requirements that underlie the information flows that support the functions. In this way, it can be used to identify the specific infrastructure requirements of the system. Furthermore, the description

TABLE 3.1
A Summary of ATIS/CVO Functional Characteristics

IRANS:	IVSAWS:
Trip planning	Immediate hazard warning
Multimode travel coordination and planning	Road condition information
Predrive route and destination selection	Automatic aid request
Dynamic route selection	Manual aid request
Route guidance	Vehicle condition monitoring
Route navigation	CVO-specific (cargo and vehicle monitoring)
Automated toll collection	
CVO-specific (route scheduling)	
IMSIS:	CVO-specific:
Broadcast services/attractions	Fleet resource management
Services/attractions directory	Dispatch
Destination coordination	Regulatory administration
Message transfer	Regulatory enforcement
ISIS:	
Roadway guidance sign information	
Roadway notification sign information	
Roadway regulatory sign information	
CVO-specific (road restriction information)	

of the system through its physical features identifies potential human needs and limits that depend on the particular interface characteristics, such as electronic maps, icons, and synthetic voice commands.

Table 3.2 illustrates three levels of this analysis by showing the relation between selected subsystems, functional characteristics, and physical features. Table 3.2 does not contain a comprehensive listing, but only a sample of the complete description. A broader description of ATIS/CVO, which links functions to objectives, subsystems, and physical features, is included in Lee et al. (1993). The remainder of this chapter focuses on ATIS/CVO functions and a network analysis of the information flows that link these functions.

HUMAN FACTORS ISSUES ASSOCIATED WITH ATIS/CVO FUNCTIONS

One useful application of the functional description of ATIS/CVO is to identify potential human factors issues that may affect these systems. The five subsystems (IRANS, IMSIS, ISIS, IVSAWS, and CVO) have been used to organize the following description of ATIS/CVO functions and their associated human factors issues.

TABLE 3.2
Physical Features Associated with a Sample of
Various Subsystems and Functional Characteristics

Subsystem	Functional Characteristic	Physical Features
IRANS	Route guidance	Directional guidance with icons
		Synthesized voice for turn-by-turn guidance
		Electronic map with highlighted route
	Predrive route and	Incident alert icons
	destination selection	Wake-up signal based on a.m. arrival time
		Real-time congestion data
		Smart card with driver preferences
IMSIS	Services/attractions	"Yellow pages" database touchscreen interface
	directory	Electronic map location display
		GPS position information for relative distance
IMSIS	Destination coordination	Parking availability data
		Text messages for restaurant reservations
		Real-time restaurant, hotel availability
		Travel time advice
ISIS	Roadway guidance sign	Smart card with driver preferences
	information	Head-up display
	Roadway regulatory sign	Icon on speedometer showing legal limit
	information	Head-up display
		Roadside beacon
IVSAWS	Immediate hazard warning	Tone, buzzer, or chime
		Digital voice
		Emergency vehicle broadcast beacon
	Automatic aid request	Roadside emergency request button
		Car theft tracking device
CVO	Fleet resource	Automatic vehicle location beacon
	management	Electronic logbook using smart cards
		Real-time two-way communication
	Regulatory administration	Vehicle classification using bar codes
		Electronic credentials
		Hazardous materials database

In-Vehicle Routing and Navigation Systems (IRANS)

IRANS provide drivers with information about potential destinations, travel modes, and directions to guide them from one place to another. When integrated with traffic management centers, IRANS provide information on recurrent and nonrecurrent traffic congestion, and is capable of calculating, selecting, and displaying optimum routes based on real-time traffic data. Table 3.3 provides a summary and definition of IRANS functions.

The most salient human factors concerns associated with the IRANS functions include factors influencing driver interpretation and attitudes regarding route and navigation information. Driver reaction to inaccurate

TABLE 3.3
ATIS/CVO Functions Associated with IRANS

IRANS

Trip planning—concerns the coordination of long, multiple-stop/destination journeys. Coordination of these journeys may involve identifying routes based on scenery, historical sites, hotel accommodations, restaurants, and vehicle services.

Multimode travel coordination and planning—facilitates decisions regarding different modes of transportation (e.g., buses, trains, and subways) that offer alternatives to single mode transportation. This information might include real-time updates of actual bus arrival times and car pool opportunities.

Predrive route and destination selection—facilitates destination and route selection choices while the car is parked. These choices include entering and selecting the destination, selecting a departure time, and a route to the destination. The information provided by this system might include real-time or historical congestion information, estimated travel time, and routes that optimize travel on a variety of parameters.

Dynamic route selection—encompasses any route selection capabilities that the driver engages in, or those that engage automatically while the car is moving. Dynamic route selection includes presentation of updated traffic and incident information that might affect driver route selection.

Automated toll collection—would allow a vehicle to travel through a toll roadway without the need to stop for tolls. Tolls would be deducted from the driver's account automatically as the vehicle passes toll collection areas. This will facilitate traditional toll collection and congestion pricing schemes.

CVO-specific (route scheduling)—coordinates numerous destinations to minimize travel time or to minimize lateness on deliveries. This function takes the driver's destinations as input and provides an optimal order for traveling between destinations. Although primarily an aid to commercial drivers, private drivers can also benefit by using this function to coordinate multiple errands or appointments.

or counterintuitive information may be a strong determinant of driver acceptance and use of many IRANS functions. Driver attitudes regarding the capabilities of the computer might lead the driver to reject computer-generated trip plans even when they are appropriate (Green & Brand, 1992; Rillings & Betsold, 1991). Drivers' ability to understand and respond to route guidance information is another important concern. Specifically, the functions of route guidance and route navigation require the system to present drivers with information that can be quickly assimilated so that it does not interfere with the primary task of driving.

In-Vehicle Motorist Services Information Systems (IMSIS)

IMSIS provide motorists with electronic yellow pages, commercial logos and signing for motels, eating facilities, service stations, and other signing displayed inside the vehicle directing motorists to recreational areas and

TABLE 3.4
ATIS/CVO Functions Associated with IMSIS

IMSIS

Broadcast services/attractions—provides travelers with information that might otherwise be found on commercial roadside signs. This information may be tailored to drivers' particular needs by using the ATIS to filter the information based on a profile of the driver's interests.

Services/attractions directory—provides information similar to that contained in a yellow pages directory with the flexibility of a computer database to facilitate a wide variety of search options. The information provided by this function may span a wide spectrum, from location of the nearest gas station to the cost of meals at the area's most authentic Indian restaurant.

Destination coordination—passes information between the vehicle and the final destination. This may include restaurant and hotel reservations or information about parking availability and location.

Message transfer—facilitates drivers' ability to communicate with others. Currently this function is accommodated with cellular telephones and CB radios; however, future ATIS systems may improve upon this technology by automatically generating preset messages at the touch of a button and receiving messages for future use.

historical sites. Table 3.4 provides a summary and definition of IMSIS functions. Examining IMSIS functions helps to identify several important human factors concerns. Most importantly, several functions require the division of attention between the primary task of driving and attending to an in-vehicle display that presents potentially extraneous information. With *broadcast services/attractions,* drivers could be exposed to substantial amounts of distracting information that would need to be filtered by the driver to identify useful information. Depending on the density of broadcast information, this could be an arduous task. To lighten the attentional demands, these systems could be designed to minimize extraneous information by filtering the information using a profile of driver needs and preferences.

In-Vehicle Signing Information Systems (ISIS)

ISIS provide noncommercial routing, warning, regulatory, and advisory information that is currently depicted on external roadway signs. ISIS functions are distinguished from IVSAWS by the relative permanence of the information displayed by this system. ISIS provide information that could be displayed on permanent roadway signs (not including changeable message signs). Table 3.5 provides a summary and definition of ISIS functions.

The most significant human constraint concerning ISIS functions is a tendency for ISIS to focus driver attention to in-vehicle displays and away

TABLE 3.5
ATIS/CVO Functions Associated with ISIS

ISIS
Roadway guidance sign information—brings information normally found outside the vehicle (e.g., street signs, interchange graphics, route markers, and mileposts) into the vehicle and displays it to the driver. Supplementary guidance information might include route-specific signs that are tailored to a specific driver's journey.
Roadway notification sign information—notifies drivers of potential hazards or changes in the roadway. This information will include merge signs, advisory speed limits, chevrons, and curve arrows. In addition, notification information may include temporary or dynamic notification information such as road closures, road maintenance, or road construction.
Roadway regulatory sign information—provides regulatory information through in-vehicle displays. Examples of this information include speed limit signs, stop signs, yield signs, turn prohibitions, and lane use control (e.g., left turn only).
CVO-specific (road restriction information)—provides CVO-specific regulatory sign information (e.g., truck speed limits, weight limits). In addition, regulation information influences commercial vehicle routes (e.g., avoidance of no-truck routes).

from the roadway. The ease of processing ISIS information may compensate for this shift in attention. In particular, ISIS displays will not be subject to environmental factors (rain, snow, and fog) that can obscure roadway signs. However, a greater proportion of the driver's attention will now be in-vehicle, potentially leaving insufficient attention for environmental scanning. The potential distractions associated with the in-vehicle display could further reduce attention to external events (potentially during inopportune circumstances).

In-Vehicle Safety Advisory and Warning Systems (IVSAWS)

IVSAWS provide warnings of unsafe conditions and situations affecting the driver on the roadway ahead. IVSAWS provide advanced warnings that give the driver time for remedial actions. IVSAWS messages are often related to relatively transient conditions, requiring modifications to the messages at irregular intervals. *Manual* and *automatic aid request* (Mayday) systems have been subsumed under IVSAWS for the purposes of the present discussion. IVSAWS do not encompass in-vehicle warnings of imminent danger requiring immediate action (e.g., collision avoidance devices). Table 3.6 summarizes IVSAWS functions.

Because IVSAWS often present time critical warning information, icons (both auditory and visual) will be obvious display options. The facile interpretation of these icons presents a human factors challenge, as the driver's ability to respond to warning signals governs the performance of

TABLE 3.6
ATIS/CVO Functions Associated with IVSAWS

IVSAWS

Immediate hazard warning—provides information regarding immediate hazards. This information may include the relative location of a hazard and the type of hazard. This information might also include warning the driver of an accident immediately ahead, the approach of emergency vehicles, or even of a stopped school bus. Thus, this information focuses on the location of specific localized incidents.

Road condition information—provides information that may include traction, visibility, congestion, construction activity, or weather conditions. Compared to the information conveyed by the immediate hazard information system, this function provides general information that could cover a wider geographic area and a longer time span.

Automatic aid request—provides a "mayday" signal in circumstances requiring emergency response. This signal will activate in circumstances where manual aid requests are not possible and where immediate response is essential (e.g., severe accidents). The signal will provide location information and could provide severity information to the emergency response personnel.

Manual aid request—allows the driver to request emergency services without leaving the vehicle. This function provides the driver with immediate access to a wide range of roadside assistance (e.g., police, ambulance, towing, and fire department) without the need to locate a phone, know the appropriate phone number, or even know the current location. This function might also include feedback after it has been used to notify the driver of the status of the response (e.g., expected arrival time).

Vehicle condition monitoring—is used to track the overall condition of the vehicle and to inform the driver of current and potential problems. Vehicular monitoring might range from reminding the driver to perform certain services (e.g., oil change) to warning the driver about current problems (e.g., engine overheating or broken fan belt). This system could also be interactive and allow the driver to interrogate the system to obtain more details about the situation.

CVO-specific (cargo and vehicle monitoring)—goes beyond *Vehicle condition monitoring*, because a commercial vehicle might have more detailed and diverse information that might include a more precise indication of engine performance. Also, commercial vehicles carry sensitive cargo, which requires careful monitoring of temperature, humidity, and vibration. Furthermore, this information may need to be distributed to dispatchers and company managers, where it could help coordinate maintenance and ensure the timely delivery of goods.

these functions. Driver attitudes may play an important role as well. If the computer has supplied erroneous information in the past, the driver may disregard the computer's information as unreliable (Lee & Moray, 1992; Sorkin, Kantowitz, & Kantowitz, 1988). Likewise, warning adherence and risk perception may also influence whether drivers respond to warning information. Unless these systems are introduced while considering the broader issues of driver acceptance, drivers may sabotage the system and undermine its potential.

Commercial Vehicle Operations (CVO)

CVO includes fleets of trucks, buses, vans, taxis, and emergency vehicles. Potential functions include both in-vehicle and out-of-vehicle CVO applications. In-vehicle CVO applications include the use of the basic ATIS subsystem capabilities and additional supporting capabilities specific to CVO requirements. Out-of-vehicle applications center on various support functions for commercial vehicle dispatchers and regulatory personnel. Table 3.7 provides a summary and definition of CVO-specific functions.

Several CVO-specific functions address the coordination of large and diverse numbers of resources in a dynamic and uncertain environment. These demands stretch the capabilities of human decision making. As such, decision-making biases may limit human effectiveness (Fischhoff, 1982; Kahneman, Slovic, & Tversky, 1982). Using computerized management support systems may help mitigate these biases. To be successful, designers will need to convey the capabilities and the inherent limitations of CVO-specific functions to the users. The *regulator enforcement* function requires personnel to review a complex set of criteria and quickly arrive at a decision. Therefore, an important consideration is how regulatory information can be displayed to enforcement personnel so that they can evaluate the information effectively and decide whether to stop the vehicle.

TABLE 3.7
CVO-Specific Functions

CVO-Specific

Fleet resource management—facilitates the coordination of resources (e.g., people, money, and equipment) within the company by enhancing the communication between the dispatchers, the drivers, and company management. As such, this function would allow the company to know the status of vehicles (e.g., location, condition of cargo, and maintenance needs). This function has particular importance in coordinating public transport where it is necessary to provide continuous service to customers, while maintaining the lowest possible rates.

Dispatch—coordinates individual CVO activities, whether they are the movement of a shipment from one part of the country to another, or the response of an emergency vehicle to an accident location. The dispatch function directly supports the vehicle operator and serves as the liaison between the driver and customer.

Regulatory administration—facilitates compliance with various regulatory administrative requirements, including taxes, licenses, permits, and coordinating the transport of hazardous material. In addition, this function may be involved with checking the required training programs and other administrative functions required of a CVO company by law.

Regulatory enforcement—facilitates the enforcement of various regulatory and legal requirements to which commercial drivers must adhere. These include traffic enforcement, marking of hazardous cargo, vehicle condition, driver fitness for duty, and driver operating hours. This function also supports *Regulatory administration* by ensuring compliance with licenses, permits, and payment of road and fuel taxes.

A likely scenario uses leading indicators (e.g., mileage since last stop, cargo type, gross vehicle weight, and historical experience with the carrier) to reduce enforcement personnel workload.

NETWORK ANALYSIS OF FUNCTIONAL RELATIONS

Beyond identifying general human factors concerns, another product of a functional description is a catalog of information flows that link ATIS/CVO functions. The information flows are important because they identify links that must be established between system components. To establish these links the driver must transform, code, and enter information. Alternatively the designer must integrate the functions so that the information flows from one function to another automatically. Whether functions are linked automatically, or by driver actions, can have a large effect on driver tasks, workload, and system usability. Systems that force drivers to enter information because functions are poorly integrated could degrade safety and undermine driver acceptance. In particular, a driver entering complex information when the vehicle is in motion presents a significant safety hazard. As an example of typical information flows, Fig. 3.3 shows information flows associated with the *immediate hazard warning* function. One

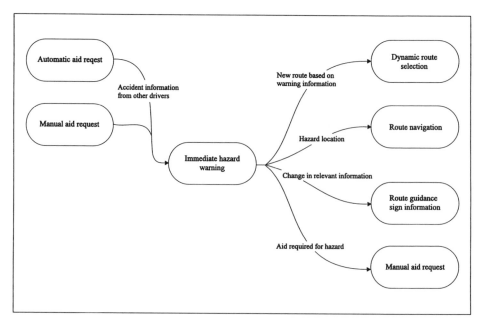

FIG. 3.3. Information flow between "Immediate Hazard Warning" and other ATIS/CVO functions.

purpose of analyzing information flows is to guide designers to integrate groups of functions that are tightly linked by information flows.

Examining links between functions can help focus a task analysis. A detailed task analysis of each ATIS/CVO function and their potential interrelations would generate an enormous amount of data that could obscure important relations between ATIS/CVO functions and associated tasks. To avoid obscuring these relations and to focus the task analysis, an innovative method was used to identify central functions and their critical interactions. Network analysis was used to analyze interrelations by examining the information flows that link ATIS/CVO functions. This method, routinely used to examine social networks, provides a systematic tool to guide the task analysis of a system (Borgatti, Everett, & Freeman, 1992). This approach enables the analyst to quantitatively determine which functions are the most central, thus reducing the scope of the analysis while enhancing its value. This method also identifies meaningful clusters of functions that are linked by information flows, which should be considered as a whole in the task analysis. Thus, network analysis provides a quantitative evaluation of information flows that focuses a task analysis on important functions and their critical interactions.

Using Quantitative Methods to Go Beyond Graphical Analysis of Information Flows

Graphical and quantitative network analysis techniques support analysis of relations in social networks and groups. Similarly, human factors professionals use graphical network representations when performing task and system analyses. Frequently, these relations are characterized by link diagrams, task network structures, and flow charts. Although graphical representations of this information are useful, complex relations may make intuitive analyses of these representations difficult. For example, in a complex and extensive system, such as ATIS/CVO, a flow chart representation of one particular subsystem might be relatively easy to understand and conceptualize. However, when many subsystems are joined, the resulting flow chart can become excessively complex, and critical links or functions may be overlooked.

Just as social scientists use network analysis techniques to examine interactions among people, human factors professionals can use network analysis to examine complex systems. The matrix algebra formalisms of network analysis provide objective and repeatable results (e.g., Luce & Perry, 1949). These methods quantify the grouping structures present in a complex network that do not appear obvious because of system complexity. Network analysis highlights important functions and groupings embedded in the overall representation.

Information flows between functions of the ATIS/CVO systems provided the data for this analysis. Using a description of each function, it was possible to identify the network of information flows that link the ATIS/CVO functions (Lee et al., 1993). Figure 3.3 shows the information flows associated with one ATIS/CVO function, immediate hazard warning. Similar descriptions of information flows associated with the other 25 functions were combined for this analysis. These functions, and the information flows that connect them, can be considered as a network, with functions (boxes) representing the nodes of the network and information flows (arrows) representing the links between nodes. This network can also be considered as a matrix of input–output relations. Such a matrix was created to summarize the information flows between all the ATIS/CVO functions.

Network Analysis Techniques

Network analysis techniques can generate two types of information: measures of *centrality* and measures of *clustering*. Measures of centrality can be defined in various ways and no single definition exists. Intuitively, it is accepted that, in social networks, the person with the most direct contacts with others is in the most central position. A person who is strategically located on the communication paths that link others, has also been suggested to be a central element (Freeman, 1979). In other words, measures of centrality indicate which elements have the most direct contact with other elements, or which elements have the greatest number of paths between themselves and other elements. Various measures of centrality exist and can be calculated using frequency counts of information flows entering and leaving a node, or using more complicated metrics of information transmission based on the overall network (Freeman, 1979).

Measures of clustering identify group structures (i.e., groups of tightly coupled functions). Several measures exist; however, only two are addressed here. One measure groups the functions into *cliques*, and the other groups them into *clusters*. Each measure uses different criteria to group functions, but they provide converging measures that reflect how information flows link functions. A clique represents a formal description of the density of links between network nodes. With a clique, each member must have at least as many connections to other functions in the clique as there are functions in the clique. Specifically, in a clique of three nodes, each node must have links to each of the other nodes in the clique, plus any links to other nodes not part of the clique. Thus, a clique shows a group of functions that share information directly between each other. Clusters do not have the same strict, formal definition that a clique does. A cluster analysis optimizes a cost function that is based on the extent that a group of functions consist of linked cliquelike structures (Borgatti et al., 1992).

Thus, the cluster analysis identifies groups of functions in a manner vaguely analogous to factor analysis in traditional statistics.

Results of the Network Analysis of ATIS/CVO Functions

Centrality Analysis. A simple measure of the importance of a function is the number of times it either receives from or provides information to other functions. Adding the row and column totals in a matrix provides a crude estimate of function centrality. However, it is possible to go beyond this simple count and use measures that provide a more accurate estimate of centrality. For example, simple counts estimate centrality by considering the input and output of each function separately. In many cases, this may not be appropriate. For instance, the output from a function that has nine inputs should be weighted more heavily than from a function that has only one input. To overcome the limits of simple counts, a network analysis measure was used that normalized counts based on the number of input and output flows.

Freeman's measure of centrality estimates the centrality of the functions based on their output and input (Freeman, 1979). This estimate of centrality reflects the percentage of adjacent vertices to a given vertex relative to the maximum possible vertices. Table 3.8 summarizes this measure of

TABLE 3.8
Centrality Measures for ATIS Private Driver Functions

ATIS/CVO Function			Output Centrality	Input Centrality
IRANS	1.1	Trip Planning	16	44
	1.2	Multimode Travel Coordination	27	16
	1.3	Predrive Route and Destination Selection	27	38
	1.4	Dynamic Route Selection	33	27
	1.5	Route Guidance	16	0
IRANS	1.6	Route Navigation	*44*	11
	1.7	Automated Toll Collection	0	16
IMSIS	2.1	Broadcast Services/Attractions	16	16
	2.2	Services/Attractions Directory	11	22
	2.3	Destination Coordination	*44*	5
	2.4	Message Transfer	16	0
ISIS	3.1	Roadway Sign Guidance Information	27	11
	3.2	Roadway Sign Notification Information	16	5
	3.3	Roadway Sign Regulatory Information	0	0
IVSAWS	4.1	Immediate Hazard Warning	11	22
	4.2	Road Condition Information	16	*50*
	4.3	Automatic Aid Request	5	11
	4.4	Manual Aid Request	16	22
	4.5	Vehicle Condition Monitoring	0	27

centrality and shows that destination coordination and route navigation are very central in terms of the inputs they receive from other functions, whereas road condition information is very central based on the input it provides to other functions.

Focusing a task analysis is an important practical benefit of the centrality measures. Specifically, this analysis indicated that route navigation and destination coordination deserve particular scrutiny concerning tasks involving entering information into the system. In contrast, road condition information deserves analysis to ensure that the information it generates gets distributed to the other functions without stretching drivers' abilities. By identifying central functions, this analysis can focus a task analysis on important elements of the system. Specifically, by highlighting central functions, this analysis helped guide the ATIS task analysis so that it captured a representative set of tasks associated with the functions critical to system success. This analysis identified several highly central ATIS functions that might have otherwise been obscured by the system's complexity before network analysis.

Grouping Analysis. In addition to estimates of function centrality, an estimate of how information flows link functions into groups was performed. This information is valuable because it shows which functions, or groups of functions, share information. Because the complexity of large systems may make flow charts and other graphical representations of information flow uninterpretable, a mathematical procedure for identifying groups is valuable.

Several measures of functional grouping exist and each captures a slightly different element of the relation among functions. A clique is a simple measure that is defined as a group of functions where each function is directly linked to all the other functions in the clique. For the private driver applications of ATIS, the network analysis identified 20 cliques of three functions or more, as shown in Table 3.9. The analysis of cliques and clusters have several practical applications. In particular, these measures can be used with the measures of centrality to guide a task analysis. Because the information flow helps specify the drivers' role within the system, the task analysis should consider these groups of functions as a primary point of analysis. Another useful measure of the links among functions is a cluster analysis. Cluster analysis generated four clusters, as shown in Table 3.10. A task analysis not considering these groups will fail to address important interactions among functions and the need to support drivers who may need to integrate information from several functions.

The results of the network analysis can be used to address important human factors issues, such as helping designers integrate related functions and focusing a task analysis. This chapter has outlined an analysis that

TABLE 3.9
Cliques for ATIS Private Driver Functions

Clique	Functions (from Table 3.8)				
1	1.1	1.2	1.3	1.6	2.2
2	1.1	1.2	1.3	1.6	4.2
3	1.1	1.3	1.6	2.1	
4	1.1	1.3	1.6	3.1	4.2
5	1.1	1.3	1.6	3.2	
6	1.1	1.2	1.3	1.7	
7	1.1	1.2	1.3	2.3	4.2
8	1.1	1.3	2.1	2.3	
9	1.3	1.5	3.1		
10	1.4	1.6	3.1	4.1	
11	1.4	1.6	3.1	4.2	
12	1.4	1.6	2.1		
13	1.2	1.4	1.6	4.2	
14	1.4	1.6	3.2		
15	1.4	1.5	3.1		
16	1.2	1.4	2.3	4.2	
17	1.4	2.1	2.3		
18	1.4	2.1	4.5		
19	2.3	2.4	4.2	4.4	
20	2.2	4.4	4.5		

TABLE 3.10
Cluster Analysis for ATIS Private Driver Functions

Clusters	Functions		ATIS/CVO Systems
1. Planning and Navigation	1.1	Trip planning	IRANS
	1.2	Multimode travel coordination and planning	IRANS
	1.3	Predrive route and destination selection	IRANS
	1.4	Dynamic route selection	IRANS
	1.6	Route navigation	IRANS
	3.1	Guidance sign information	ISIS
	4.2	Road condition information	IVSAWS
2. Miscellaneous Functions	This cluster is not meaningful. It is composed of two functions and each interacts only with itself.		
3. Aid & Emergency Services	2.2	Services/attractions directory	IMSIS
	4.1	Immediate hazard notification	IVSAWS
	4.3	Automatic aid request	IVSAWS
	4.4	Manual aid request	IVSAWS
	4.5	Vehicle condition monitoring	IVSAWS
4. Travel Coordination	2.1	Broadcast services/attractions	IMSIS
	2.3	Destination coordination	IMSIS
	2.4	Message transfer	IMSIS

helped guide a task analysis of ATIS/CVO functions. In general, network analysis of the information flows that link ATIS/CVO functions can provide a quantitative approach for selecting functions, and their combinations, for a detailed task analysis. Measures of centrality identify important functions that are critical to information flow through the network, thereby reducing the complexity of the task analysis by focusing on important functions. Network analysis also identifies groups of functions, linked by information flows. Task analysis should address these groups to capture important interactions between functions.

By adopting network analysis techniques, it was possible to focus the ATIS/CVO task analysis on central functions and on critical groupings of these functions. The centrality analysis identified highly central functions for which a detailed tasks analysis was performed, whereas the cluster analysis identified groups of functions that should be considered as integrated units in the ATIS design. For example, as part of the task analysis, scenarios were conceptualized to illustrate how future drivers interact with groups of functions identified with the network analysis. Thus, the findings from the network analysis identified meaningful combinations of functions for these scenarios that would have been overlooked by traditional analysis.

In summary, network analysis has the advantage of illuminating important information that may be obscured by numerous interconnections that are shown in flow charts or link diagrams. However, the network analysis requires an adequate listing of functions and information flows linking these functions. Without an accurate representation of the system's functions and their interrelations, results produced by network analysis are meaningless. If these basic concepts are clearly defined, network analysis will capture highly coupled groups of functions and reveal particularly important functions. Detecting functions central to ATIS/CVO, and identifying those that form highly coupled groups, provides a strong basis for a task analysis. A task analysis focused by network analysis will provide a more accurate description of cognitive demands and provide better design guidance.

CONCLUSIONS

The functional analysis of ATIS/CVO provides a starting point for considering the task demands associated with accessing and responding to ATIS/CVO information. A qualitative analysis of ATIS/CVO functions identifies potential human factors issues associated with future systems. The functional analysis can also support quantitative analyses, such as a network analysis of information flows. The network analysis techniques described in this chapter show how a quantitative analysis of information flows among functions can help focus a more detailed analyses, such as task analysis.

Because the function analysis provides a description independent of any particular physical implementation, it provides useful insights that are independent of any particular system, as demonstrated by the discussion of human factors issues and network analysis. The multilevel approach to a functional description places the functional description in the context of system objectives and the concrete physical features of potential systems. This highlights the variety of design alternatives. This may provide a useful representation of the designer's problem space, which could be a basis for the effective presentation of human factors design guidelines. Thus, the functional description can provide a common language that links an engineering description of ATIS/CVO with a psychological description of driver needs, limits, and capabilities.

ACKNOWLEDGMENTS

This research was supported by contract DFTH61-92-C-00102 from the U.S. Federal Highway Administration. I would like to thank Mireille Raby, William Wheeler, Thomas Dingus, and Mellisa Hulse for their contributions to the research referred to in this chapter.

REFERENCES

Borgatti, S. P., Everett, M. G., & Freeman, L. C. (1992). *UCINET IV version 1.0 reference manual.* Columbia, SC: Analytic Technologies.

Fischhoff, B. (1982). Debiasing. In D. Kahneman, P. Slovic, & A. Tversky (Eds.), *Judgement under uncertainty: Heuristics and biases* (pp. 421–444). New York: Cambridge University Press.

Freeman, L. C. (1979). Centrality in social networks: Conceptual clarification. *Social Networks, 1,* 215–239.

Frey, P. R., Rouse, W. B., & Garris, R. D. (1992). Big graphics and little screens: Designing graphical displays for maintenance tasks. *IEEE Transactions on Systems, Man, and Cybernetics, 23,* 10–20.

Green, P., & Brand, J. (1992). *Future in-car information systems: Input from focus groups* (SAE Tech. Paper Series SAE No. 920614, 1-9). Warrendale, PA: Society of Automotive Engineers.

Kahneman, D., Slovic, P., & Tversky, A. (1982). *Judgement under uncertainty: Heuristics and biases.* New York: Cambridge University Press.

Lee, J., & Moray, N. (1992). Trust and the allocation of function in the control of automatic systems. *Ergonomics, 35*(10), 1243–1270.

Lee, J. D., Morgan, J., Wheeler, W. A., Hulse, M. C., & Dingus, T. A. (1993). *Development of human factors guidelines for Advanced Traveler Information Systems and Commercial Vehicle Operations. Task C working paper: Description of ATIS/CVO functions.* Seattle, WA: Battelle Seattle Research Center.

Luce, R. D., & Perry, A. D. (1949). A method of matrix analysis of group structure. *Psychometrika, 14,* 95–117.

Rasmussen, J. (1986). *Information processing and human–machine interaction: An approach to cognitive engineering.* New York: North-Holland.

Rasmussen, J., Pejtersen, A., & Goodstein, L. P. (1994). *Cognitive systems engineering.* New York: Wiley.

Rillings, J. H., & Betsold, R. J. (1991). Advanced driver information systems. *IEEE Transactions on Vehicular Technology, 40*(1), 31–40.

Sorkin, R. D., Kantowitz, B. H., & Kantowitz, S. C. (1988). Likelihood alarm displays. *Human Factors, 30,* 445–459.

Wheeler, W. A., Lee, J. D., Raby, M., Kinghorn, R. A., Bittner, A. C., Jr., & McCallum, M. C. (1993). *Development of human factors guidelines for Advanced Traveler Information Systems and Commercial Vehicle Operations. Task E working paper: Task Analysis of ATIS/CVO functions.* Seattle, WA: Battelle Seattle Research Center.

Chapter 4

Analysis of Driving a Car With a Navigation System in an Urban Area

Motoyuki Akamatsu
Matsutaro Yoshioka
National Institute of Bioscience and Human Technology, Tsukuba, Japan

Nobuhiro Imacho
Public Works Research Institute, Tsukuba, Japan

Tatsuru Daimon
Hironao Kawashima
Keio University, Yokohama, Japan

To arrive at a destination, a driver requires information such as the current location of the vehicle and the route from the current location to the destination. When driving in an unfamiliar area, it is thought that the driver arrives at the destination by recognizing the current location and following the directions to the destination, with the aid of both maps and road signs. An onboard navigation system may facilitate these tasks.

In Japan, more than 20 different types of car navigation systems are now on the market. Some are built-in by the car manufacturer (factory fitted), whereas others can be bought in shops and installed by the car owner (aftermarket). Most systems show the destination, current location of the car, and distances and directions to the destination on a digital map; some also provide recommended routes.

Several aspects of ergonomic interface design are of particular interest, including the optimal placement of the display, the color of display markings, and the size of character that best facilitates navigation while minimizing distraction. One important consideration is the kind of information that should be presented on the display. There have been several studies on interface design for car navigation systems that have examined the effectiveness of the system or the manner in which information is presented to the driver (Faeber & Popp, 1991; Labiale, 1989, 1990; Parkes, 1989; Streeter, Vitello, & Janssen, 1985). Most of these studies were conducted on highways or low traffic roads; however, the car navigation system can also be useful for driving in large cities with complex road networks. Although driving simulators can be used to replicate driving in an urban environment, their ability to reproduce complex traffic conditions and road environments is limited. Consequently, field experiments were conducted to explore driver behavior and the processing of information when navigation systems are used in real urban areas.

Sufficient information by which to infer cognitive processes cannot be obtained by simply observing a driver's behavior, so a method based on subjective observation—called the *thinking-aloud* or *verbal protocol method*— can be used to gather additional data in order to better understand the cognitive process. With the thinking-aloud method all thoughts are vocalized, and the comments are recorded while the subject performs the task (Anzai, 1984). In driving research, the verbal protocol is obtained from the driver while driving. Although several studies have used this method, there are no standardized procedures for collecting and analyzing the data (Noy & Zaidel, 1991). Schraagen (1993) used this method to analyze cognitive processes during navigation in a city using a paper map, and categorized the verbalized words into several areas of knowledge. Such categorization is the simplest and most common way of analyzing the data obtained by this method.

In this study, driver behavior when using a navigation system in the central area of Tokyo was recorded by means of small video cameras, and the landmark information used by drivers was analyzed using the thinking-aloud method. In the analysis, verbalized words were categorized into several types of landmark information.

METHOD

Experiment

Eight cars equipped with navigation systems were used in this experiment, which was conducted in the central area of Tokyo. The distance from the starting point to the final destination was about 2.75 km, and three subdestinations to large, well-known buildings were selected by the experimenter. The traffic was heavy, and there were many buildings of varying sizes along two- or three-lane streets (Fig. 4.1). The experiments were carried out between 10 a.m. and 5 p.m. on weekdays.

The navigation task consisted of driving the car from the starting point to each of the subdestinations and then to the final destination. The drivers were shown the locations of the subdestinations on a paper map just prior to the start of the trip and were allowed to choose their own route. There was no time limitation for selecting the route. After the subjects selected the route and programmed the navigation system, the journey to the destination began. When the chosen route was not followed to a subdestination, the driver selected a new route to the next one.

FIG. 4.1. An example of the road environments in the experiments.

Data Acquisition

Four small video cameras of 17.5 mm in diameter were installed inside the car to record the driver's behavior without restricting the subject's movements or field of view. The cameras monitored the following: the driver's face and eyes, the view through the front windshield, the driver's upper body movements, and the status and driver's operation of the navigation system interface (Fig. 4.2). Using a video wiper, the images obtained from the cameras were recorded onto a video recorder. The images of Camera 1 allowed researchers to determine when the driver looked at the navigation system interface. In order to establish a reference, the driver was asked to look at the display at the beginning of each session, and the images of the direction of the face and eyes recorded by Camera 1 at this time were used in the analysis to identify when the subject glanced at the display. From the images of Cameras 3 and 4, the driver's operation of the system was recorded.

The thinking-aloud method requires that the subject's thoughts be spoken aloud; thus they were given a practice session in driving and verbalizing their thoughts on the day before the experiment. The drivers' comments were recorded with a small microphone for later analysis. The travel time ranged between 32 and 96 minutes ($M = 67.1$ min), and the travel distance ranged from 9.8 km to 14.9 km ($M = 12.0$ km).

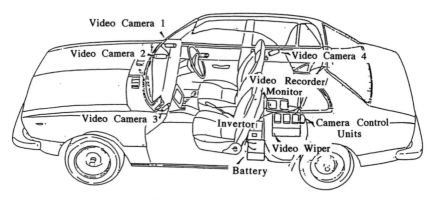

FIG. 4.2.   Experimental setup.

The Car Navigation Systems Used in the Experiment

This study aimed to observe the driver's behavior while using a navigation system; eight different systems were used and the results were analyzed together in order to avoid focusing on the characteristics of one particular system. All the systems indicated the vehicle's current location and the destinations on digital maps; however, the landmarks such as narrow roads, buildings, and the names of streets and places shown on the maps differed from one system to another. The systems used in the experiment did not provide a route guidance function; consequently, the drivers determined the routes by themselves using either a digital or a paper map and followed the chosen route on the digital map. Because the navigation systems used were still under development at the time of this experiment, they were not available commercially.

Subjects

Eight male drivers participated in the experiments, half of whom were familiar with the Tokyo area and half of whom were not. Their ages ranged from 25 to 35 years, and all had more than 5 years of driving experience. All eight subjects knew how to use the car navigation systems because each had been involved in developing the system installed in the car.

RESULTS

When Does a Driver Use Information From the Navigation System?

The camera images of the driver's face and the view through the windshield were used to determine the circumstances under which the drivers looked at the navigation system. The number of glances per minute at the display was used as a measure of the frequency of use of the navigation system. The glancing frequency was found to vary along the route; both the familiar and unfamiliar driver groups gradually increased their glances at the display as they approached an intersection at which they intended to turn and decreased their glances after completing the turn. Figure 4.3 shows the mean frequency of glances around the time of turning at an intersection. Analysis

a: Drivers who were not familiar with the area

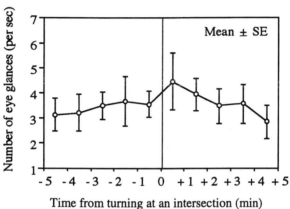

b: Drivers who were familiar with the area

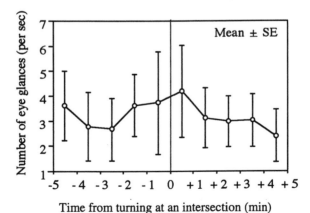

FIG. 4.3. The means and the standard errors of the number of glances at the navigation system interface per minute, for 5 minutes before and 5 minutes after turning at intersections, as observed for unfamiliar subjects (a) and familiar subjects (b).

of the audio recording of the drivers' comments corroborated that they tended to use the navigation system to identify a designated intersection and to confirm that the correct turn had been made. This observation suggests it is important to design the interface and the digital map of the navigation system to enable the driver to easily identify the intersection that is seen through the windshield.

The recordings of the drivers' comments indicated that the subjects used landmarks and other information along the route to identify their location. Consequently, when designing the display and the map's database, it is necessary to know which landmarks are likely to be used as reference points. *Landmarks*, as used here, means information used to identify a location. The next section analyzes the types of landmarks used to identify intersections and pinpoint the location of the car along the route using the recording of the drivers' comments.

The Landmarks Used

In accordance with the thinking-aloud method, words relating to landmarks were extracted from the recorded comments. These words referred to structures or large land features such as buildings, bridges, tunnels, parks, and towers; road geometry; the names of streets, intersections, and places; and distances and directions to the destination provided by road signs or maps. Thorndyke and Hayes-Roth (1982) called this information "survey knowledge" because it is related to the topographic characteristics of an environment.

Words alluding to landmarks were selected from the driver's recorded comments; in total, 246 spoken words related to landmarks. Table 4.1 shows the number of words for each landmark, and the "Others" category includes words related to traffic signals and road geometry such as slope, curve, or width. The values for each subject group are frequencies of the use of words. The values in parentheses in the "Road:System" columns are comparisons between the number of words spoken when the driver looked outside the car at the road environment and those spoken when the driver looked at the navigation system display. The result for the total in the first column was obtained by summing the values for the eight drivers. The relative frequencies of the use of words are also shown.

The totals of the number of words showed that the most frequently reported landmarks were structures and street names, followed by intersection names, distance information, and place names. In addition, road signs, directions, and railway tracks were used. More than half of the 246 spoken words were related to structures, street names, and intersection names.

Information about structures was obtained from both the navigation system and the road environment. About two thirds of the words (30 of

TABLE 4.1

Frequency of Landmarks Used by Subjects for Route Navigation (Total Number of Words = 246)

Type of Landmark	Total (8 Subjects)		Drivers Not Familiar With Tokyo (4 Subjects)		Drivers Familiar With Tokyo (4 Subjects)	
	Frequency	(Road:System)	Frequency	(Road:System)	Frequency	(Road:System)
Structure	49 (19.9%)	(27: 22)	30	(14: 16)	19	(13: 6)
Name of street	44 (17.9%)	(14: 30)	22	(7: 15)	22	(7: 15)
Name of intersection	35 (14.2%)	(24: 11)	12	(4: 8)	23	(20: 3)
Distance	27 (11.0%)	(1: 26)	12	(1: 11)	15	(0: 15)
Name of place	25 (10.2%)	(11: 14)	22	(9: 13)	3	(2: 1)
Road sign	14 (5.7%)	(14: 0)	9	(9: 0)	5	(5: 0)
Direction	13 (5.3%)	(1: 12)	6	(0: 6)	7	(1: 6)
Railway	12 (4.9%)	(11: 1)	9	(9: 0)	3	(2: 1)
Expressway	8 (3.3%)	(4: 4)	4	(3: 1)	4	(1: 3)
Railway station	8 (3.3%)	(4: 4)	4	(0: 4)	4	(4: 0)
Entrance to expressway	4 (1.6%)	(2: 2)	3	(1: 2)	1	(1: 0)
Others	7 (2.9%)	(2: 5)	7	(2: 5)	0	(0: 0)
Total	246 (100%)	(115:131)	140	(59: 81)	106	(56: 50)

44) relating to street names were taken from the navigation system display and the other third came from the road environment, H_0: $p = 115/246$; $B(44, 115/246) \cong N(20.5, 10.9)$, $z = 2.88$, $p < .01$. These data indicate that the drivers obtained more information about street names from the navigation system than from the road and traffic environment. This tendency was more pronounced for distance and direction information, most of which was obtained from the navigation system: 26 of 27 words for distance—$B(27, 115/246) \cong N(12.6, 6.7)$, $z = 5.17$, $p < .001$—and 12 of 13 words for direction—$B(13, 115/246) \cong N(6.1, 3.2)$, $z = 3.29$, $p < .001$. Conversely, information about the names of intersections and road signs was obtained mainly from the infrastructure: 24 of 35 words for intersection names—$B(35, 115/246) \cong N(16.3, 8.7)$, $z = 2.59$, $p < .005$—and 14 of 14 words for road signs—$B(14, 115/246) \cong N(6.5, 3.5)$, $z = 2.59$, $p < .005$.

The familiar and unfamiliar driver groups differed in the number of words they used for place names. The drivers unfamiliar with Tokyo used place names as landmarks more frequently than those who were familiar: 22 of 140 words versus 3 of 106 words, $\chi^2(df = 1) = 11.04$, $p < .005$. Two thirds of the place names were spoken when the unfamiliar driver looked at the navigation system, which may be explained by the fact that place names were often presented on the display but not in the road environment. Whereas this finding suggests that unfamiliar drivers tended to rely on the navigation system to identify their location, comparisons of numbers of words showed that the drivers who were familiar with Tokyo obtained most of the intersection names from the road environment (20 of 23 words).

Various structures were used by the drivers as landmarks. In total, 49 words referred to 25 different structures, including government offices, hotels, parks, and department stores.

DISCUSSION

Even when a driver's current location and the intended destination are provided by a navigation system, the driver must match the current location indicated on the display with the view through the windshield in order to follow the route. Because the system's estimate of the car's location can be in error and due to the driver's uncertainty as to the exact distance to an intersection, external landmarks must also be used to identify the current location. In addition, the map presents the streets and structures from the perspective of a bird's-eye view, which differs from the view seen through the windshield. Frequent use of the navigation system near intersections suggested that drivers used landmarks to decide whether to turn at the approaching intersection and to confirm whether a correct turn had been made.

As mentioned earlier, Schraagen (1993) performed a navigation task experiment in a city using a paper map using the thinking-aloud method. His verbalization categories were street names; road signs; landmarks such as schools, churches, and railways; topological knowledge, including road characteristics and road types; and metric knowledge such as compass direction, distance, and angle. His "landmarks" corresponded to "structures" in this study. Schraagen found the most frequently used information was street names, followed by topological knowledge and landmarks (structures). Compared to his findings, buildings and other structures were used more frequently, but topological knowledge (the "Others" category) was used less frequently by the subjects in this study. Although both studies were conducted in urban traffic environments, these differences may be due to differences in the road environment. In central Tokyo, large and small buildings are close to each other and are easily seen by drivers; in addition, there is little consistency in building design. Because of the large variety in building types, they serve as good landmarks. Schraagen found that the relative frequency of words relating to metric knowledge (direction and distance) was 4% as compared to more than 15% in this study. This difference may be related to the use of a navigation system as opposed to a paper map. Most of the words related to distances and directions were spoken in this study when drivers consulted the navigation system display. The navigation system presented direction and distance to the destination on its display, which may have facilitated the use of metric knowledge to identify the current location.

The choice of landmarks used by the drivers when navigating depended on the road environment; the results obtained here apply only to landmarks in the central urban area of Tokyo. Analysis of the structures used as landmarks revealed that they possessed specific characteristics. They were well-known buildings or public facilities, had a distinctive appearance with easily recognized names, and were large or tall. Their characteristics can be summarized in the following way: They were visible from a distance, possessed distinctive features, their names were visible from the intersection, and they were situated close to the roadside or to an intersection. In this experiment, the buildings of several government offices and hotels were used by the subjects as landmarks; however, it should be noted that one designated subdestination was a ministry building and two were hotels, which suggests that the choice of landmark may be influenced by the nature of the destination and the navigation task.

Other landmarks often used by the subjects were road signs and the names of streets, intersections, and places, which are under the control of road administrators (such as the Ministry of Construction and Tokyo Metropolis). Comparisons of the number of words for landmarks taken from the display with those taken from the road environment indicate that street names were obtained from the navigation system, whereas road sign information was obtained from the road and traffic infrastructure. For drivers who were familiar with Tokyo, most street names were also obtained from the road infrastructure. It is usually easier to identify intersection names on the road because the signs showing their names are attached to highly visible traffic lights, whereas not all intersection names were displayed by the navigation systems. The same was true for road signs; they were fixed above streets and easily seen by drivers, but the navigation systems used here did not show road signs on the display. On the other hand, although street names were displayed on the navigation system's display, the panels showing the street names were sometimes difficult to see from the car because the drivers did not know their location or because heavy traffic hid the panels. The difference in the frequency of use of street and intersection names between the navigation system and the road environment may be due to inconsistencies between the information displayed by the navigation system and that of the road environment. The use of navigation systems could be facilitated by ensuring that street names are easily seen, which is the responsibility of the administrators at the ministry or the metropolis. Whenever possible, navigation systems should display road sign information and intersection names in the same location as the actual signs in the road environment. Unlike the names of streets and intersections, structures are arbitrarily chosen by drivers as landmarks; therefore, the interface should show structures that are likely to be useful as landmarks (i.e., that have all the characteristics already described). The information on the navigation system should also be updated to reflect changes in the road and traffic environment.

Comparisons of the frequency of word use showed several differences between the familiar and unfamiliar driver groups. Both the navigation system's interface and the road traffic infrastructure should be designed to facilitate navigation by unfamiliar drivers. For example, the signs for place names should be clearly visible in order to help unfamiliar drivers to identify the place. A small number of subjects were used in this experiment, thus further analysis of a large number of subjects would be necessary to confirm the effects of familiarity on navigation.

CONCLUSIONS

Subjects often used the navigation system in order to identify where to turn. Commonly used landmarks included buildings, street names, intersection names, distances, place names, and road signs. In order to make using navigation systems more efficient, the signs showing street names should be clearly visible and intersection names and road signs should be displayed by the navigation system in a manner that is compatible with their location in the real traffic environment.

ACKNOWLEDGMENTS

This study is part of the research project of the working group for man–machine interface in the Individual Communication Group, the Route/Automobile Communication System of Japan, and has been supported by companies that are members of the group.

REFERENCES

Anzai, Y. (1984). Cognitive control of real-time event-driven systems. *Cognitive Science, 8,* 221–254.

Faeber, B., & Popp, M. M. (1991). Route guidance systems: Technological constraints and user needs. In Y. Queinnec & F. Daniellou (Eds.), *Designing for everyone* (pp. 1480–1482). London: Taylor & Francis.

Noy, Y. I., & Zaidel, D. (1991). Methodological framework for evaluating the ergonomics and safety of advanced driver information systems. In Y. Queinnec & F. Daniellou (Eds.), *Designing for everyone* (pp. 1607–1609). London: Taylor & Francis.

Labiale, G. (1989). Influence of in-car navigation map displays on drivers' performances. *SAE Technical Paper Series 891683, Future Transportation Technology Conference and Exposition.* Vancouver: Society of Automotive Engineers.

Labiale, G. (1990). In-car road information: Comparisons of auditory and visual presentations. In *Proceedings of the Human Factors Society 34th Annual Meeting* (pp. 623–627). Orlando: Human Factors Society.

Parkes, A. M. (1989). Changes in driver behaviour due to two modes of route guidance information presentation: A multi-level approach (Rep. No. 21). *DRIVE Project,* V1017.

Schraagen, J. M. C. (1993). Information presentation in in-car navigation systems. In A. M. Parkes & S. Franzen (Eds.), *Driving future vehicles* (pp. 171–185). London: Taylor & Francis.

Streeter, L. A., Vitello, D., & Janssen, W. H. (1985). How to tell people where to go: Comparing navigation aids. *International Journal of Man–Machine Studies, 22,* 549–562.

Thorndyke, P. W., & Hayes-Roth, B. (1982). Differences in spatial knowledge acquired from maps and navigation. *Cognitive Psychology, 14,* 560–589.

Chapter 5

Effect of In-Vehicle Route Guidance Systems on Driver Workload and Choice of Vehicle Speed: Findings From a Driving Simulator Experiment

Raghavan Srinivasan
Dowling College, Oakdale, NY

Paul P. Jovanis
University of California, Davis

Recent innovations in microcomputer and display technology have led to the feasibility of sophisticated route guidance systems that can help drivers in choosing and maintaining efficient routes. Human factors professionals have been concerned about the degree and severity of driver distraction resulting from provision of guidance information in the vehicle. The objective of this study was to explore how the characteristics of route guidance systems affect the attentional demand and efficiency of the driving task. Specifically, this study was conducted to understand how drivers react to complex route guidance systems under varying task demands resulting from driving in different types of roads.

There are three primary issues that motivate this study. In order to place these issues in the proper context, consider a brief discussion of the literature.

MODE OF INFORMATION PRESENTATION

Route guidance information may be presented using visual displays, audio messages, or both. Most of the information needed for the driving task is obtained visually, so it has been argued that audio route guidance systems will lead to less distraction from the driving task. However, at the same time, audio messages can be considered intrusive (Stokes, Wickens, & Kite, 1990). A recent study by Parkes and Burnett (1993) indicated that subjects spent more time looking, and made more glances toward a visual route guidance display when visual guidance information alone was presented, compared to when both visual and auditory information were provided. However, the study did not explore whether this difference in driver scanning patterns was associated with changes in driving performance: changes in driving speed, reaction times to external events, lane position, and so on. Further research is needed concerning how visual and audio route guidance systems affect driving performance and driver preferences.

INFORMATION CONTENT AND FORMAT

In the case of visual displays, there are two common ways of providing route guidance information: a route map or a turn-by-turn display. Dingus and Hulse (1993) argued that because a turn-by-turn display is less complex, it will require less attention. However, the study also mentioned that a route map will be useful in complex routes to recall and plan for upcoming maneuvers. Burnett and Joyner (1993) conducted an experiment with an electronic map-based route guidance system and a baseline method. Half of the subjects used instructions given by a passenger as the baseline measure; the other half used a set of maps (from which subjects could make notes if preferred). The results indicated that although subjects perceived their workload in the electronic map to be lower than with maps/notes, the electronic map was associated with large amounts of time with eyes off the road. However, data were again not collected on whether this difference in driver scanning patterns was associated with changes in driving performance. In summary, although studies seem to indicate an electronic route map to be associated with large attentional demand, research is needed to study how electronic maps and turn-by-turn displays affect driving performance.

Audio messages, because of their very nature, should not be very long and complex. The information provided in the messages should probably be limited to turn street name, distance/number of blocks to the turn, and the direction of turn (Means et al., 1993). Walker, Alicandri, Sedney, and Roberts (1990) confirmed this observation: Complex audio messages were associated with larger number of navigation errors compared to simpler audio messages.

LOCATION OF VISUAL DISPLAYS

It could be argued that displays located close to the driver's forward field of view would lead to less distraction from the driving task. Based on this argument, a display that is projected in front of the driver (head-up display) is better than a display located in the dashboard. Little research has been conducted using head-up route guidance displays for automotive applications. However, a few studies have been conducted comparing head-up and head-down displays in aircraft, although the results have been inconclusive (Stokes et al., 1990). Research is clearly needed in this area.

STUDY APPROACH AND OBJECTIVES

The literature review helped in the development of the objectives of the study. A common limitation of previous research was that they did not include head-up displays. Some of the studies also have the limitation that they did not collect data on how the different route guidance systems affected driving performance: speed, reaction times to external events, lane deviation, and so on (data collected was limited to eye scanning patterns). This research addresses the previous shortcomings by using a high fidelity driving simulator to collect detailed driving performance data. The specific objectives of the study were to determine the following:

1. Do electronic route guidance devices lead to better driving performance compared to paper maps?
2. Do audio route guidance systems lead to better driving performance and lower workload compared to visual electronic route guidance devices and paper maps?
3. Does a head-up turn-by-turn display in combination with a head-down electronic route map lead to better driving performance and lower workload compared to a head-down electronic route map?

The study was conducted using a high fidelity fixed-base driving simulator that has been developed by the Hughes Aircraft Corporation (Hein, 1993). The simulator is equipped with three screens that result in a total field of view of 170 degrees. Computer-generated images (such as roadway segments, traffic control devices, roadway traffic, etc.) are projected on the screens. The movement of these objects is synchronized with the vehicle movement generated by the driver as in a typical car. The focus of the experiments were on route following using four route guidance systems. Subjects were asked to drive from an origin to a destination using a predetermined route. All subjects used all four route guidance devices.

DRIVING ENVIRONMENT

The basic study area was a simulated network (2 mile × 2 mile section of Los Angeles) with three types of roadway:

1. Urban two-lane undivided arterials—10-ft lanes and 30 mph speed limit. This was the most difficult driving situation in the experiment.

2. Urban four-lane undivided arterials—12-ft lanes and 40 mph speed limit.

3. Parkway type—four-lane divided road with 12-ft lanes and wide shoulders; a rural environment with 55 mph speed limit. This was the least difficult driving situation.

Each driving trial (scenario) consisted of roadway segments from all the three roadway types, but corresponded to a different route in the network. In addition, each trial had a different set of street names. Each scenario was designed so that it would approximately match the other in terms of time to complete and overall complexity. The traffic environment consisted of three types of vehicles: vehicles that would use the cross street at intersections, vehicles that traveled on adjacent and opposing lanes, and lead vehicles introduced in some segments of the trial. The speed of the lead vehicles varied from as low as 5 mph up to the speed limit. The average speed of other vehicles in the driving scene were close to the speed limit of the particular section. The traffic density was kept low to maintain visual scenes with a high update rate (typically 30 Hz). An extended discussion of the experimental set up is contained in Srinivasan et al. (1992).

ROUTE GUIDANCE DEVICES

Four route guidance systems were tested in the simulator: a head-down electronic route map, a paper map, head-up turn-by-turn guidance display with head-down electronic map, and voice guidance with head-down electronic map. In each case, the systems were the result of numerous trials and pilot tests to desire designs that worked well, at least for the experimenters. Members of the team included human factors design engineers from Hughes Aircraft Company with many years of interface design experience. Electronic maps are included as a common component in all electronic systems because of a desire to test the performance of displays in combination.

Head-Down Electronic Route Map

The electronic map was a 6-in. color liquid-crystal display located in the instrument panel to the right of the driver (Srinivasan, Yang, Jovanis, Kitamura, & Anwar, 1994). The network was shown in green with the intended route highlighted in red. The thickness of the lines in the map was used to represent the three types of roadway segments. Before the subject started driving, the complete route network was shown (full scale). Once the subject started driving, the display changed to a half-mile scale (Fig. 5.1).

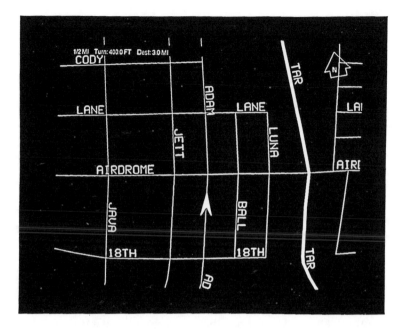

FIG. 5.1. Head-down electronic map (half-mile scale).

The position of the driver's vehicle was shown by an icon (arrow) in the center of the map. The location of the destination was shown by a "star." The top left of the map showed the distance to the turn and the distance to the destination. The orientation of the map was always "heading up" (i.e., the map rotated in such a way that the driver was always heading up the display).

Paper Map

The basic design of the paper map was similar to the full-scale electronic map, with the obvious difference being that vehicle position was not tracked. The size of the map was 11″ × 17″ (Srinivasan, Yang, et al., 1994). Each driver was instructed to use the map in a way that was most convenient.

Head-Up Turn-by-Turn Guidance Display (HUD) with Head-Down Electronic Map

The HUD (Fig. 5.2) was projected directly in front of the driver just above the hood of the car. It consisted of a vertical line and street name indicating the street on which the driver was traveling, and a horizontal line and street name indicating the street onto which the driver has to make a turn, and the direction of turn. The distance to the decision point (shown in tenths of a mile until the driver was 500 feet from the intersection, after

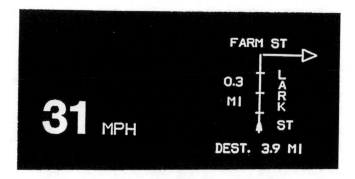

FIG. 5.2. Head-up turn-by-turn guidance display.

which it is shown in feet) and the distance to the destination were also shown. The left side of the HUD showed the speed of the vehicle.

The horizontal bars on the vertical line indicated the distance to the decision point. The distance between two consecutive bars represented 25% of the distance between the decision point and the previous turn. Each of these bars disappeared after the driver crossed that particular point on the roadway. The vertical line thus represented a variable distance to the next distance point, the value of which was displayed numerically (Srinivasan, Yang, et al., 1994).

Voice Guidance With Head-Down Electronic Map

A prerecorded female voice was used for providing guidance information (Srinivasan, Yang, et al., 1994). Two messages were given for each turn. The distance (from the next turn) at which the first message was given, depended on the type of road: 1,200 ft for parkways, 700 ft for four-lane undivided roads, and 400 ft for two-lane roads. The second message (in all the three types of roads) was given 200 ft before the turn. An example of the content of the first message was: "In 400 feet, turn right onto Zuma." An example of the content of the second message was: "Turn right onto Zuma."

EXPERIMENTAL DESIGN

Each subject drove two trials (scenarios) using each of the four route guidance systems, in turn. Each trial was 3 to 4 miles long, had 9 to 10 turns, and took 10 to 15 min to complete. The order of presentation of the route guidance systems was counterbalanced across 4 × 4 Latin squares, to reduce order effects. Similarly, 8 × 8 Latin squares were obtained for the 8 scenarios, and were randomly matched with 4 × 4 Latin squares to obtain the sequence for each subject (Srinivasan et al., 1992; Srinivasan, Yang, et al., 1994).

Each subject had a training session to get them accustomed to the simulator. The training was conducted in stages, over a 1- to 2-hr period. Once the subjects were comfortable in driving at the posted speed limit, making turns and reacting to the external events, then this was end of the training session. At this time, a final decision was made on the suitability of an individual as a test subject based on whether the subject exhibited any symptoms of simulator sickness. Each subject also drove a practice trial with each route guidance device before the two data collection runs.

Subjects were recruited by a market research firm to satisfy experimental and subject adequacy criteria. Eighteen subjects between the ages of 30 and 40 completed the experiments. Due to limited project resources, a decision was made to recruit subjects who had a higher probability of completing the experiments. Older subjects have been found to have an increased propensity to simulator sickness, and hence were excluded from the study. Half of the subjects were in the high experience group (driving more than 15,000 miles a year) and half were in the low experience group (driving less than 12,000 miles a year) (Srinivasan et al., 1992).

DEPENDENT MEASURES

A variety of performance measures were collected during the study. They included:

1. *Driving speed.* This data was collected at 10 Hz during the experiments. Because subjects were asked to drive as close to the speed limit as comfortable for them, it was not expected that large differences in speed would be observed. It was postulated that subjects would drive slightly faster with a route guidance system that required lower attentional demand.

2. *Workload.* The NASA TLX subjective workload test (Hart & Staveland, 1988) was used to obtain ratings on workload. This test assumes workload to be a function of six factors, namely: mental demand, physical demand, temporal demand, performance, frustration level, and effort. The first step in this method involved rating each route guidance system on these dimensions on a scale of 0 to 100. The second step involved a weighting procedure consisting of assigning weights to the different dimensions based on their importance to the route-following task. These weights in combination with the ratings is used to develop a weighted workload. A lower workload value meant a more favorable route guidance device.

3. *Number of navigation errors.* Defined as the number of times the subject deviated from the intended route either by turning the wrong direction

or by missing a turn. It was postulated that a more complex route guidance system would lead to a larger number of errors.

4. *Reaction times.* Reaction times to external events in the simulator (e.g., pedestrians, crossing vehicles, change in traffic signal, left turning vehicles, and obstacles) were recorded whenever an event occurred in the event. It was hypothesized that subjects would be able to react faster if they spend more of their time looking at the roadway scene rather than the route guidance system.

Apart from these measures, subjects were asked to fill in a pretest questionnaire that queried them on demographic information and their familiarity and use of technologies like cellular phones, VCR, computers, and so on (Srinivasan et al., 1992). After completing trials with each route guidance system, the subjects were asked to comment on specific attributes of the different route guidance devices. Detailed modeling of reaction times is in Srinivasan, Yang, et al. (1994).

ANALYSIS AND RESULTS

Driving Speed

Analysis of variance (ANOVA) models were developed for mean speed over a decision segment. A decision segment was defined as the roadway segment between two consecutive turns. Separate models were developed for urban two-lane, urban four-lane, and parkway roads. The models included route guidance system, period (order) effect, repetition effect and scenario effect (trial effect) as fixed factors, and effect due to individual subjects as a random factor. The period effect represents a combination of effects such as fatigue and practice in the simulator. The scenario effect was included to account for differences between the scenarios. The repetition effect was included to test whether subjects performed differently in the second trial with a device compared to the first trial. Models were estimated by specifying the different categories within each fixed factor as dummy variables. The maximum likelihood method was used to fit the ANOVA models. Chi-square tests were conducted to test for the collective significance of each of the individual factors that were included. Table 5.1 gives a desciption of the variables and abbreviations used in the ANOVA models.

The ANOVA model assumes that the variance of the random error component is not a function of the dependent variable or the different levels of the independent variables. To test this assumption, the residuals obtained from the model were plotted against the predicted value of the mean speed and also the independent variables. If the residual plots

TABLE 5.1
Description of Variables and Abbreviations Used in the ANOVA Models

Factor	Categories	Dummy Variables
Route Guidance Type	Paper Map	P
	Electronic Map	E
	HUD with Electronic Map	HE
	Voice with Electronic Map	VE
Scenario Type	Scenario 3	Scen3
	Scenario 4	Scen4
	Scenario 5	Scen5
	Scenario 6	Scen6
	Scenario 7	Scen7
	Scenario 8	Scen8
	Scenario 9	Scen9
	Scenario 10	Scen10
Period (Order) Effect	Period 1	P1
	Period 2	P2
	Period 3	P3
	Period 4	P4
Repetition Effect	Second trial with a particular device	Repeat

showed any evidence of violation of the assumptions of the ANOVA procedure, then a Box–Cox power transformation was applied to the dependent variable (Montgomery, 1991).

Model for Two-Lane Roads. Table 5.2 shows the results of the model that was developed for the two-lane roads. The chi-square tests indicated that scenario effect and route guidance effect were statistically significant ($p <$ .05). The parameter estimates indicate subjects drove fastest with the voice and electronic map combination (VE), followed by the electronic map alone option (E). They drove slowest with the paper map (P). The p values indicate that the difference in speed between P and VE, and between P and E are statistically significant ($p < .05$). However, the difference in speed between HE and P is marginal ($p = .086$). Further models were developed with E as a base case, and HE as a base case, to statistically test the differences between the electronic devices. Results from these models are summarized in Table 5.3. The table indicates that the subjects drove significantly faster with the voice and electronic map combination not only compared to the paper map, but also compared to the other two electronic route guidance systems.

The model was used to calculate the mean predicted value of the mean speed for the four route guidance systems (Fig. 5.3). The predicted mean speeds are between 15 and 17 mph. The predicted mean speeds for VE, E, and HE are higher than the paper map (P), by 11.62%, 5.55%, and 4.51%, respectively.

TABLE 5.2
ANOVA Model for Mean Speed (Two-Lane Roads)

Parameter	Estimate	Standard Error	Est/S.E.	P Value
scen3	−0.003	0.052	−0.067	0.947
scen4	0.145	0.056	2.599	0.009
scen5	−0.013	0.050	−0.256	0.798
scen6	−0.181	0.052	−3.503	0.000
scen7	−0.076	0.051	−1.477	0.140
scen8	−0.056	0.051	−1.105	0.269
scen9	0.158	0.054	2.923	0.003
P2	0.005	0.036	0.140	0.889
P3	0.021	0.037	0.577	0.564
P4	0.080	0.036	2.192	0.028
Repeat	0.005	0.026	0.205	0.838
E	0.083	0.037	2.261	0.024
HE	0.068	0.039	1.715	0.086
VE	0.172	0.039	4.445	0.000
Constant	3.392	0.056	61.106	0.000
Subj(random)	0.014	0.006		

Note. Dependent Variable = (Mean Speed)$^{0.45a}$
Number of observations = 684

[a]Box–Cox transformation
Results of χ^2 test:
  Subj(random) Effect   (χ^2 = 45.938, df = 1, p = .000)
  Repeat   (χ^2 = .042, df = 1, p = .838)
  Scenario Effect   (χ^2 = 54.015, df = 7, p = .000)
  Period (Order) Effect (χ^2 = 6.035, df = 3, p = .110)
  Guidance Type   (χ^2 = 19.788, df = 3, p = .000)

The results have indicated that drivers perform better in the driving task with electronic devices compared to the paper map. It seems that the paper map forced the subjects to drive slower because of its complexity and the absence of real-time information on the location of the driver in the network. It is also clear that the subjects performed the best in the driving task with the audio systems. The audio system permitted the subjects

TABLE 5.3
P Values for Comparison of Mean Speed Between
Route Guidance Systems (Two-Lane Roads)

	P	HE	E	VE
P		0.086	0.024	0.000
HE			0.702	0.009
E				0.021
VE				

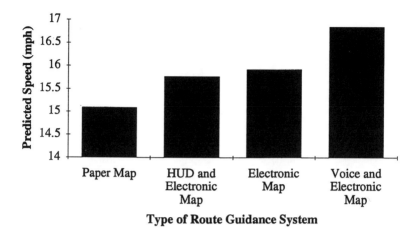

FIG. 5.3. Predicted value of speed (two-lane roads).

to look at the driving scene more often than the visual systems, and hence allowed the subjects to drive faster.

It was surprising that subjects drove slightly faster with the electronic map than the head-up display/electronic map combination. The HUD was expected to do better because it was located closer to the forward field of view and was less complex compared to the electronic map. Subjects comments during the exit interview, identified two design flaws with the head-up display (HUD): First, the distance represented by the vertical countdown bars was not constant but depended on the decision segment length. Second, there were some differences between the position of the cursor and the vehicle position in the simulated world (the top of the cursor was a fraction of an inch below the intersection in the HUD when the vehicle was in the simulated intersection). Because of software differences, the electronic map cursor did not contain this positional error.

Model for Four-Lane Roads. Table 5.4 shows the results of the ANOVA model that was developed for the four-lane roads. The chi-square tests indicated that scenario effect and route guidance effect were statistically significant ($p < .05$). The voice guidance and electronic map combination (VE) were again associated with the highest mean speed. The parameter estimates indicated E and VE to be associated with higher speeds compared to P and HE, similar to the two-lane roads. The difference between VE and P (base case) was statistically significant ($p = .043$), whereas the difference between P and E was marginal ($p = .070$). It was also found that the difference between VE and HE, and the difference between E and HE were statistically significant (Table 5.5).

Figure 5.4 shows the predicted value of the mean speeds for four-lane roads. It can be seen that although VE and E are associated with higher

TABLE 5.4
ANOVA Model for Mean Speed (Four-Lane Undivided Urban Roads)

Parameter	Estimate	Standard Error	Est/S.E.	P Value
scen3	−9.821	2.197	−4.469	0.000
scen4	−4.985	2.224	−2.241	0.025
scen5	0.722	2.482	0.291	0.771
scen6	4.600	2.595	1.772	0.076
scen7	12.287	2.185	5.624	0.000
scen8	−0.707	2.402	−0.294	0.769
scen9	−1.753	2.156	−0.813	0.416
P2	1.452	1.559	0.931	0.352
P3	2.042	1.589	1.285	0.199
P4	2.706	1.560	1.735	0.083
Repeat	1.102	1.102	1.000	0.317
E	2.931	1.620	1.810	0.070
HE	−0.272	1.639	−0.166	0.868
VE	3.416	1.688	2.024	0.043
Constant	52.620	2.435	21.612	0.000
Subj(random)	24.780	10.078		

Note. Dependent Variable = (Mean Speed)$^{1.33}$[a]
        Number of cases = 461
[a]Box–Cox transformation
Results of χ^2 test:
    Subj(random) Effect   ($\chi^2 = 43.215$, $df = 1$, $p = .000$)
    Repeat   ($\chi^2 = .999$, $df = 1$, $p = .318$)
    Scenario Effect   ($\chi^2 = 123.751$, $df = 7$, $p = .000$)
    Period (Order) Effect   ($\chi^2 = 3.247$, $df = 3$, $p = .355$)
    Guidance Type   ($\chi^2 = 8.265$, $df = 3$, $p = .041$)

speeds in comparison to HE and P, the percentage difference between the audio route guidance system and the paper map (calculated to be approximately 5%) is smaller compared to that obtained in the two-lane roads. It should be noted that the two-lane roads had narrow lanes (10-ft lanes) and more intersection traffic control devices, which resulted in a more difficult driving environment. Hence, these results seem to indicate

TABLE 5.5
P Values for Comparison of Mean Speed Between Route
Guidance Systems (Four-Lane Undivided Urban Roads)

	HE	P	E	VE
HE		0.868	0.050	0.025
P			0.070	0.043
E				0.774
VE				

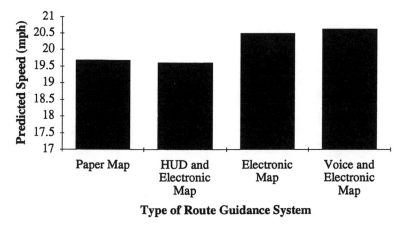

FIG. 5.4. Predicted value of speed (four-lane undivided roads).

that the largest benefit due to the audio system could be obtained in relatively complex driving situations.

Model for Parkways. The results showed that the audio system was again associated with the highest speed followed by the electronic map. The paper map was associated with the slowest speed. The percentage difference between the guidance systems was smaller to those obtained in the two-lane roads and close to those obtained from the four-lane models. This result confirms the previous observation that larger benefits due to the electronic devices (especially the audio system) could be obtained in relatively complex driving situations (in this experiment: two-lane roads).

Subjective Workload

Weighted workload was obtained for each of the four route guidance systems based on the NASA TLX procedure (Hart & Staveland, 1988). The nonparametric Friedman's procedure was used to perform statistical tests, because the workload data did not conform to a normal distribution (Neter, Wasserman, & Kutner, 1990). Friedman's test ranks the four devices based on its workload values, for each subject separately. The lowest workload value in each subject is given Rank 1, the highest workload value is given Rank 4, and so on. These ranks are then summed over the 18 subjects to obtain the rank sum for each route guidance device. Figure 5.5 shows that the voice/electronic map combination has the lowest mean workload. The paper map has the highest rank sum and the highest mean workload. The electronic map alone option has slightly lower mean workload and slightly lower rank sum compared to the HUD/electronic map combination.

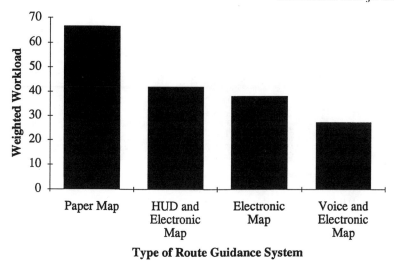

FIG. 5.5. Weighted workload.

The Friedman's test rejected the null hypothesis that all the four route guidance systems have the same workload ($p < .0001$). A multiple comparison test was conducted to find out which of the route guidance devices were different from each other (Table 5.6). The multiple comparison test calculated a parameter called *ZSTAT* based on the rank sum values. The null hypothesis was rejected if *ZSTAT* was larger than the critical value *ZC*, where

$$1 - \Phi(ZC) = \alpha/[K(K-1)],$$

where Φ is the cumulative standard normal distribution function, α is the desired overall significance level, and K is the number of route guidance systems (in this case, 4) compared (Neter et al., 1990). Table 5.6 confirms the qualitative observations that were made based on Fig. 5.5.

The workload results reaffirm the conclusions that were made based on the speed data. The audio system was the most favorable option followed by the electronic map alone option. The paper map was the least favorable option. The electronic map alone option does slightly better than the HUD/electronic map combination, probably due to the design deficiencies in the HUD.

Navigation Errors

The number of navigation errors is a direct indicator of ability to follow guidance information. The voice/electronic map combination had the least number of errors (.037 per trial) followed by the electronic map

TABLE 5.6
Results of Multiple Comparison Test (*ZSTAT* Values)

	P	HE	E	VE
P		2.58	2.97	5.29
HE			0.39	2.71
E				2.32
VE				

Note. The critical value (*ZC*) is 2.39 for $\alpha = .10$; 2.64 for $\alpha = .05$.

alone option (.111 per trial) and the HUD/electronic map combination (.192 per trial). The paper map had the largest number of errors (.696 per trial). These results also reaffirm the results obtained from the subjective workload and driving speed (i.e., the voice/electronic map combination is the most favored and the paper map is the least favored). The electronic map alone option was second best to the voice/electronic map combination.

Reaction Times to External Events

ANOVA models were developed with reaction time as a dependent variable (Srinivasan, Yang, et al., 1994). Separate models were developed for five different external events. The independent variables included route guidance type and order effects. The effect due to subjects was included as a random factor. It was not consistently clear from the results on which route guidance device was the best (i.e., shortest reaction time). The best device varied depending on the type of external event. For example, in the case of crossing vehicles, the electronic map was the best. In the case of left turning vehicles, the HUD/electronic map combination was the best. In the case of traffic signal events, the HUD/electronic map combination and the voice guidance/electronic map combination were the best. Work is now in progress to modify these reaction time models to include predicted driving speed as a covariate. This will explicitly account for the fact that subjects might react differently to external events depending on their driving speed.

CONCLUSIONS AND FUTURE RESEARCH

The analysis has clearly demonstrated that, for the particular devices tested, driving performance is improved with the electronic devices compared to the paper map. It is also clear that the voice guidance/electronic map is associated with better driving performance compared to the visual electronic devices. This is consistent with preliminary results that have been obtained from the TravTek field experiments (Perez, Fleischman, Golembiewski, & Dennard, 1993).

The results from workload, navigation errors, and driving speed are consistent. Subjects had the highest workload when using the paper map and the lowest workload when using the voice guidance/electronic map combination. Subjects had slightly lower workload when using the electronic map alone option compared to the HUD/electronic map combination.

The paper map was associated with the largest number of navigation errors and the voice guidance/electronic map combination was associated with the least number of errors. The HUD/electronic map combination had slightly more errors than the electronic map alone option.

Subjects drove fastest with the voice guidance/electronic map combination and slowest with the paper map. The percentage difference in speed between the paper map and the voice guidance/electronic map combination was larger in the two-lane roads (with 10-ft lanes) compared to the four-lane urban roads and parkways. This indicates that larger benefits due to audio route guidance messages could probably be obtained under complex driving conditions. The driving speed associated with the electronic map alone option was slightly higher than that of the HUD/electronic map option.

Exit interviews revealed some interesting observations about the route guidance devices. Although the audio system had the least workload and was most preferred, some subjects found the audio messages to be irritating and wanted to have the option to turn them off. This option was included in TravTek (Means et al., 1993). The interviews also revealed two deficiences in the design of the head-up display: First, the distance represented by the vertical countdown bars was not constant but depended on the decision segment length. Second, there were some differences between the position of the cursor and the vehicle position in the simulated world (the top of the cursor was a fraction of an inch below the intersection in the HUD when the vehicle was in the simulated intersection). Because of software differences, the electronic map cursor did not contain this positional error. It is postulated that these deficiencies probably led to its performance being slightly inferior to that of the electronic map alone option. Further experiments with an alternative head-up turn-by-turn display have received more favorable comments from subjects (Landau, Laur, Hein, Srinivasan, & Jovanis, 1994; Srinivasan, Landau, Hein, & Jovanis, 1994).

Future Research

The simulator experiments contained in this study are typical of current capabilities in high-fidelity simulators: A relatively large number of dependent variables can be used to assess driving performance in the context of an experimental design reflecting subject attributes and trial randomization. A major improvement would be the development and application of multivariate statistical models that allow the study of interrelations between variables. An example of such a model is the use of predicted speed as an instrumented variable in reaction time models. This formulation captures the notion that speed is endogenous to reaction time—that is, it is an outcome of a driver decision-making process, as is reaction time. The proper model would not treat speed as an independent variable, but as a variable chosen by the driver and conditions the reaction time observed. Although drivers are asked to drive close to the speed limit, the relative fidelity of the simulation may lead them to choose speeds affecting reaction times. A model of the type described would more precisely control for that effect.

Research needs to be conducted using older subject populations. This study used relatively young subjects (between ages 30 and 40). Further research is needed with alternative formats of turn-by-turn displays for both head-up and head-down modes. This would help in determining whether a head-up display leads to better driving performance compared to a head-down display, given a particular display format. Research is also needed to explore whether a simpler electronic map leads to better driving performance compared to a complex electronic map. Finally, experiments need to be conducted on-the-road to validate results from simulator experiments.

ACKNOWLEDGMENTS

The authors would like to thank the Office of Competitive Technology of the California Department of Commerce for their financial support during this study. The authors are indebted to Cheryl Hein, Barry Berson, Francine Landau, Mike Laur, Craig Lee, and many other employees of the Hughes Aircraft Company whose contribution was invaluable in the development of the driving scenes and data collection protocols.

REFERENCES

Burnett, G. E., & Joyner, S. M. (1993). An investigation of man machine interfaces to existing route guidance systems. *Proceedings of the 4th International Conference on Vehicle Navigation and Information Systems*, 395–400. Piscataway, NJ: IEEE.

Dingus, T. A., & Hulse, M. C. (1994). Some human factors design issues and recommendations for automobile navigation information systems. *Transportation Research, 1C*(2), 119–131.

Hart, S. G., & Staveland, L. E. (1988). Development of NASA-TLX (Task Load Index): Results and theoretical research. In P. A. Hancock & N. Meshkati (Eds.), *Human mental workload* (pp. 139–183). Amsterdam: Elsevier Science.

Hein, C. M. (1993). Driving simulators: Six years of hands-on experience at Hughes Aircraft Company. *Proceedings of the 37th Annual Meeting of the Human Factors and Ergonomics Society, 1,* 607–611. Santa Monica, CA: Human Factors and Ergonomics Society.

Landau, F., Laur, M., Hein, C. M., Srinivasan, R., & Jovanis, P. (1994). *A simulator evaluation of five in-vehicle navigation aids.* Institute of Transportation Studies, University of California, Davis. (UCD-ITS-RR-94-19)

Means, L., Carpenter, J. T., Szczublewski, F. E., Fleischman, R. N., Dingus, T. A., & Krage, M. K. (1993). Design of the TravTek auditory interface. *Transportation Research Record 1403,* Transportation Research Board, National Research Council, Washington, DC.

Montgomery, D. C. (1991). *Design and analysis of experiments* (3rd ed.). New York: Wiley.

Neter, J., Wasserman, W., & Kutner, M. H. (1990). *Applied linear statistical models: Regression, analysis of variance, and experimental designs.* Homewood, IL: Irwin.

Parkes, A. M., & Burnett, G. E. (1993). An evaluation of medium range 'advance information' in route-guidance displays for use in vehicles. *Proceedings of the 4th International IEEE Vehicle Navigation and Information Systems Conference,* 238–241. Piscataway, NJ: IEEE.

Perez, W. A., Fleischman, R., Golembiewski, G., & Dennard, D. (1993). TravTek field study results to data. *Proceedings of the 3rd Annual Meeting of IVHS America,* 667–673. Washington, DC: IVHS America.

Srinivasan, R., Landau, F., Hein, C. M., & Jovanis, P. (1994). Effect of in-vehicle driver information systems on driving performance: Simulation studies. *Proceedings of the First ATT/IVHS World Congress "Towards an Intelligent Transport System," 4,* 1717–1725.

Srinivasan, R., Yang, C. Z., Jovanis, P., Kitamura, R., & Anwar, M. (1994). Simulation study of driving performance with selected route guidance systems. *Transportation Research, 2C(2),* 73–90.

Srinivasan, R., Yang, C. Z., Jovanis, P., Kitamura, R., Anwar, M., Hein, C. M., & Landau, F. (1992). *California advanced driver information system.* Institute of Transportation Studies, University of California, Davis. (UCD-ITS-RR-92-20)

Stokes, A., Wickens, C., & Kite, K. (1990). *Display technology: Human factors concepts.* Warrendale, PA: Society of Automotive Engineers.

Walker, J., Alicandri, E., Sedney, C., & Roberts, K. (1990). *In-vehicle navigation devices: Effects on the safety of driver performance* (FHWA-RD-90-053). Office of Safety and Traffic Operations, Research and Development, Federal Highway Administration, McLean, VA.

Chapter 6

An Assessment of Moving Map and Symbol-Based Route Guidance Systems

Gary Burnett
Sue Joyner
HUSAT, Loughborough University, Leics., UK

Route guidance systems are in-vehicle electronic devices that aid drivers in choosing and maintaining efficient routes. Since the initial development of these systems, human factors practitioners have expressed concern over possible safety-related implications (e.g., Barrow, 1990; Dewar, 1988; Parkes, 1991). Whereas route guidance systems should support drivers in the navigation task and reduce workload, they also carry the potential for information overload and distraction, with the subsequent consequences of poorer driver performance and decreased safety.

Large programs, such as Prometheus and DRIVE in Europe and IVHS in North America, have supported research into a wide range of human factors issues associated with the design of route guidance systems. Given the imminent widespread introduction of these systems, it is important that research results are rapidly translated into usable design recommendations and guidelines.

HUMAN FACTORS ISSUES FOR THE DESIGN OF ROUTE GUIDANCE SYSTEMS

Previous human factors work regarding the presentation of information by route guidance systems has typically been of two types. First, researchers have conducted experimental work in laboratories, in simulators, or on roads, examining specific design issues. Such issues can be classified into three broad areas of concern:

1. What route guidance information is required to minimize navigational uncertainty, either during the pretrip/planning stage (e.g., traffic conditions, estimated journey time, availability of parking places, and avoidance of certain classes of road—see Joint, Bonsall, & Parry, 1990; Wallace & Streff, 1993), or en route at the microlevel (e.g., road layout, distance, landmarks, road signs—see Alm, 1990; Bengler, Haller, & Zimmer, 1994; Schraggen, 1991).

2. When route guidance information should be presented to drivers (see Ross, Nicolle, & Brade, 1994; Kishi & Suiguira, 1993).

3. How route guidance information should be presented to drivers (head-up or head-down; verbal, spatial, or combination; visual, auditory or combination—see Burnett & Parkes, 1993; Todoriki, Fukano, Obabayashi, Sakata, & Tsuda, 1994; Van Winsum, van Knippenberg, & Brookhuis, 1989; Verwey, 1993).

The second type of typical human factors work comprises road-based evaluations of existing and prototype route guidance systems. For such research, a wide range of issues with respect to all three of the previous questions can be addressed within a single study, because it is the whole interface style that is under investigation. However, such an approach does not permit the study of any one issue in depth. The work reported here falls within this second category of study.

Several authors have evaluated moving map-based systems and have found that such systems can lead to large amounts of time with drivers' eyes off the road. For example, Wierwille, Antin, Dingus, and Hulse (1988) investigated the effects of three navigation methods (memorized route, paper map, and electronic map-based navigator) on driver behavior. The electronic navigator was associated with increased visual distraction; on average, 33% of the total journey time was spent looking at the device, compared with none for the memorized route and 7% for the paper map. However, no differences were found in lane exceedences and brake accentuations between the three conditions; it was therefore concluded that the electronic navigator could be used "effectively" by the driver. This was based on the view that drivers demonstrate appropriate adaptation in their visual scanning patterns for high demands in the driving task (also suggested by Rockwell, 1988).

Labiale (1989) investigated the effects of different electronic map displays (map alone, map plus auditory message, and map with written message) on driver behavior. In contrast to Wierwille et al., he found that use of the map-based systems did influence driving performance: Drivers strategically reduced their speed when consulting such displays, presumably to cater for increases in mental workload due to the introduction of the in-vehicle display. Such a result has also been found by Van Winsum et al. (1989).

Increasingly, there are systems that present the driver with immediate guidance instructions (using symbols and often simple voice messages) en route, with little other information; herein they are referred to as *symbol-based systems*. There is emerging evidence that such systems result in less visual workload than their map-based counterparts. For example, Ashby, Fairclough, and Parkes (1991) compared 24 subjects using two electronic route guidance systems, one giving map information to drivers (Bosch TravelPilot) and the other directional symbols and auditory guidance instructions (LISB/Ali-Scout). They found that subjects using the LISB system attended to the display for significantly less of the total time in motion (8%) than with TravelPilot (14.4%). However, the LISB system was associated with increased temporal demand, a reflection on the pacing aspect of symbol-based systems, particularly their voice output.

METHODOLOGY ISSUES

There are several methodology issues regarding previous route guidance studies. First, a number of studies have incorporated a control or baseline condition of "normal" driving and navigating behavior (e.g., Parkes, 1989; Pauzie & Marin-Lamellet, 1989; Schraggen, 1990; Streeter, Vitello, & Wonsiewicz, 1985; Verwey & Janssen, 1988). However, closer inspection of these studies reveals that experimenters have imposed a number of different restrictions on such normal navigational methods. These restrictions include the type of map(s), the position in which the map(s) were located within the vehicle, enforced pretrip planning, and the required use of memory. It is quite likely that, for many of the subjects within these trials, the strategies emposed on them were not their preferred methods, and given the choice they would have chosen alternative means of navigating.

Second, previous studies have often looked at one or two dependent variables in isolation. In contrast, Parkes (1991) argued for a multilevel approach to usability evaluation in order to provide a rich view of driver behavior and performance. This advocates the use of a wide range of vehicle control, behavioral, and subjective measures at different levels of the driving task.

Third, previous studies have commonly used gross measures of visual distraction (e.g., mean glance duration or glance frequency across a route). Although these are undoubtedly useful when comparing conditions, they lack context (i.e., what was happening to the control of the vehicle and its interaction with other road users within the road network?). Reduced glance duration and/or frequency does not necessarily mean a system is safer. Safety will depend on when a system requires a driver to take their eyes off the road, and the resulting effect on the control of the vehicle.

BACKGROUND TO THE FIELD STUDIES

This chapter reports on the conduct of two field studies designed to evaluate the man–machine interfaces (MMIs) of three existing route guidance
systems. Both studies had the following aims: to provide generic recom-
mendations/guidelines for the design of MMIs to route guidance systems,
and to establish the critical issues of concern for such systems so that
controlled experiments could then be conducted in the future.

At the time of the studies, the three systems were either commercially
available outside the United Kingdom within mainland Europe, or were
under development. Table 6.1 provides a summary of the characteristics
of each of the route guidance systems evaluated, including a schematic of
the visual information presented.

METHOD: FIELD STUDY 1

In this study, one experimental condition and two conditions of "baseline"
driving were used:

> *MM.* Driving to the destination using route guidance provided by the
> moving map-based system.
>
> *Maps/notes.* Subjects were provided with a 1:50,000 sheet map of the
> whole area and street-level maps of the start and end destination zones.
> The maps had a highlighted route to follow and blank paper and pens
> were provided. Subjects were instructed to drive to the destination using
> their preferred method of navigation (e.g., paper maps, prewritten in-
> structions, sketches, etc.).
>
> *Instructions.* Driving to the destination with an experimenter sitting in
> the passenger seat providing verbal instructions (e.g., "turn right at the
> church," "take the next left"). Instructions were standardized and were
> only given when the experimenter felt it was most appropriate to do
> so, given the current road and traffic conditions. The messages were
> designed to make the navigation task as simple as possible; this condition
> aimed therefore to simulate the ideal route guidance system.

Sixteen male and eight female drivers from 40 to 60 years old were
selected for the study. The subjects were all experienced drivers and were
unfamiliar with the trial area.

There were two experimental routes, matched, as far as possible, by the
approximate journey time, the number and types of representative
roads/junctions, and probable traffic densities. Route 1 was 12.1 miles

TABLE 6.1
Summary of Systems Evaluated

	System MM	System S1	System S2
Display	A separate unit near the center of the cockpit (5.75″ monochrome CRT).	A separate unit near the center of the cockpit (5.75″ color LCD).	In the dashboard (4″ color LCD).
Schematic of visual information presented by system while driving			
Description of visual information presented by system while driving	Heading-up map of the road network with the route to be taken highlighted. Choice of map scale. Names of current road and destination—available on request. Direct straight line distance to the destination in miles to one decimal place—available on request.	Simple graphical representation of the junction (from small library of symbols). Miles to the next turn to two decimal places (e.g., .46). Names/numbers of current/next road. Countdown blocks each representing one quarter of the distance between the last and next maneuver.	Accurate graphical representation of the junction. Meters to the next maneuver from 3,500 m away (the countdown increments reduce from 500 m to 10 m as the distance to the junction reduces).
Description of voice information	No voice information.	A simple single instruction (e.g., turn left) supplemented by complementary visual information.	A series of voice messages, each supplemented by updated visual information (e.g., take 2nd turning left, take 1st turning left).
Mechanism for dealing with two close maneuvers	Two maneuvers near to each other shown within map view.	Instructions not linked for maneuvers close to each other.	Instructions linked for maneuvers close to each other (e.g., take next left then turn right).

long and contained 14 decision points (four *T* junctions, five turn off roads, four roundabouts, and one slip road exit). Route 2 was 9.5 miles long (there was less dual carriageway) and contained 14 decision points (three *T* junctions, four turn off roads, six roundabouts, and one slip road exit). Without navigational errors, Route 1 took approximately 24 min to drive, whereas Route 2 took about 21 min. Both routes contained a mixture of rural and urban roads, and were driven during off-peak daylight hours with low to medium traffic densities.

A repeated-measures design ensured each subject completed one route using the moving map-based system and the other route using a baseline measure of navigation (half used the Maps/Notes, the other half Instructions). The ordering of routes and experimental conditions was counterbalanced across all subjects.

An instrumented vehicle and observational techniques allowed a wide range of objective measures to be taken. Those reported here are as follows:

Journey time. Defined as the total time taken by drivers to reach their destination minus nonvoluntary stoppage time (i.e., due to traffic incidents but not voluntary stoppages to look at a map or notes).

Navigational errors. Defined as having occurred when a subject strayed from the designated route.

Three aspects of visual attention. The mean duration of a single glance, glance frequency, and the percentage of journey time in motion spent glancing toward a given area of the visual scene. Glance durations were measured using video editing equipment. A single glance was defined as the time from the moment the eyes left the road ahead to view another area in the visual scene until the moment they returned to the road ahead.

Vehicle performance parameters. Vehicle speed, steering wheel variability, indicator use, and lane-changing behavior.

Subjective data comprised the measurement of perceived physical and mental workload and driver opinion. Perceived workload was measured using a specially tailored version of the NASA—Raw Task Load Index (RTLX) (Fairclough, 1991; Hart & Staveland, 1988). In completing this, each subject made a rating on six discrete components of perceived workload (mental demand, mental effort, physical demand, time pressure, distraction, and stress). The sum of the component values was divided by six to calculate each subject's overall perceived workload score.

A questionnaire was developed to determine subjective opinions on aspects such as ease of use, display quality, perceived safety, and acceptability. For most questions, subjects gave ratings on a semantically labeled, 6-point scale from "very easy" to "very difficult."

Prior to driving the experimental route with the route guidance system, subjects were given three periods of training. First, they were given a static demonstration of the information presented by the system. Second, they drove a quiet route in a residential area in which the experimenter pointed out, en route, how the system worked, including use of the different map scales available within the display (during the experimental route they were free to change these scales as perceived appropriate). Finally, they drove a more complex route during which they were only given help if they requested it. After driving this route, if the experimenter felt the subject was experiencing learning difficulties with the system, or if the subject was not completely confident, then another route was driven. The total time for familiarization was approximately 1 hr.

RESULTS: FIELD STUDY 1

Journey Time

Unpaired t tests showed that subjects took significantly longer to complete a route with MM (M: Route 1 = 1,425 sec; Route 2 = 1,150 sec) than with Instructions (M: Route 1 = 1,325 sec; Route 2 = 1,010) ($p < .05$ for both of the test routes). Although not significant, the Maps/Notes condition generally resulted in longer journey times (M: Route 1 = 1,450 sec; Route 2 = 1,325 sec) than both MM and Instructions conditions.

Navigational Errors

Navigational errors were classified according to the type of maneuver at which the error occurred and the action taken by the subject. An error rate value was calculated accounting for the variation in the numbers of the different maneuver types within the routes, and the differences in subject numbers across the MM and Map/Notes conditions. Therefore, these values represent the likelihood (in percentage form) of a maneuver of a particular type resulting in a navigational error within this study. This analysis was speculative, and was conducted to allow examination of gross differences between the different error types. Therefore, it was only considered appropriate to carry out statistical testing for the overall error rate. The maneuvers resulting in at least one error are shown in Table 6.2. The

TABLE 6.2
Navigational Error Rates (%) for MM and Maps/Notes

Type of Maneuver	Type of Error	MM	Maps/Notes
Dual carriageway exit	Turned too soon	2.1	0
	Missed turning	4.2	0
Side turn off	Turned too soon	0.9	1.8
	Missed turning	8.8	1.8
	Wrong direction	0	0
Roundabout	Wrong exit	2.3	0.9
Total error rate (across all maneuvers)		4.3*	1.5

*significant at $p < .05$.

total error rate at the foot of the table also incorporates those maneuvers (i.e., *T* junctions) where no errors were made. Subjects made no errors in the Instructions condition.

Gross Measures of Visual Attention

The initial analysis required the calculation of glance duration, frequency, and percentage of time in motion across the routes, as used in a number of other studies (e.g., Ashby et al., 1991; Wierwille et al., 1988). The results of this analysis for glances toward the sources of route guidance information are shown in Table 6.3; to account for differences in journey time, glance frequency has been expressed as a value per unit time.

Similar analyses of the glance variables were conducted for other areas of the visual scene. This analysis revealed that the duration of a glance to the road ahead was significantly less in the MM condition ($M = 2.94$ sec) than for the baseline conditions (*M*: Instructions = 9.35 sec; Maps/Notes = 6.85 sec) ($p < .001$; $p < .05$, respectively). Furthermore, the percentage of time in motion spent glancing toward the road ahead was significantly less for MM ($M = 72.4\%$) than for the baseline conditions (*M*: Instructions = 91.0%; Maps/Notes = 85.8%) ($p < .001$; $p < .01$, respectively).

Subjects spent a significantly reduced proportion of their journey time glancing toward the rearview mirror for the MM condition ($M = 2.1\%$) than for the Instructions condition ($M = 3.3\%$) ($p < .05$). Furthermore, the number of glances per unit time made toward the rearview mirror was significantly less for MM ($M = .031$) than for the Instructions condition ($M = .059$) ($p < .01$). There was no significant difference between the MM and Maps/Notes conditions for these variables.

Approaching Junctions: Visual Attention

As mentioned earlier, in order to make results more meaningful, it is necessary to provide additional context to glance data. Therefore, an analysis was carried out to see whether glance patterns toward MM changed as subjects progressed from one junction to another (e.g., Are there particular points on a road where more information is required from the route

TABLE 6.3
Visual Attention Allocation Toward the MM Display and Maps/Notes

	MM	Maps/Notes
Mean glance duration (sec)	1.12	1.20
Mean glance frequency (number/sec)	0.182***	0.045
Mean percentage of time in motion (%)	20.40***	5.30

***significant at $p < .001$.

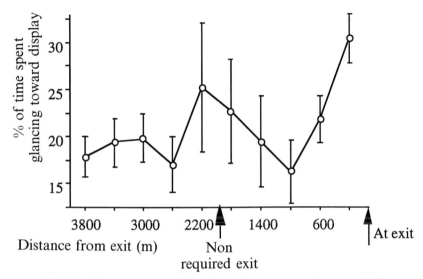

FIG. 6.1. Mean percentage of time in motion spent glancing toward MM along a section of dual carriageway (standard deviation bars are given), n = 12.

guidance system than others? Are there particular events which cause the route guidance system to be sampled more?). A series of approaches to different junctions were isolated for analysis where, based on the error data, it was known that difficulties were arising. Figure 6.1 shows the results for one such stretch of road on Route 1, which was the approach to an exit from a dual carriageway road. The stretch of road was split into equal distance portions and the percentage of time spent glancing toward MM for each of those sections was calculated. Similar glance patterns were found for the equivalent stretch of road on Route 2.

Approaching a Dual Carriageway Exit: Steering Wheel Variability

Steering wheel variability was analyzed for the final minute approach to the dual carriageway exit on both routes, where it was known that large amounts of time were spent glancing toward the route guidance system (see Fig. 6.1). The purpose of this analysis was to examine whether subjects made compensatory adjustments to the steering to correct any path deviations caused by looking away from the road ahead. Statistical analysis revealed that there was significantly more variability in steering wheel movements for the MM condition than for either baseline condition on Route 1 ($p < .05$). However, this result was not replicated for the final minute approach to the dual carriageway exit on Route 2.

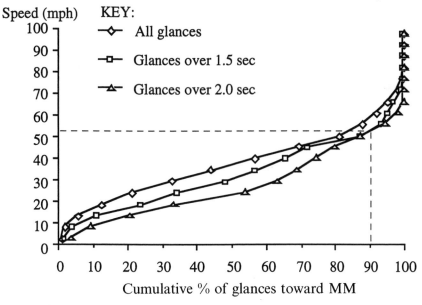

FIG. 6.2. Cumulative glance frequency curves for MM by duration and speed.

Visual Attention and Vehicle Speed

To investigate the speeds at which longer glances toward MM were made, cumulative frequency plots were made of the speeds at which glances of certain durations were made (Fig. 6.2). This method of assessing the visual cost of a system places the emphasis on the likelihood of long glance durations at high speeds.

Perceived Workload

Mean NASA–RTLX scores associated with each condition are shown in Table 6.4. The results of statistical analysis are also shown. A value of 0 corresponds to low perceived workload, whereas 100 corresponds to high perceived workload.

Subjective Opinion

For this study, the questionnaire revealed that subjects found it easier to get to their destination with Instructions or MM than with Maps/Notes ($p < .01$ in both cases). Indeed, 80% of subjects found it at least "easy" to get to their destination with MM. Furthermore, the questionnaire revealed that subjects found it easier to read the moving map-based display rather than to read Maps/Notes ($p < .05$). Although all subjects felt safe using

TABLE 6.4
Field Study 1: Results From NASA–RTLX Workload Questionnaire

Component	MM	Maps/Notes	Instructions	Significance Testing
Mental demand	37	60	29	*/+
Mental effort	45	49	39	
Physical demand	15	30	11	+
Time pressure	16	36	6	*
Distraction	33	50	13	*/+
Stress	16	39	14	+
Overall score	27	44	19	*/+

*significant difference between MM and Instructions, $p < .05$.
+significant difference between MM and Maps/Notes, $p < .05$.

either MM or Instructions, two felt unsafe driving in the Maps/Notes condition.

DISCUSSION: FIELD STUDY 1

A number of negative aspects of the moving map-based system were revealed by the study. For instance, MM affected driver performance by reducing the time taken to reach a destination when compared with instructions given by a well-informed passenger. The most apparent reason for this difference relates to the difference in navigational errors between the two conditions, because each mistake contributed to an increase in total journey time. However, it is also possible that drivers generally reduced their vehicle speed, perhaps as a consequence of the increased task demands associated with using MM.

The primary reason for the errors with MM was the fact that the system's displayed vehicle location lagged behind the vehicle's real road position. As a consequence, subjects often felt the correct turning was further away than it actually was; this explains why the majority of errors occurred when subjects missed the correct side turning. Such problems undoubtedly would have reduced as drivers gained experience of the system, but the overhead of permanent calculation and road matching will place an additional cognitive load on the driver.

The rotating aspects of the map display also caused some difficulties, which were most apparent when traveling on roundabouts. These occurred partly because the display was rotating in the opposite direction to that in which subjects were progressing, and hence was considered distracting and confusing, and partly because of the strategies employed by subjects. For instance, those subjects who performed best on roundabouts were those who knew exactly what they had to do before entering a roundabout, and then made few glances at the display while actually on it.

The most apparent negative aspect of MM was the increased time spent with the eyes off the road as compared with Maps/Notes. However, the glance data obtained from those subjects who used the Maps/Notes for navigation were confounded to a large extent by the strategies employed. Indeed, if one defines an individual's strategy within the Maps/Notes condition in terms of whether maps or notes were the primary information sources used, the extent to which the information sources was used on the move, and the main position of the information source during the journey, then only 2 of the 12 subjects adopted the same overall strategy within this study. Therefore, large subject numbers would be required to account fully for the navigational styles and abilities of a specific user population. For this pragmatic reason, it was decided that Instructions would be the more appropriate baseline condition for Field Study 2.

The analysis of drivers' visual behavior along the dual and single carriageway stretches of road revealed particular difficulties with this style of interface. Along the dual carriageway, a number of subjects increased the time spent glancing toward the MM display when passing an exit (i.e., checking to ensure they were not required to leave at that point). Not all subjects experienced this difficulty, as borne out by the measure of variability. On the final approach to the dual carriageway exit, drivers consistently spent more time glancing toward the MM display as new information scrolled onto the display. A similar result occurred on the single carriageway stretch of road. This reflects the visual demands placed on the driver by a system that requires visual checking for new information. Subjects were taught how to change the scale of the map display while moving, and encouraged to do so in order to gain advance warning of the oncoming maneuvers. However, not all of the subjects attempted to use the scale facilities when on the experimental routes. Several of those who did not make such attempts commented that they did not want to press the scale buttons while moving, either for safety reasons, or in case they could not return to the preferred scale. In most cases, these are problems resolved with increased experience.

Subjects were found to exhibit greater variability in their steering wheel movements when traveling on the dual carriageway and approaching the required exit on Route 1, where it was known that large amounts of time were spent with the eyes off the road glancing toward the moving map-based system. This gives some cause for concern, because this result is evidence that, while traveling along the dual carriageway and glancing toward MM, drivers were making more compensatory adjustments to the steering to correct any path deviations caused by looking away from the road ahead. In other words, they were wavering within their lanes. However, this result was not observed for the approach to the dual carriageway exit on Route 2, and it is not clear as to why such route differences may have arisen.

As expected, the plot of vehicle speed and glance duration shows longer glances tend to be made at slightly slower speeds, although 10% of 2-sec glances were made at speeds in excess of 50 mph. A similar distribution was found for Maps/Notes. This may reflect a driver's increased desire (perceived or otherwise) to watch the road ahead, and not the navigation information source, when traveling at high speeds. It is felt that such plots may be useful when comparing conditions in future studies.

METHOD: FIELD STUDY 2

The method employed for the evaluation of the two symbol-based systems was similar to that for system MM, although (as discussed earlier), it was decided to use only one baseline condition. The conditions used were:

S1. Driving to a destination using route guidance provided by the symbol-based system S1.

S2. Driving to a destination using route guidance provided by the symbol-based system S2.

Instructions. As for Field Study 1.

It was hoped that all 24 of the subjects from the first study could be used in this study. However, only 18 were available, so 6 new subjects were chosen; these new subjects were selected to be similar (in terms of age and driving experience) to those from the first study.

Field Study 2 also used a repeated measures design, with each subject completing all three conditions (S1, S2, and Instructions). Each subject attended two sessions. During their first session, subjects used one of the route guidance systems on one of the two routes. During their second session, they used the second route guidance system on the other route, and then completed their Instructions run (baseline measure) using the route driven during their first session. As Instructions is a measure of normal driving, it was felt that preknowledge of the route would not adversely affect the results. Given the differences in the route-finding algorithms of the three systems, it was impossible to use the same two routes as those used within the first study. However, it was possible to select two matched routes that were similar to those used in Field Study 1. Route 1 was 10.3 miles long and contained 12 decision points (three T junctions, three turn off roads, five roundabouts, and one slip road exit). Route 2 was 8.7 miles long (there was less dual carriageway) and also contained 12 decision points (three T junctions, four turn off roads, four roundabouts, and one slip road exit). Without navigational errors, Route 1 took approximately 17 min to drive, whereas Route 2 took about 16 min. As for Field Study 1, both routes contained a mixture of rural and urban roads, and were driven during off-peak daylight hours with low to medium traffic densities.

The measures taken were very similar to those within Field Study 1. However, due to technical difficulties, steering wheel variability could not be measured in this study.

The training procedure used in this study was as for Field Study 1.

RESULTS: FIELD STUDY 2

Journey Time

Subjects took significantly longer to complete a route with S1 (M: Route 1 = 884 sec; Route 2 = 1,063 sec) than with Instructions (M: Route 1 = 793 sec; Route 2 = 965 sec) ($p < .01$ for both of the test routes). Furthermore, subjects took significantly longer to complete a route with S2 (M: Route 1 = 881 sec; Route 2 = 998 sec) than with Instructions ($p < .05$ for both of the test routes). However, there were no significant differences between S1 and S2 for either of the routes.

Navigational Errors

A similar analysis was conducted for this study as for Field Study 1, and the results are shown in Table 6.5.

TABLE 6.5
Navigational Error Rates (%) for S1 and S2

Type of Maneuver	Type of Error	S1	S2
Dual carriageway exit	Turned too soon	2.1	4.2
	Missed turning	12.5	0
Side turn off	Turned too soon	3.2	0
	Missed turning	1.4	0.5
	Wrong direction	0	0.9
Roundabout	Wrong exit	1.7	3.4
Total error rate (across all maneuvers)		3.1	1.9

TABLE 6.6
Visual Attention Allocation Toward the S1 and S2 Displays

	S1	S2
Mean glance duration (secs)	0.99	0.96
Mean glance frequency (number/sec)	0.182	0.169
Mean percentage of time in motion (%)	18.06*	15.91

*significant at $p < .05$.

Gross Measures of Visual Attention

Table 6.6 shows mean values for glance duration, glance frequency, and the percentage of time in motion spent glancing toward the S1 and S2 displays across both routes.

Analysis of the other areas in the visual scene provided very similar results to Field Study 1. For example, the duration of a glance to the road ahead was significantly less in the system conditions (M: S1 = 2.80 sec; S2 = 2.76 sec) as compared with the Instructions condition (M = 6.65 sec) ($p < .001$ for both systems). Furthermore, the percentage of time in motion spent glancing toward the road ahead was significantly less in the system conditions (M: S1 = 72.4%; S2 = 70.8%) than for the Instructions condition (M = 87.3%) ($p < .001$ for both systems).

Subjects spent a significantly reduced proportion of their journey time glancing toward the rearview mirror for the system conditions (M: S1 = 2.3%; S2 = 2.8%) than for the Instructions condition (M = 4.3%) ($p < .01$ for both systems). The number of glances per unit time made toward the rearview mirror was also significantly less in the system conditions (M: S1 = .033; S2 = .037) than for the Instructions condition (M = .056) ($p < .01$ for both systems).

Approach to a Complex Junction: Visual Attention

As for Field Study 1, drivers' visual behavior was examined on the approach to a number of different junctions. Table 6.7 presents the results for the approach to a complex junction (a large roundabout where the driver had to take the fourth exit), where an interesting disparity was found between the glance duration and frequency variables. Examination of the table shows how the mean duration of glances toward both symbol-based displays generally increased over the initial approach (1,000 m to 200 m from the maneuver) and then reduced markedly over the last 200 m. However, for the glance frequency variable, a distinct system difference was observed. Over the last 200 m, the mean number of glances dropped considerably with S2, but remained high with S1.

TABLE 6.7
Driver Visual Behavior on the Approach to a Roundabout

		1000 m–800 m	800 m–600 m	600 m–400 m	400 m–200 m	200 m–junction
S1	Mean glance duration (sec)	0.72	0.90	1.02	1.10	0.87
	Mean glance frequency (no./sec)	0.15	0.20	0.18	0.28	0.26
S2	Mean glance duration (sec)	0.80	0.81	0.83	0.99	0.82
	Mean glance frequency (no./sec)	0.10	0.15	0.26	0.25	0.08

Approach to a Complex Junction: Indicating and Lane-Changing Behavior

To investigate the result within Table 6.7, three aspects of additional context data were added. First, the mean points at which each of the two systems presented guidance information on the approach to the junction; second, the mean points at which subjects changed lanes; and third, the mean points at which subjects employed their indicators (see Table 6.8).

Perceived Workload

Mean RTLX scores associated with each condition are shown in Table 6.9.

Subjective Opinion

Within this study, the questionnaire revealed a wide range of more positive opinions toward S2 than S1. For example, subjects felt that it was easier to get to their destination, felt more confident, felt safer, and found the position of the display more satisfactory than with S1 ($p < .01$, $p < .001$, p

TABLE 6.8
Indicating and Lane-Changing Behavior
on the Approach to a Roundabout

	Instructions	S1	S2
Mean distance before junction that information presented	600 m	71 m	445 m
Mean distance before junction that subjects moved into right-hand lane	303 m	−60 m (i.e., on roundabout)	210 m
Mean distance before junction that subjects used indicators	203 m	10 m	149 m

TABLE 6.9
Field Study 2: Results From NASA–RTLX Workload Questionnaire

Component	S1	S2	Instructions
Mental demand	42**	45**	15
Mental effort	43*	46*	22
Physical demand	19**	16*	7
Time pressure	26**	22**	9
Distraction	37***	34***	4
Stress	24**	24**	8
Overall score	32**	31**	11

*significant difference between S1/S2 and Instructions, $p < .05$.
**significant difference between S1/S2 and Instructions, $p < .01$.
***significant difference between S1/S2 and Instructions, $p < .001$.

$< .01$, $p < .01$, respectively). Subjects were also asked about their strategies for using the various visual components of S1 and S2. This revealed that, for system S1, subjects felt they relied more on the distance and road name/number information as compared with the graphical representation. The converse result was found for S2.

DISCUSSION: FIELD STUDY 2

There were a number of reasons for errors made while using the symbol-based systems. For S1, the poor design of the countdown bar (each block representing a quarter of the distance to the next maneuver) meant subjects found it difficult to use this source of information effectively. This was because they could not relate the bar to a prior expectation of distance, it was difficult to quantify the distance represented, and there were too few bars to give the driver a clear impression of relative distance changes. Presenting the supporting alphanumeric distance in miles to two decimal places was generally considered useful when there was a large distance to the next turn (e.g., 5.52 miles), but was extremely difficult to interpret on the approach to a maneuver (e.g., .07 miles).

S1 also tended to present information late on the approach to certain maneuvers. The consequence of this poor timing was especially evident along the dual carriageway where subjects were not able to get across to the inside lane in time, and so missed the exit. Errors also occurred for side turns and, once again, poor distance representation may be to blame. However, the errors may also be attributable to the system's limited presentation of road layout information. There was no information (either visual or verbal) regarding side turns before the actual turn; many subjects therefore turned into the first available side turn after the message appeared.

Based on subjects' comments, it is apparent that reduced road layout information (as given by S1) promotes greater reliance on distance and road/street name information. This may be inappropriate for four reasons. First, any system that relies heavily on distance information is likely to require more visual checking of the display on approach to a maneuver. Second, any system where the user relies on text information (such as street names/road numbers) will promote greater amounts of time with the eyes off the road (Dicks, Burnett, & Joyner, 1995). Third, road sign information can be poorly visible, misleading, or missing. Finally, road sign information is often to the side of the road, requiring the driver to look away from the road ahead.

Subjects made less errors with S2, and most of these occurred for technical reasons rather than interface deficiencies. The main interface problem arose where two close turns were grouped together into a single presentation (both visual and voice). Following completion of the first stage of a maneuver, the symbol remained the same and did not turn to represent the driver's view. This caused several subjects to make errors on the second stage of the maneuver, highlighting the need always to match the image to the view outside the windscreen.

Less time was spent glancing toward system S2 than S1. It is felt this result reflects the reduced information within the S2 display (just road layout and text distance), and the suitability of S2's distance and road layout representations in reducing navigational uncertainty.

Indicating and lane-changing behavior were analyzed within Field Study 2 in order to investigate the effect of poor timing of messages on the driving task. The most apparent consequence of late timing was found on the approach to a complex roundabout where information was presented very late by S1. The mean point at which subjects began to indicate to move into the right-hand lane with S1 was 10 m before the junction. In fact, many subjects did not indicate at all with this system and, on average, subjects did not move into the right-hand lane until they were actually on the roundabout. Given the complexity of this maneuver and the amount of traffic generally present, this was a potentially dangerous maneuver. Indeed, late and sometimes sudden lane-changing resulted in other road users sounding their horns at two subjects.

Analysis of drivers' visual behavior when approaching the aforementioned roundabout showed that, when the demands of the driving task are high (slowing down, checking mirrors, indicating, getting into lane, etc.), drivers will compromise between the conflicting demands of driving and navigating by making a relatively large number of short duration glances toward the route guidance display. This result supports the work of Rockwell (1988), who concluded glance duration is sensitive to the demands of the environment, whereas glance frequency is a better measure of poor display/control design. Obviously, the late message timing associated with S1 was largely to blame; however, incompatibility between the timing of the visual and voice components was also at fault. For this maneuver, and several others, the voice messages were presented after their visual counterpart, and it is felt this encouraged subjects to increase their visual checking of the display.

DISCUSSION

Some tentative comparisons can be made between the three systems on the basis of the results from the two studies.

The most apparent negative aspect of all three route guidance systems was the increased time spent with the eyes off the road as compared with the baseline conditions. Prior to the second trial it was hypothesized that the presence of voice instructions with the symbol-based systems would markedly reduce the "visual cost" of their displays, compared with the map-based system. The results have indicated some support for this. For example, there was a reduction in the mean duration of a glance toward the symbol-based systems, as compared with the moving map-based system (see Tables 6.3 and 6.6), which probably reflects the reduced complexity of the displays. However, there was little difference in the mean number of glances per second made toward the three systems. Deficiencies in the design of the symbol-based interfaces (as discussed earlier) are likely to be the main cause of this result, although it is possible that route and subject differences, and system novelty and reliability may have also influenced this.

Use of all three route guidance systems significantly reduced the visual attention allocation toward the road ahead and the rearview mirror, as compared with the Instructions condition. These results replicate previous experimental findings (Ashby, Fairclough, & Parkes, 1991; Fairclough, Ashby, & Parkes, 1993) and form further evidence for Rumar's (1988) descriptions of the side effects of visual workload (i.e., that "spare" visual resources are allocated to the display at the expense of other areas in the visual scene).

Based on the response of those subjects who experienced all three systems across the two studies, S2 was by far the most preferred system. Subjects' comments confirmed that it offered increased functionality, overall better performance, a superior input mechanism, greater reliability, and better message timing.

However, favorable comments were also made regarding the ease of use and safety of the other two systems. There were several subjects who felt particularly positive about the moving map-based system over the symbol-based systems, and it is possible that future systems should adopt a level of adaptation to cater for such individual preferences.

It must be noted that, although subjects may genuinely hold such opinions, there could be a novelty factor associated with such new technology. Longitudinal studies would overcome such effects but they are inevitably associated with high financial costs; indeed, it could be argued that a snap-shot study such as this has the advantage of allowing the worst case safety scenario (i.e., the initial use of a system) to be tested. However, it is likely that as the technology matures, such longitudinal studies may become more feasible.

A number of issues arose in the two studies that highlight the problems of evaluating first generation systems whose technology is, as yet, not 100% proven. First, several preferred routes had to be rejected for the studies, given differences in route selection algorithms between the systems, and errors with mapping data that resulted in the systems recommending drivers to make dangerous or illegal maneuvers. Second, insufficient processing speed meant that there was a lag in the moving map-based system and this caused a number of navigational errors, as discussed earlier. Finally, the tracking system for S1 was not totally reliable and, on a few occasions, the system would "drift" off the designated route, thus failing to provide guidance. This necessitated the rejection of some data in Field Study 2.

CONCLUSIONS AND FUTURE WORK

The trials revealed a number of deficiencies in the interfaces to all three route guidance systems. For instance, with the moving map-based system the lag and rotating aspects of the display meant that driver uncertainty was sometimes high and a number of errors subsequently occurred. With the symbol-based systems, the representation of distance to a maneuver caused some problems and, although recommendations for change can be made, further work is required to ascertain the optimum means of representing such information.

The approach adopted by S1 (i.e., simple standard symbols together with distance and street name/number) was found to produce a number of difficulties for subjects. However, such a technical solution lends itself to the presentation of information on a simple, low-cost display and so may be desirable for system designers/manufacturers. In addition, good human factors practice, such as the use of aerial view symbols (Green & Williams, 1992) and the integration of distance countdown bars within the road layout representation, can enhance the usability of standard symbols. Therefore, if such an approach is to be adopted, careful consideration of the various components of the interface should take place to ensure their optimization; this is particularly important for the distance representation and the content/structure of voice messages.

All three route guidance systems resulted in significant eyes-off-the-road time. However, the difficulty lies in deciding what constitutes "unacceptable" or "unsafe" behavior. The techniques introduced here have attempted to address this by adding context to visual behavior, thus aiding in the examination of systems in terms of potential safety-related issues.

As an example, there is some evidence that, for high visual demand situations, a moving map-based system results in increased steering wheel variability, which indicates that subjects were wavering within their lanes. However, this result was not conclusive and more controlled research is required to test this hypothesis fully.

It was clear that, for particular maneuvers, the information presented by the symbol-based systems (especially S1) was too late. Furthermore, it is clear that the resulting driving behavior has strong implications for driver safety. However, it is impossible, based on the results of this analysis, to be sure what the optimum timing of route guidance messages is. Even in the system conditions, where no obvious unsafe driving occurred, drivers did not indicate or change lanes as early as for the "ideal" (i.e., the Instructions condition). Again, further research is needed to determine guidelines for the timing of route guidance messages.

ACKNOWLEDGMENTS

This work was carried out as part of the Basic Research in Man–Machine Interaction (BRIMMI) project, conducted within the EUREKA-supported PROMETHEUS program. It was supported in part by the U.K. Department of Trade and Industry and the motor manufacturing partner was the Ford Motor Company Limited.

REFERENCES

Alm, H. (1990). Drivers' cognitive models of routes. *DRIVE 2 Project VI041 (GIDS)*, Contribution to deliverable GIDS/NAV2. Linköping, Sweden: VTI.

Ashby, M., Fairclough, S. H., & Parkes, A. M. (1991). A comparison of two route information systems in a urban environment. *DRIVE 1 Project VI017 (BERTIE)*, Report no. 47. Loughborough, UK: HUSAT Research Institute.

Barrow, K. (1990). *Human factors issues surrounding the implementation of in-vehicle Navigation and Information systems* (SAE Tech. paper 910870).

Bengler, K., Haller, R., & Zimmer, A. (1994). Experimental optimisation of route guidance information using context information, "Towards an Intelligent Transport System," *Proceedings of the First World Congress on Applications of Transport and Intelligent Vehicle-Highway Systems, 4*, 1758–1765.

Burnett, G. E., & Parkes, A. M. (1993). The benefits of "Pre-Information" in route guidance systems design for vehicles. In E. J. Lovesey (Ed.), *Contemporary ergonomics* (pp. 397–402). London: Taylor & Francis.

Dewar, R. E. (1988). In-vehicle information and driver overload. *International Journal Vehicle Design, 9*(4/5), 557–564.

Dicks, L., Burnett, G. E., & Joyner, S. M. (1995). An evaluation of different types and levels of route guidance information. In S. A. Robertson (Ed.), *Contemporary ergonomics* (pp. 193–298). London: Taylor & Francis.

Fairclough, S. H. (1991). Adapting the TLX to assess driver mental workload. *DRIVE 1 Project V1017 (BERTIE)* (Rep. No. 71). Loughborough, UK: HUSAT Research Institute.

Fairclough, S. H., Ashby, M. C., & Parkes, A. (1993). In-vehicle displays, visual workload and usability evaluation. In A. G. Gale (Ed.), *Vision in vehicles* (Vol. 4). Amsterdam: Elsevier Science.

Green, P., & Williams, M. (1992). Perspective in orientation/navigation displays: a human factors test. In *Proceedings of the Third International Conference on Vehicle Navigation and Information Systems* (pp. 221–226). Piscataway, NJ: IEEE.

Hart, S. G., & Staveland, L. E. (1988). Development of NASA-TLX (Task Load Index): Results and theoretical research. In P. A. Hancock & N. Meshkati (Eds.), *Human mental workload* (pp. 139–183). North Holland: Elsevier Science.

Joint, M., Bonsall, P. W., & Parry, T. (1990). *Drivers' basic requirements for route guidance information* (Technical Note 271). United Kingdom: Institute for Transport Studies, University of Leeds.

Kishi, H., & Sugiura, S. (1993). *Human factors considerations for voice route guidance* (SAE Tech. paper 930553).

Labiale, G. (1989). In-car navigation VDUs—A psycho-ergonomic study. In A. G. Gale et al. (Ed.), *Vision in vehicles* (Vol. 3). Amsterdam: Elsevier Science.

Parkes, A. M. (1989). Changes in driver behaviour due to two modes of route guidance information presentation: A multi-level approach. *DRIVE 1 Project V1017 (BERTIE)* (Report No. 21). Loughborough, UK: HUSAT Research Institute.

Parkes, A. M. (1991). Data capture techniques for RTI usability evaluation. In *Advanced telematics in road transport, proceedings of the DRIVE Conference* (Vol. 2, pp. 1440–1456). Amsterdam: Elsevier Science.

Pauzie, A., & Marin-Lamellet, C. (1989). Analysis of aging drivers behaviours navigating with in-vehicle visual display systems. In *Proceedings of the First IEE/IEEE Vehicle Navigation and Information Systems Conference* (pp. 224–249). Piscataway, NJ: Institute of Electrical and Electronic Engineers.

Rockwell, T. H. (1988). Spare visual capacity in driving—revisited. In A. G. Gale et al. (Eds.), *Vision in vehicles* (Vol. 2, pp. 317–324). Amsterdam: Elsevier Science.

Ross, T., Nicolle, C., & Brade, S. (1994). An empirical study to determine guidelines for optimum timing of route guidance instructions. *DRIVE 2 Project V2008 HARDIE*, Deliverable 13.2.

Rumar, K. (1988). In-vehicle information systems. *International Journal of Vehicle Design, 9*(4–5), 548–556.

Schraggen, J. M. C. (1990). *Strategy differences in map information use for route following in unfamiliar cities: Implications for in-car navigation systems* (Rep. No. IZF 1990 B-6). Soesterberg, the Netherlands: TNO Institute for Perception.

Schraggen, J. M. C. (1991). *An experimental comparison between different types of in-car navigation information* (Rep. No. IZF 1991 B-1). Soesterberg, the Netherlands: TNO Institute for Perception.

Streeter, L. A., Vitello, D., & Wonsiewicz, S. A. (1985). How to tell people where to go: Comparing navigational aids. *International Journal of Man–Machine Studies, 22,* 549–562.

Todoriki, T., Fukano, J., Obabayashi, S., Sakata, M., & Tsuda, H. (1994). Application of head-up displays for in-vehicle navigation/route guidance. In *Proceedings of the IEE/IEEE Vehicle Navigation and Information Systems Conference* (pp. 479–484). Piscataway, NJ: Institute of Electrical and Electronic Engineers.

Van Winsum, W., van Knippenberg, C., & Brookhuis, K. (1989). Effect of navigation support on drivers' mental workload. *Proceedings of European Transport and Planning 17th Annual Meeting, 1,* 69–84.

Verwey, W. B. (1993). Further evidence for benefits of verbal route guidance instructions over symbolic spatial guidance instructions. In D. H. M. Reekie (Ed.), *Proceedings of the IEE/IEEE Vehicle Navigation and Information Systems Conference* (pp. 227–231). Piscataway, NJ: Institute of Electrical and Electronic Engineers.

Verwey, W. B., & Janssen, W. H. (1988). *Route following and driving performance with in-car route guidance systems* (Rep. No. IZF 1988 C-14). Soesterberg, the Netherlands: TNO Institute for Perception.

Wallace, R. R., & Streff, F. M. (1993). Traveler information in support of drivers' diversion decisions: A survey of driver's preferences. In D. H. M. Reekie (Ed.), *Proceedings of the IEE/IEEE Vehicle Navigation and Information Systems Conference* (pp. 242–250). Piscataway, NJ: Institute of Electrical and Electronic Engineers.

Wierwille, W. W., Antin, J. F., Dingus, T. A., & Hulse, M. C. (1988). Visual attentional demand of an in-car navigational display system. In A. G. Gale et al. (Eds.), *Vision in vehicles* (Vol. 2, pp. 307–316). Amsterdam: Elsevier Science.

Chapter 7

Design Guidelines for Route Guidance Systems: Development Process and an Empirical Example for Timing of Guidance Instructions

Tracy Ross
Gill Vaughan
Colette Nicolle
HUSAT, Loughborough University, Leics., UK

During the development of route guidance or navigation systems, there are many stages in the design process requiring the input of human factors knowledge if the final system is to be usable and safe. A good review of the questions that need to be considered at each stage is provided in Ashby and Parkes (1993). These questions include: Who are the users?, What information do users need?, How should the information be presented?, and What are the consequences of interface design on the driver? Each has equal importance and the CEC DGXIII Transport Telematics Program (also referred to as DRIVE II) was instrumental in supporting the research needed to answer them. This chapter details some of the activity of a DRIVE II research project that aimed to answer the question, How should the information be presented?, by developing design guidelines in four intelligent vehicle highway systems (IVHS) application areas. The first half of this chapter describes the process used for the development of guidelines in one area, route guidance and navigation. The main stages were expert evaluations of existing systems, a review of existing guidelines, an identification of the main gaps in existing guidelines, and the development of new guidelines. A hierarchical structure of human factors design issues was developed as a framework on which to present results throughout the project. The second half of the chapter describes an example of empirical work conducted to develop new guidelines for one of these design issues. The study described aimed to develop prescriptive guidelines for the timing of guidance instructions.

PROCESS FOR THE DEVELOPMENT OF GUIDELINES

Expert Evaluation of Existing Systems

The systems evaluated were at various stages of development. They included Bosch TravelPilot™, Philips Carin™, Autoguide™/LISB™, and Travtek™. The aim of the expert evaluation was to identify the design aspects of existing systems that do and do not adhere to good ergonomic principles. These contributed to the HARDIE guidelines as illustrative examples of best practice as well as examples of what to avoid. The evaluations also identified those areas of design where human factors experts are unsure as to best practice, that is, the areas where further research toward guidelines is needed.

The expert evaluations proved useful in gathering detailed information on good and poor design aspects of existing systems. Some examples of good design aspects are: "The location of related information is consistent." "An alerting auditory signal is given if new information appears on the screen." Some examples of poor design aspects are: "Roundabout symbols do not correspond to the actual layout." "The rapid sequence of three commands close together caused difficulty." This information contributed to the development of design guidelines as illustrative "examples" for the related design issue. For certain aspects of design, it was clear that human factors experts are unsure of best practice due to lack of empirical evidence. This information helped identify the gaps in existing guidelines discussed later.

Review of Existing Guidelines

The aim for this part of the work was to discover easily accessible sources of information containing guidelines that could be applied by designers. Thus, the search excluded research findings requiring interpretation before they could be useful in the design process. Although there is a plethora of the latter information, it is, in the main, of more use to the academic world than to the design teams of vehicle and equipment manufacturers.

The availability of easily applied guidelines for route guidance and navigation systems appears to be limited. The review showed that current sources tend to be written in an informative fashion rather than for immediate application to system design. Three references provide concise guidance. Antin (1993) covered timing of information, duration of presentation, orientation of information, the use of windows, spatial relations, personalizing information, reducing information overload and distraction from the driving task, and the advantages of combining visual and auditory information. Stokes, Wickens, and Kite (1990) covered when to use different presentation types, the use of colors and their stereotypes, timing of information, orientation of information, and the benefits of presenting visual and auditory information in combination. Recommendations from Schraagen (1993) covered information density, harmonization of vehicle and road information, and orientation.

Identification of Gaps in Existing Guidelines

By taking the information from the review of guidelines, and using the hierarchy of design issues, it was possible to form a list of issues for which no design support was available. This, together with the results of the expert evaluations, helped decide the issues requiring further research. These results were supported via consultation with designers. The questions fell into five main areas.

Choice of Information Presentation Mode. Few guidelines exist for the best mode, or combination of modes, to use (i.e., visual, auditory, or more uncommonly tactile). Some guidelines do exist for generic applications, for example, visual information should be used if the message needs to be referred to later, or if it is long and complex, but they would be much easier to apply to route guidance and navigation systems if they were presented in the context of the vehicle environment and accompanied by appropriate examples. When different presentation modes are used together, guidelines are needed to avoid conflicting or redundant information. Individual preference and the requirements of drivers with special needs must also be considered.

Design of Information Presentation Mode. Once the mode of information presentation has been chosen, the designer requires guidelines on the content, structure, and layout of that information (e.g., for visual information, should text, symbols, or video be used?). Some specific questions are:

Should auditory messages be repeated automatically, or on request?

What is the best structure for the information contained in a text message?

What color-coding stereotypes exist? Are these common across Europe?

Optimum Timing of Guidance Instructions. This is particularly important to safety and efficiency of the system. If a guidance instruction is provided too early or too late, it may cause the driver to make an incorrect or dangerous maneuver or miss the turn completely. No specific guidance is available with regard to timing, probably because *optimum* timing depends on several variables: vehicle position on a multilane road, the vehicle speed, the positioning and density of surrounding traffic, and the geometry of the maneuver itself. The following are relevant questions in this area:

Should there be a pre-announcement of an instruction? If so, how long beforehand?

When time between maneuvers is short, should instructions be "stacked"?

Should the timing of in-vehicle information be linked to the related external information?

Orientation and Harmonization of Vehicle and Road Information. The nature of a route guidance or navigation system means that it must convey an idea of the driver's location and direction, for example:

How should the information describing distance to the next maneuver be presented?

Which parts of a display (i.e., the representation of the vehicle and of the road network) should be static and which dynamic? Does this depend on the type of maneuver?

What landmarks should be used to aid navigation and are they international?

Driver/System Interaction. This area covers issues of the driver's control over the system, as well as the effect of the system on the driver. Some of the issues in this area relate to the driver input to the system and the dialogue. These areas of design were outside the defined scope of HARDIE, but some still have implications for the design of the presented information, for example:

What should be the maximum complexity of information presented at any one time?

How should incoming messages (e.g., on traffic congestion) be filtered?

Development of New Guidelines

At the conclusion of the HARDIE project, guidelines for route guidance and navigation systems were published as part of a design guidelines handbook. The guidelines concentrated on those issues that are safety critical, a high priority for the designers consulted by the project, and vital to the functioning of every route guidance system. The handbook development took place at three levels.

First, guidelines from existing sources were collated to ensure that there was no unnecessary duplication of work. Second, existing research literature was reviewed to identify results that could support new guidelines. These results were not in the form of guidelines and were therefore inaccessible to designers. The task was to determine which valid empirical findings showed a consensus of results and could be used as the rationale for a new guideline. The final level of research was to conduct new empirical studies for priority design issues where, currently, there are gaps.

EMPIRICAL STUDY: TIMING OF GUIDANCE INSTRUCTIONS

This section describes one example of new empirical studies conducted within the project. One gap identified in existing guidelines was information on the optimum timing of guidance instructions. This design issue was chosen for investigation because it met the project criteria: safety critical, a high priority for designers, and vital for the function of all route guidance systems.

Previous Research

For a route guidance system to be acceptable to drivers, it must meet their expectations based on their previous navigation experience. Also, the system output to the driver should be designed in such a way as to maintain or improve current levels of safety. Most certainly, it must not be detrimental to safe driving as a result of driver distraction, lack of confidence, or erratic vehicle control.

The topic of this investigation—the appropriate timing of a route guidance instruction—is critical to system acceptability and safety. If an instruction is given too early, the driver is likely to forget the information while attending to other driving tasks. If the instruction is given too late, it will either cause the driver to brake suddenly or to maintain safe driving but miss the maneuver completely. The research literature provides some information on problems created by poor timing of instructions. Fairclough, Ashby, Lorenz, and Parkes (1991) showed that drivers reported a high temporal demand resulting from instructions being given close to a junction. However, finding detailed information on values (in terms of time or distance) resulting in appropriate and acceptable timing is more difficult.

One approach has been to relate the timing of an instruction to external cues or objects rather than specific time or distance intervals. Schraagen (1990) stated that on main roads (with signs), the next guidance instruction should be presented immediately upon completion of the last maneuver.

He also recommended that on highways, the instruction should be given as soon as the drivers can see the related road sign. In normal weather conditions, this distance is 1–1.5 km. Relating guidance to existing road signs is also supported by Kishi and Sugiura (1993) and is advisable as a way of increasing driver confidence in the system. This approach to timing is applicable to roads where signs are located, providing the route guidance system has knowledge of the placement of these signs. However, for roads without signs, or where signs are only visible once the maneuver has been reached, timing needs to be based on other factors, probably independent of any external cues.

The most obvious criteria on which to base the timing of an instruction is distance to the junction or predicted time to the junction (based on current speed). Only two sources have been found that attempt to quantify timing in this way. The first is based on time to the maneuver. Verwey et al. (1993) stated that if two maneuvers are less than 10 sec apart, then the two instructions from the route guidance system should be given together, prior to the first maneuver. Making an assumption based on this, it would seem that the advice is that any single instruction needs to be given more than 10 sec prior to a maneuver. This equates to a distance of 167 m at a speed of 60 kph or 278 m at a speed of 100 kph. Another study (Kimura, Maranuaka, & Sugiura, 1994) with a small number of subjects ($n = 6$) specified that on "surface roads" in Japan, a distance of 700 m was needed to enable a lane-change maneuver and a distance of 300 m prior to a junction where the driver was required to turn across the flow of traffic (it is not clear if oncoming traffic was in motion or stationary). The study also states that vehicle speed, passing vehicles, and number of traffic signals passed will affect appropriate timing of the instruction. However, there is no indication of how timing should be adjusted according to these varying conditions.

The evidence from the literature pointed to the fact that more work was needed on the timing of route guidance instructions and justified an investigation to produce quantifiable results.

Method

Fifteen subjects were used (age range = 24–62; $M = 42$). All had experience of driving on all types of road. The whole procedure took 2 hr.

The passenger (Experimenter 1) had a fixed list of navigation instructions that were short, precise, and contained an intrinsic prompt ("Take the next left turn" or "Take the second right turn"). The timing of the instructions was based on the experimenters subjective assessment of prevailing conditions, and were balanced to ensure an equal number of late/okay/early instructions for each subject and each junction. Auditory instructions alone were used, as this is well accepted to be the most appropriate primary source of guidance information. They were deliberately short so as not to introduce a memory effect into the study. In a working system, visual displays should be used for back-up information, and should include direction of next maneuver, current position, road layout, distance to next maneuver, landmarks, street names/relevant road signs, and distance to destination. This was not done within this study, as the chosen design of the display may have influenced the results.

The driver (subject) may have had knowledge of the roads used but did not know the route to be taken. The driver commented whether the instruction was too early/too late according to a 6-point rating scale.

Experimenter 2 was in the back of car to operate the equipment. The data logger was started when the instruction was given and stopped when the maneuver was reached.

The main consideration of the study was to identify the factors that may affect timing and, to select those that were the most important, those that current/future systems could detect/measure, and those that the current study could control, select for or measure.

The maneuver chosen was "leaving the current route," that is, turning into a side road or slip road. Timing of instructions is very important for this type of maneuver. "End of road" maneuvers (e.g., at T junctions or roundabouts) were not included, as they are of lower priority with regard to the importance of timing; drivers have to slow down anyway at these points. Lane changing is an important maneuver and should be targeted for further investigation. The experimental route was selected to include 18 maneuvers with a range of approach speeds and a range of angles.

Independent variables were:

- Angle of maneuver: $0°–60°$ ($n = 7$), $61°–120°$ ($n = 6$), $121°–180°$ ($n = 5$)
- Speed at instruction
- Right ($n = 7$) or left ($n = 11$) turn
- Traffic density: low, medium, high. This was based on subjective assessment by the experimenter. Two different experimenters took part in the trials and a pretrial ensured consistency of assessment.
- Lane changes required to complete the maneuver.

Dependent variables were:

- Time to maneuver when the instruction was given
- Distance to maneuver when the instruction was given
- Onset of indicator use

- Onset of deceleration
- Onset of brake activation
- Time of peak deceleration
- Time of peak braking
- Peak deceleration value
- Peak brake depression value

The subject was also asked to give a subjective rating of the timeliness of the instruction at each maneuver, using the following scale: 1—much too early; 2—bit too early; 3—tiny bit too early; 4—tiny bit too late; 5—bit too late; 6—much too late. The precise meaning associated with each rating was explained to the subjects and did not cause any difficulty.

A pretrial drive ensured familiarization with the vehicle to eliminate training effects. Each subject drove the same experimental route. Order effects were minimized in the analysis as, for each independent variable, variations in, for example, angle of maneuver, were spread over the whole route. The analysis was not conducted on a junction-by-junction basis, but combined data based on the independent variables.

Results

A first pass through the data showed that some junctions were eliciting different behavior from the majority. These junctions proved to be those where a slip road was provided, making the transition from the main route to the side road a more gradual process than for the majority of junctions. For the analysis, these two types of junctions were separated. Those without a slip road were termed "Side Road Maneuvers" ($n = 14$); those with a slip road were termed "Slip Road Maneuvers" ($n = 3$). One junction was removed from the analysis completely as it proved to be two maneuvers in one, rather than a single maneuver.

Throughout the analyses, a hypothesis was accepted if the following criteria were reached: a correlation $\geq \pm.3$ (in the predicted direction) and a p value of $\geq .05$.

Step 1

The detailed analysis to be conducted was based on the subject response rating scale, therefore it was imperative to ascertain that the information provided by it was valid. The rating scale was created for this study and had not been formally validated. For the purposes of this study, validity was checked by performing a correlation analysis between paired values of rating and vehicle control data.

Hypothesis 1. Later instructions (reflected by a higher rating) would definitely be accompanied by decreased distance and time to the junction and could possibly be accompanied by decreased time before the junction for onset of indication, peak deceleration, peak braking, start of braking and start of deceleration, and by increased peak braking and peak deceleration.

The main part of the hypothesis was accepted for both side road and slip road maneuvers, giving validity to the rating scale and permitting further analysis on this basis.

Step 2

The next step was to determine if there was a relation between vehicle speed and distance when the instruction was given. If this was the case, regression analyses could provide an equation for timing based on speed alone.

Hypothesis 2. For each value on the rating scale (1–6), a higher speed is accompanied by a longer distance to the junction (at the point when the instruction is given).

For side road maneuvers, the hypothesis was accepted for Responses 2–6 and rejected for Response 1. A strong linear relation exists between vehicle speed and the appropriate distance at which to give a guidance instruction. Further analysis will show the influence of the other independent variables.

For slip road maneuvers, the hypothesis was rejected for all responses and analysis ceased at this point.

Step 3

The next stage of the analysis concentrated only on side road maneuvers. Also, the analysis concentrated on Responses 3 and 4 only. This was because these two center ratings were considered to be the subjects' determination of an "okay" timing and it is this timing in which designers are interested. Once a relation had been established between speed and distance to the maneuver, it was of interest to determine whether any of the other variables under consideration had an influence on this relation. First it was necessary to identify which subgroups of data had sufficient observations to show a significant correlation between speed and distance to maneuver.

Hypothesis 3. For each subgroup within each variable, higher speed is accompanied by a longer distance to the junction (at the point when the instruction is given).

The results showed that it was only possible to look at the effect of the following subgroups (i.e., those where the hypothesis was accepted): low versus medium traffic density, left versus right turns, angles of maneuver of 61°–120° versus 121°–180°.

Regression lines were plotted for these three comparisons. The graphs showed that traffic density (low vs. medium) does not have an effect on appropriate timing; direction of turn (left vs. right) does have an effect on appropriate timing, but it is not consistent; angle of maneuver (61°–120° vs. 121°–180°) does not have an effect on appropriate timing.

Resulting Guideline

The results showed that, from this study, a prediction of optimum timing (in terms of distance) can be based on vehicle speed alone. The guideline was produced in the following way. The paired data for "distance to the junction at the instruction" and "speed at the instruction" for Responses 3 and 4 were used in the analysis. The reason for this was that these two ratings represented subjective opinion of an "okay" timing. Regression lines were plotted and associated equations were calculated for preferred minimum distance (based on data for the Subject Rating 4—tiny bit too late); preferred maximum distance (based on data for the Subject Rating 3—tiny bit too early); and ideal distance (based on combined data for Subject Ratings 3 and 4).

In the early stages of developing a route guidance system, it will be necessary to determine an algorithm to calculate at what point prior to a maneuver the driver requires the instruction. From the results of this study, the following advice can be offered.

For maneuvers meeting the criteria—leaving the current route; not requiring a lane change; not at the end of a road; without a slip road (i.e., a sudden turn); for speeds between 18 kph and 101 kph—the timing of a guidance instruction can be based on the equations provided in Table 7.1 (shown graphically in Fig. 7.1).

TABLE 7.1
Equations for Appropriate Timing of an Instruction

Equations		Calculated Values (meters) for 20 kph Steps				
		20	40	60	80	100
Preferred Minimum Distance	= (Speed × 1.637) + 14.799	48	80	113	146	178
		(22)	(14)	(8)	(14)	(22)
Ideal Distance	= (Speed × 1.973) + 21.307	61	100	140	179	219
		(25)	(16)	(11)	(14)	(25)
Preferred Maximum Distance	= (Speed × 2.222) + 37.144	82	126	170	215	259
		(44)	(27)	(16)	(23)	(40)

Note. Figures in parentheses indicate the 95% confidence intervals (±) for the calculated distances.

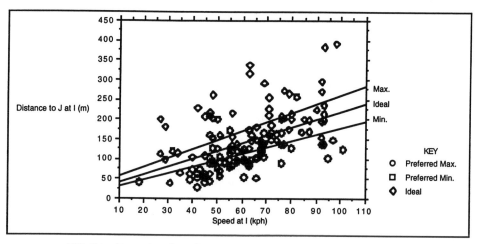

FIG. 7.1. Regression lines for appropriate timing of an instruction. J = Junction, I = Instruction.

To err on the side of caution, it may be advisable to implement the equation for preferred maximum distance. An instruction given slightly too early is preferable to one given too late, particularly if it is possible to request an auditory repeat of the last message.

As with all aspects of system design, the timing of instructions for a particular system should also be investigated through user trials to determine acceptability by drivers at several maneuvers.

DISCUSSION

The empirical study reported here is, to the authors' knowledge, the first of its kind, although important work is also taking place at the University of Michigan Transportation Research Institute that has yet to be published. The study has the limitation that only one particular type of junction was investigated, namely, turns off the current route. Although this means that, strictly, the results should only be applied to this type of maneuver, it is likely that, with some thought, they could be more widely used. Two other types of maneuver were identified earlier, namely, turning at the end of a road (this includes giving-way at a roundabout) and changing lanes. For the former maneuvers, which require the driver to stop or give way, an acceptable timing is likely to be later than for the maneuvers in this study. The guidelines given in this report could therefore be used as a "best guess" for these end-of-the-road maneuvers, in the absence of any other data. For maneuvers that include changing lanes, the guidelines can only be used as minimum criteria. A separate study needs to be carried out to determine the factors affecting these types of maneuver in isolation. The results could then be combined with those here to determine appropriate timings for any possible type or combination of maneuver.

As well as providing timing advice for single turns, the results could be used in decisions regarding when to "stack" instructions (i.e., give instructions for two subsequent maneuvers in a single message). According to the rule: When the distance between two subsequent maneuvers is less than the minimum preferred distance for that speed, the instructions should be "stacked."

Because the results of this study focused mostly on turns from a single carriageway into a side road, there was no influence of road sign information. In the case of maneuvers where a road direction sign is present, the driver must be given the guidance information in good time prior to the sign. This enables the driver to match the information from the system to that in the external environment. The exact timing will be dependent on the length of the in-vehicle guidance message and the complexity of the external information. Within this study the onset of the instruction was defined as the start of the verbal instruction and the instruction given was quite short (e.g., "Take the next left turn"). Long verbal messages are inadvisable due to their temporal nature but, should a long message be necessary in an exceptional case, the timing of the instruction should take this into account.

CONCLUSIONS

HARDIE has identified the need for a single source of guidelines covering information presentation by route guidance and navigation systems. Such literature did not exist in a form that was accessible and easily applied by designers. Research conducted within the project produced a set of guidelines based on current human factors knowledge. The empirical study reported has made an important step in the area of timing of route guidance instructions. Previous research in the area has been minimal and the detrimental effects of this can be seen in current prototypes of such systems. This study did not have the resources available to study all the maneuvers of interest, nor all the variables that could affect timing. However, it did result in a concrete guideline for a particular type of maneuver and has highlighted points that should be taken into consideration in future studies.

The guidelines produced are by no means complete, nor are they unchangeable. New, as yet unforeseen, technologies may require adaptation of the guidelines, although it is intended that the guidelines be as technology independent as possible. One of the main advantages of presenting such guidelines is that they provide a basis for discussion among experts in the area, thus enabling addition to, or refinement of, their content.

Such activity can only benefit the usability and safety of such systems and ensure that designers are aware of best practice.

The handbook also contains guidelines on traffic and road information, collision avoidance, and autonomous intelligent cruise control. The format of the handbook is designed to provide a succinct guideline with supporting examples where necessary, and additional information on the rationale behind the guideline, further relevant reading, and other guidelines of relevance.

The guidelines will provide a summary of current best ergonomic practice for IVHS information presentation. This will be a valuable source of information for designers (the prime target audience), for behavioral researchers, and for standards bodies (particularly ISO TC22 SC13 WG8 and ISO TC 204 WG13).

ACKNOWLEDGMENTS

The authors wish to acknowledge the valuable contribution of others in the HARDIE consortium: Transport Research Laboratory (UK); INRETS-LESCO (F); BAe Sowerby Research Centre (UK); Universidad Politécnica de Madrid (E); TÜV Bayern e.V. (D) and TÖI (N). The HARDIE project was supported by the CEC DGXIII Transport Telematics Programme. Some of the material contained in the section "Empirical Study: Timing of Guidance Instructions" was previously published in Ross and Brade (1995). (Inclusion in this chapter is in accordance with the copyright rules.)

REFERENCES

Antin, J. F. (1993). Informational aspects of car design: Navigation. In B. Peacock & W. Karwowski (Eds.), *Automotive ergonomics* (pp. 321–335). London: Taylor & Francis.

Ashby, M. C., & Parkes, A. M. (1993). Interface design for navigation and guidance. In A. M. Parkes & S. Franzen (Eds.), *Driving future vehicles* (pp. 295–310). London: Taylor & Francis.

Fairclough, S. H., Ashby, M. C., Lorenz, K., & Parkes, A. M. (1991). A comparison of route navigation and route guidance systems in an urban environment. *Proceedings of 24th ISATA International Symposium on Automotive Technology and Automation*, 659–666.

Kimura, K., Maranuaka, K., & Sugiura, S. (1994). Human factors considerations for automotive navigation systems. In S. McFadden, L. Innes, & M. Hill (Eds.), *Proceedings of the 12th Triennial Congress of the International Ergonomics Association: Vol. 4. Ergonomics and design* (pp. 162–165). Ontario, Canada: Human Factor Association of Canada.

Kishi, H., & Sugiura, S. (1993). *Human factors considerations for voice route guidance* (Tech. Paper No. 930553). Warrendale, PA: Society of Automotive Engineers.

Ross, T., & Brade, S. (1995). An empirical study to determine guidelines for optimum timing of route guidance instructions. In *Proceedings of the Colloquium on Design of the Driver Interface* (Digest No. 1995/007, pp. 1/1–1/5). London: Institution of Electrical Engineers.

Schraagen, J. M. C. (1990). Use of different types of map information for route following in unfamiliar cities. In *CEC DRIVE I Project V1041, Deliverable GIDS/NAV2*, 49–66.

Schraagen, J. M. C. (1993). Information presentation in in-car navigation systems. In A. M. Parkes & S. Franzen (Eds.), *Driving future vehicles* (pp. 171–185). London: Taylor & Francis.

Stokes, A., Wickens, C., & Kite, K. (1990). *Display technology—Human factors concepts.* Warrendale, PA: Society of Automotive Engineers.

Verwey, W. B., Alm, H., Groeger, J. A., Janssen, W. H., Kuiken, M. J., Schraagen, J. M., Schumann, J., van Winsum, W., & Wontorra, H. (1993). GIDS functions. In J. A. Michon (Ed.), *Generic intelligent driver support: A comprehensive report on GIDS* (pp. 113–144). London: Taylor & Francis.

Chapter 8

Human Factors Considerations for Automotive Navigation Systems— Legibility, Comprehension, and Voice Guidance

Kenji Kimura
Kenji Marunaka
Seiichi Sugiura
Toyota Motor Corporation, Aichi, Japan

An in-vehicle navigation system provides current and local area road information and route guidance to a destination. This can reduce driver stress and improve route efficiency, thereby contributing to safe and economic driving. Providing road information and route guidance is complicated by the fact that the driver's attention is divided between many simultaneous tasks: visual scanning of navigation system display/controls, motor and visual tasks involved with primary vehicle controls (steering, pedal, controls, turn signals), and visual road and traffic information perception in the direct and indirect field of view. Accordingly, the human interface of a navigation system must be developed so that the driver may perceive guidance information as unobtrusively as possible while driving.

Zwahlen, Adams, and Debald (1988) proposed that a 1–2 sec average duration per look at car displays or controls is the guide to be used when designing sophisticated in-vehicle displays, or CRT touch panel controls and/or applications. Wierwille, Antin, Dingus, and Hulse (1988) investigated glance time for drivers to observe and operate in-vehicle displays and controls (conventional and navigation systems).

In order to ensure that the driver can assimilate route guidance information from the navigation system within the available glance time, character size, color combinations of display characters, map display, and amount of route information should be considered. There are a few studies on the legibility and comprehension of displayed information in the field of automotive navigation systems. Shekhar, Coyle, Shargal, and Kozak (1991) investigated the relation between the response time and the map orientation (north, south, east, and west).

Voice route guidance may reduce the visual workload of the driver compared to a map display. Streeter, Vitello, and Wonsiewicz (1986) showed that the performance of drivers with voice guidance was better than that with a route map. In order to make voice guidance systems effective, it is important to consider the information required for drivers to find directions (Freundschuh, 1989), the timing of the information (Ross & Brade, 1995), and expression for audibility and comprehension.

In Japan, automotive navigation systems with map display, route guid-ance display, and voice route guidance have been developed. This chapter summarizes previous studies on the software specifications for automotive navigation systems (Kimura, Osumi, & Nagai, 1990; Kishi, Sugiura, & Kimura, 1992; Kishi & Sugiura, 1993).

AVAILABLE ATTENTION DIVERSION TIME

Depending on traffic conditions, the driver visually monitors the forward field of view, and periodically checks the rear field through the rearview mirrors. When there is enough available time, the driver may glance at in-vehicle displays and controls in Fig. 8.1 (speedometer, radio, air-condi-tioning control panel, navigation system, etc.). Wierwille (1993) proposed a visual time-sharing model (on driving/in-vehicle). This *available attention diversion time* has been investigated for navigation system software and interface design.

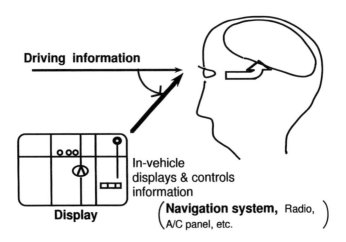

FIG. 8.1.   Driver model.

Fixation Time for Conventional In-Vehicle Displays and Controls

A study investigated the *fixation time* of drivers as they observe and operate conventional in-vehicle displays and controls on a rural expressway and urban streets through video monitoring of the driver's eyes. The fixation time was defined as the interval between the initial and final time in which the driver's eyes are on the displays and controls. Five subjects participated in this study with two repetitions. The results are shown in Fig. 8.2 (Kishi et al., 1992). The fixation time on a rural expressway was longer than that on urban streets. The fixation time was mostly distributed from 0.5 sec to 1.8 sec, in spite of the difference between tasks and locations.

Allowable Visual Distraction Time

The allowable visual distraction time that a driver can comfortably tolerate during driving was measured. Target marks that simulated in-vehicle displays and controls were set in two locations: at the center of vehicle meters and at the center of an instrument panel. While driving on rural expressways and urban streets, the subjects were instructed to gaze steadily at one of the target marks as long as they could comfortably tolerate it psychologically before restoring visual attention to the road ahead. The electrooculogram (EOG) of the subjects was measured. The glance time to the target marks (the time interval between when the subject began to gaze at it and when the subject resumed attention to the road) was determined by EOG signal voltage analysis. Eight subjects participated in this study with five repetitions. The results are shown in Fig. 8.3 (Kimura et al., 1990). The time for which 95 (50)% of the subjects were able to gaze steadily at the target marks without feeling uncomfortable was 1.0 (2.0) sec.

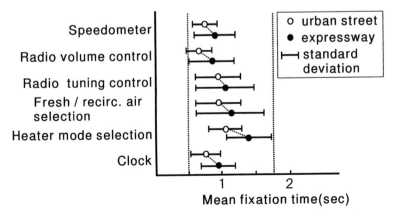

FIG. 8.2. Fixation time for conventional in-vehicle displays and controls.

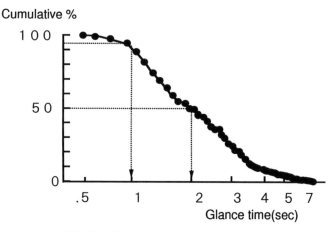

FIG. 8.3. Allowable visual distraction time.

Attention Diversion Time for a Navigation System

It was found that the driver spends .5–1.8 sec fixation times toward conventional in-vehicle displays and controls. Based on the maximum visual distraction time study results, it was found that the driver can look away from the road for 1–2 sec. The glance time consists of the transition interval between in-vehicle and road view and fixation on in-vehicle controls and displays. This interval is about .1 sec (Hayes, Kurokawa, & Wierwille, 1989). To summarize, the fixation time plus .2 sec is approximately equal to the glance time. Accordingly, the driver can see in-vehicle controls and displays within 1–2 sec interval of available glance time. Comprehension of information from the navigation system controls and displays must be completed within 1–2 sec. The legibility and operability of the controls and displays should be designed accordingly.

OPERABILITY OF USER CONTROLS

The navigation system controls must be easily identified and used within the available driver diversion time described previously. Controls that cannot be operated within the prescribed time limit should be made unavailable for use while the vehicle is being driven. The following are examples of fast and slow operation controls: *fast*—display zoom, map display orientation, road map display selection; *slow*—ten key numeric input, destination programming.

LEGIBILITY AND COMPREHENSION OF DISPLAYED INFORMATION

In order to ensure that the driver can assimilate route guidance information from the navigation system within the available glance time limits, the following specifications for the visual display were considered (Kimura et al., 1990; Kishi et al., 1992): character size, color combinations of display characters (foreground and background), background luminance for reflection and glare suppression, map display orientation, and amount of route information necessary for easy comprehension.

Character Size

Vehicle vibration makes reading small characters on the navigation display difficult. Large characters and surrounding figures are difficult to read in a single glance for a given display size. The most generally recommended character size is .27–.75 degrees (16 min–45 min) in previous studies on VDT (e.g., ANSI/HFS100-1988). The range of appropriate character size for automotive map displays may need to be smaller than other applications for reasons of vehicle vibration and short glance.

Combinations of Color for Legibility

The physical attributes of color are commonly represented by the luminance and chromaticity. In order to read the displayed information, the luminance and chromaticity of the character should contrast sufficiently with the VDT screen background. This contrast is expressed as luminance contrast and chromaticity difference, respectively, in Fig. 8.4. The luminance contrast and chromaticity difference were calculated by the following equation (Yoroizawa, Hasegawa, & Murasaki, 1986):

$$C = L_1/L_2, \quad D = ((u_1' - u_2')^2 + (v_1' - v_2')^2)^{0.5}$$

$$\textbf{Luminance contrast} \quad = \quad \frac{\text{Luminance of letter (L1)}}{\text{Luminance of background (L2)}}$$

Chromaticity difference = the distance between chromaticity of letter and of background

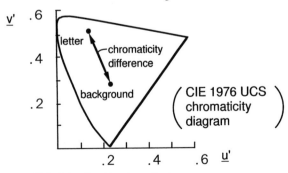

FIG. 8.4.   Combination of colors.

where C = luminance contrast
        D = chromaticity difference
        L_1 = luminance of characters
        L_2 = luminance of background
        u_1', v_1' = chromaticity of characters
        u_2', v_2' = chromaticity of background (CIE1976 UCS chromaticity
        diagram)

A study was carried out to investigate the relation of luminance contrast and chromaticity difference by subjective evaluation in reading four Chinese characters ("easy to read" or "difficult to read"). The visual distance was 800 mm, the character size was .36 degrees and the display time was 1 sec. The test covered 240 combinations of color. Ten subjects participated in this study. The results were analyzed using discriminant analysis (a kind of multivariate analyses) and appear in Fig. 8.5. The boundary between easy- and difficult-to-read conditions is represented by Line A in Fig. 8.5 (rate of misjudgment: "easy to read"—0%, "difficult to read"—20%) (Kishi et al., 1992). The appropriate range for reading characters is the area to the right of Line A. By the same methods, the appropriate range for recognition of a road on the display (line width displayed as a road: .8 mm) is the area to the right of Line B (rate of misjudgment: "easy to recognize"—4.5%, "difficult to recognize"—14%). The range necessary to distinguish road type color is the area to the right of Line C (rate of misjudgment: "easy to distinguish"— 3%, "difficult to distinguish"—32%). Finally, when the chromaticity differ-

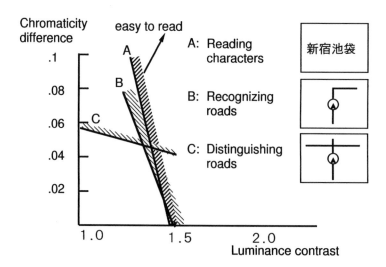

FIG. 8.5.   Relation between luminance contrast and chromaticity difference on legibility.

ence between the characters (roads) and background is zero, the luminance contrast should be greater than 1.5.

Background Luminance

The background luminance on automotive navigation displays must be considered for reflection suppression from interior materials during daytime and glare suppression at night. The upper limit of background luminance at night were investigated by pupillary diameter measurement (Kimura et al., 1990). It was found that when the rate of change of pupillary diameter exceeded .1, subjects felt dazzled. The luminance of background for which subjects' rate of change of pupillary diameter was .1 is shown in Fig. 8.6. Accordingly, the upper limit of background luminance at night should be 1.0 cd/m² in green; .6 cd/m² in red, white, and cyan; and .2 cd/m² in blue.

Map Orientation for Easy Judgment of Turn Direction

In order to judge right or left turns from the navigation display, it is necessary for the driver to correlate the map display orientation with the orientation of the crossroads ahead of the vehicle. The relation between the map orientation angle and the reaction time of judging right/left turns was investigated (Kishi et al., 1992). A test using 20 subjects and a display time of 1 sec was performed. The subjects looked at the map display and were instructed to judge right/left turns in order to reach a point on the forward crossroad as soon as possible. The results are shown in Fig. 8.7. This is coincident with "mental rotation" phenomena (Shepard & Metzler,

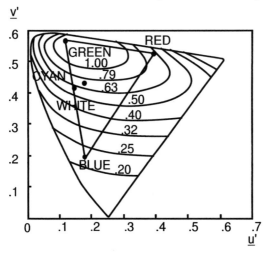

FIG. 8.6.   Upper limit luminance of background at night.

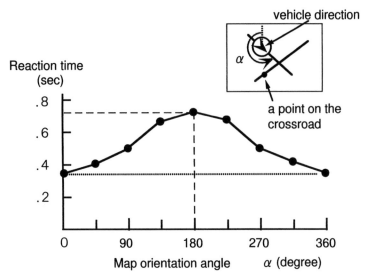

FIG. 8.7.   Map orientation.

1971). The reaction time for a map orientation that is 180 degrees different from crossroad orientation was twice as long as with a map orientation of 0 degree. For example, the map display should have a south up orientation when the vehicle is pointed south.

Amount of Information for Comprehension

Figure 8.8 shows the relation between the number of crossings on the route to a destination and the correct guidance response ratio (answer ratio) (Kishi et al., 1992). The subjects looked at the map display and were instructed to answer the routes to the destination. The number of subjects participating in the test was six, the display time was 1.0 sec, and the display frequency was twice. The correct answer ratio is reduced in proportion to the number of crossings on the route. The ratio rises (to 80% with five crossings) when route guidance points are added to the map display in Fig. 8.8. Therefore, for effective route guidance, the route crossings displayed should be minimized and route guidance points added to enhance the legibility of the correct route on the display. Roads displayed on the map should be limited to primary roads in order to reduce the number of crossings. Figure 8.9 shows the distribution of the number of route crossings within a 6″ map display (map scale: 1/20,000; map area: Tokyo, Osaka, Nagoya). This suggests that the number of route crossings can be limited to five when only arterial roads larger than 5.5 m in width are displayed.

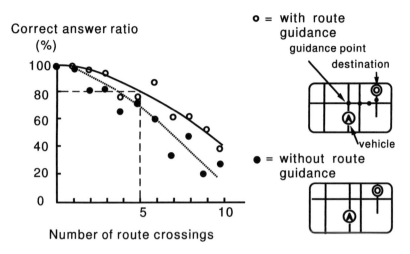

FIG. 8.8. Relation between number of crossings and correct answer response.

HUMAN INTERFACE FOR VOICE ROUTE GUIDANCE

Route guidance by both a map display and voice prompting reduces the visual workload of the driver compared to a map-only system. The following auditory interface aspects for voice route guidance were studied (Kishi & Sugiura, 1993): the information required for voice guidance on unfamiliar roads and the timing of this information, expressions for voice guidance, and effectiveness of voice guidance (confirmed by driver fixation behavior and change in heart rate).

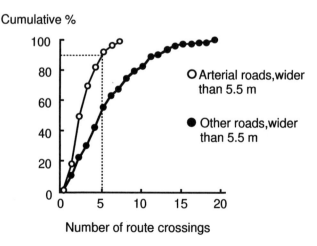

FIG. 8.9. Distribution of the number of route crossings.

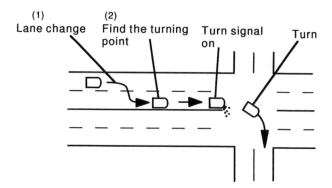

FIG. 8.10. Driving actions for right turn.

Information Required for Voice Guidance and Timing

The necessary information for a driver to make a right-hand turn involving a lane change on surface roads is the lane change direction and the information to identify a proper turning point (Fig. 8.10) (Kishi & Sugiura, 1993). The lane change direction information should be provided when the driver is able to select a proper lane before making the turn. The distance required for a lane change is related to the number of lane changes, the vehicle speed, the number of passing vehicles, and the number of signals passed (Fig. 8.11). The empirical survey into the proper timing to provide lane change information was done in Tokyo and Nagoya (number of subjects: 6, number of experimental cases: 60). The distance between the point where the subject was instructed to begin to change a given lane and the point where the subject finished lane change was measured as the

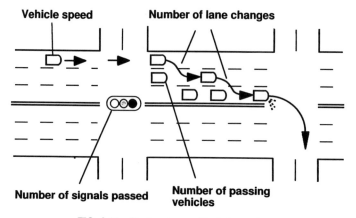

FIG. 8.11. Timing to provide information.

TABLE 8.1
Information Content and Timing on Surface Roads

No.	Information	Timing
1	Lane change direction right or left	When a driver can select a proper lane before making turn (700 m before intersection)
2	Identification of turning point landmarks (e.g., intersection sign, traffic signal, pedestrian bridge, configuration of intersection)	When the landmarks become recognizable (300 m before intersection)

distance required for a lane change. The distance for executing a lane change was found to be highly correlated to the vehicle speed, the number of lane changes, the number of signals passed, and the number of vehicles that passed by multiple regression analysis (multiple correlation: .82). The distance required for a lane change was defined about 700 m as a combination of upper level of each factor in order for drivers to execute a lane change without anxiety. Accordingly, a driver needs lane change information 700 m before the intersection. The information to identify a turning point should be provided when a landmark becomes recognizable, generally 300 m before the intersection (Table 8.1). It should be noted that on expressways, guide signs for each entrance, junction, and exit are set at 300 m, 2 km, and 2 km ahead, respectively (Fig. 8.12, Table 8.2).

Expressions of Voice Guidance

The proper expressions of voice guidance should consider information overload, the discrimination of similar sounding words, the expression for

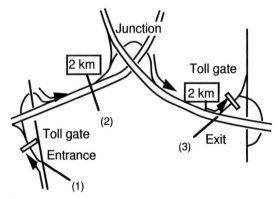

FIG. 8.12.   Locations of required route guidance signs on expressways.

TABLE 8.2
Information Content and Timing on Expressways

No.	Information	Timing
1	Direction to the expressway entrance and bearings at the ramp	Prior to the guide sign indicating the entrance (300 m before intersection)
2	Running direction at a junction	Prior to the guide sign indicating the direction of the junction (2 km before junction)
3	Name of the exit intersection	Prior to the guide sign indicating the direction of the exit (2 km before exit)

right/left directions, and the guidance for consecutive turns. Kishi and Sugiura (1993) described on the empirical findings as follows:

Information overload: The length of the voice messages should be less than 5–7 sec.

Discrimination of similar sounds: Similar pronunciation such as *usetsu* (meaning "right turn") and *sasetsu* (meaning "left turn") in Japanese are difficult to distinguish.

Expression for "right" or "left" directions: A voice guidance direction phrase should coincide with the driver's own sense of direction. The voice guidance should announce right or left direction changes relative to the driver's sense of the current road staying straight on.

Guidance for consecutive turns: When the vehicle is on a road with consecutive turns, the necessary guidance information for the first turn and the second turn must be provided simultaneously and in advance.

Examples of voice guidance on surface roads and expressways are shown in Figs. 8.13 and 8.14, respectively.

Effectiveness of the Voice Route Guidance

An experiment was carried out comparing a voice plus display guidance system with a display only guidance system (number of subjects: 4, 3 male, 1 female, subject age: 20s = 2, 30s = 1, 40s = 1, test route: surface road with the total distance of 30 km involving 4 route guidance zones, test vehicle: Toyota Celsior). Kishi and Sugiura (1993) described the detailed methods and results. The broad results are shown in Fig. 8.15. The average display fixation duration per minute with voice guidance was significantly shorter than without voice guidance during route guidance zones, $F = 26.48$, $df = 1/8$, $p < .01$. The average fixation frequency per minute with voice guidance was also significantly lower than without voice guidance during route guidance

(1)"About 700 meters ahead, turn to the right."
(2)"About 300 meters ahead, turn to the right at Toyota-cho."
(3)"About ### meters ahead, turn to the left, and then turn to the right."
(1) Lane change
(2) Find the turning point
Turn signal on
Turn
Toyota-cho
(3)

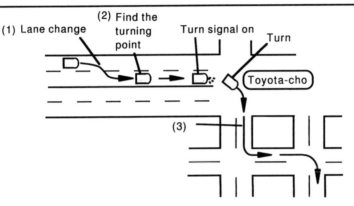

FIG. 8.13. Examples of voice guidance on surface roads.

zone, $F = 27.76$, $df = 1/8$, $p < .01$. The heart rate change was calculated by $(HR_{data} - HR_{base})/HR_{base}$ (HR_{data}: the average heart rate for 30 sec during route guidance zone before each turning intersection, HR_{base}: the average heart rate for 30 sec during straightforward vehicle running). The heart rate change with voice guidance was lower than without voice guidance, $F = 2.92$, $df = 1/36$, $.05 < p < .1$.

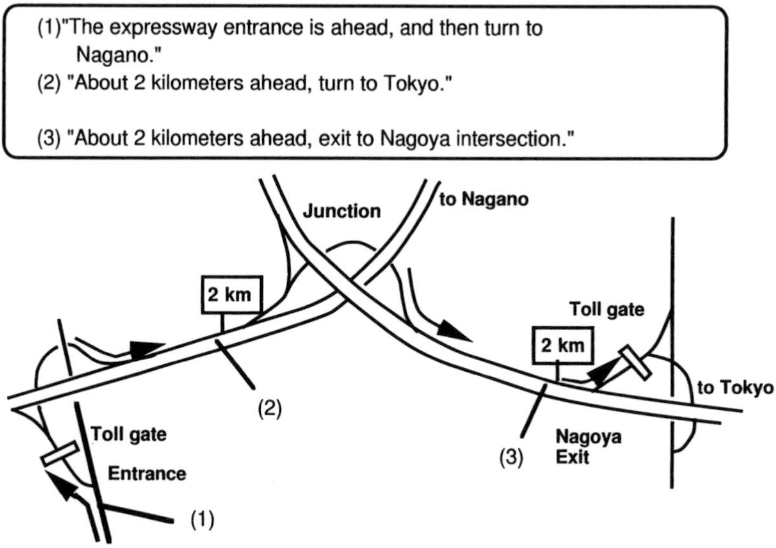

(1)"The expressway entrance is ahead, and then turn to Nagano."
(2) "About 2 kilometers ahead, turn to Tokyo."
(3) "About 2 kilometers ahead, exit to Nagoya intersection."
Junction
to Nagano
2 km
Toll gate
2 km
to Tokyo
(2)
Toll gate
Entrance
Nagoya Exit
(3)
(1)

FIG. 8.14. Examples of voice guidance on expressways.

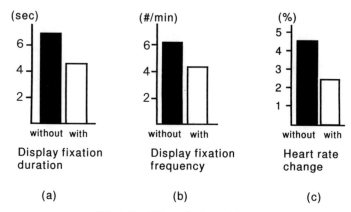

FIG. 8.15.   Effect of voice guidance.

CONCLUSIONS

1. The available attention diversion time is about 1–2 sec.
2. The operation of user controls should be completed within 1–2 sec.
3. Legibility and comprehension of displayed information:
   - Character size: within .27–.75 degrees
   - Combination of colors: luminance contrast (character to background) ≥ 1.5 when the chromaticity difference between the characters and background is zero.
   - Background luminance: reflection and glare should be minimized.
   - Map orientation: south-up orientation when the vehicle is pointed south.
   - Amount of information: the number of crossings ≤ 5.
4. Human interface specifications for voice guidance and their effectiveness:
   - The timing to provide information: surface roads—700 m before lane change, 300 m before intersection; expressways—300 m before intersection, 2 km before junctions and exits.
   - Effectiveness of voice guidance: Reduction of visual and mental workload was confirmed.

REFERENCES

ANSI/HFS 100-1988, American National Standards for Human Factors Engineering of Visual Display Terminal Workstations.

Freundschuh, S. M. (1989). Does "anybody" really want (or need) vehicle navigation aids? *1989 IEEE*, 439–442.

Hayes, B. C., Kurokawa, K., & Wierwille, W. W. (1989). Age-related decrements in automobile instrument panel task performance. *Proceedings of the Human Factors Society 33rd Annual Meeting, 1,* 159–163.

Kimura, K., Osumi, Y., & Nagai, Y. (1990). CRT display visibility in automobiles. *Ergonomics, 33*(6), 707–718.

Kishi, K., & Sugiura, S. (1993). *Human factors considerations for voice route guidance* (SAE Paper 930553). Dearborn, MI: Society of Automotive Engineers.

Kishi, K., Sugiura, S., & Kimura, K. (1992). Visibility considerations for automobile navigation display. *Journal of the Society of Automotive Engineers of Japan, 46*(9), 61–67.

Ross, T., & Brade, S. (1995). *An empirical study to determine guidelines for optimum timing of route guidance instructions: Colloquium on "Design of the driver interface."* London: IEE Professional Group C12.

Shekhar, S., Coyle, M. S., Shargal, M., & Kozak, J. J. (1991). Design and validation of headup displays for navigation in IVHS, *VNIS912795,* VNIS'91 Conference.

Shepard, R. N., & Metzler, J. (1971). *Science, 701,* 171.

Streeter, L. A., Vitello, D., & Wonsiewicz, S. A. (1985). How to tell people where to go: Comparing navigational aids. *International Journal of Man–Machine Studies, 22*(5), 549–562.

Wierwille, W. W. (1993). An initial model of visual sampling of in-car displays and controls. In *Vision in vehicles* (Vol. 4, pp. 271–280). Amsterdam: Elsevier Science.

Wierwille, W. W., Antin, J. F., Dingus, T. A., & Hulse, M. C. (1988). Visual attentional demand of an in-car navigation display system. *Vision in vehicles* (Vol. 2, pp. 307–316). Amsterdam: Elsevier Science.

Yoroizawa, I., Hasegawa, T., & Murasaki, K. (1986). Legibility evaluation of displayed characters. *Journal of the Institute of Television Engineers of Japan, 40*(4), 298–304.

Zwahlen, H. T., Adams, C. C., & Debald, D. P. (1988). Safety aspects of CRT touch panel control in automobiles. In *Vision in vehicles* (Vol. 2, pp. 335–344). Amsterdam: Elsevier Science.

Chapter 9

Cognitive Ergonomics and Intelligent Systems in the Automobile

Guy Labiale
Université Paul Valéry, Montpellier, France

The objective of this chapter is, on the one hand, to present certain categories of *adaptable* and *intelligent* systems likely to be introduced in the driving compartment by describing in particular the functions and architectures of an intelligent interface and, on the other hand, to outline consideration on certain problems related to cognitive ergonomics.

New electronic and computerized systems for handling information, communication, and automatic control are currently being developed and introduced into the driving compartment. The aim is to provide the driver with better information and assist in making the driving task more comfortable, more efficient, and safer. Among these systems are navigation and route guidance, traffic information (traffic conditions, traffic incidents, etc.) and perturbations due to the weather (slippery road, drifting snow, etc.), speed regulation and obstacle detection (AICC: Autonomous Intelligent Cruise Control), in-vehicle warning (engine problems, underinflated tires, etc.) and maintenance, cellular telephone, and motorist information systems (parking lots, hotels, weather information, etc.).

Until now, the addition of installations has consisted of juxtaposing these different systems on the dashboard (the addition of several displays and commands to those already on the traditional instrument panel), which risks making the driving compartment look dangerously like the cockpit of an aircraft. The task of driving an automobile consists of a number of complex tasks and subtasks that include controlling trajectory, avoiding obstacles, controlling the vehicle's drive line, reading road signs and signals, following the route, complying with the highway code, and so on. Therefore, carrying out these operations requires the driver to call on complex perceptual, attentional, cognitive, and motivational processes almost continuously (as opposed to the task of the aircraft pilot which, excluding taking-off and landing, does not require this continuous process of perceptual–cognitive regulation). Consequently, it becomes clear that a certain number of problems concerning mental overload, errors, and accidents may occur (Noy, 1990). Paradoxically, the introduction of these new information systems in the driving compartment risks increasing the problems of selective attention and mental load for the driver, thus running counter to the initial objective of increasing the comfort and safety of automobile driving.

However, to solve these problems, there are several methods of simplifying and integrating the presentation of these new items of information and control in the driver's compartment. Rouse, Geddes, and Curry (1987) distinguished three procedures for aiding human operators to perform their task: to make the task easier, to take over part of the task to be performed, and to carry out the task completely.

In view of this, a certain number of technological and ergonomic solutions can be proposed to aid the operators in performing their task. The following sections outline five main categories of solutions: *the integrated interface, the programmable interface, the intelligent interface, the automatic specific command,* and *the electronic copilot.*

PRESENTATION OF ADAPTABLE AND INTELLIGENT SYSTEMS

Several operational implementations concerning adaptable and intelligent systems in the driving compartment can be envisaged.

The Integrated Interface

This implementation groups all the information of the driving compartment on a screen (or possibly on a head-up display) and on a loudspeaker. It is only of limited interest on an ergonomic level, because it requires numerous and prolonged in-vehicle visual explorations. It also results in mental overload for the driver, because all the main information windows are displayed simultaneously on the same screen (speed + fuel level + road information + navigational map + alarm system, etc.). Although certain secondary information (adjustment for auditory diffusion, vehicle status, etc.) can be consulted by the driver by selecting other screen displays, this requires visual search and manual control, which are potentially dangerous to perform when driving.

The Programmable Interface

This is a programmable, integrated interface. It integrates the new informative and communicative functions on a single screen (possibly loudspeakers, and an HUD) installed on the automobile instrument panel. The driver can select the modes and complexity of information display using certain options offered by the interface. The basic option displays the essential information on a single page according to an ergonomic display format; the interrogation of additional information, requiring the selection of other screen pages, must only be performed when the vehicle is at standstill to avoid disrupting the driving task. With the custom option, drivers can program the presentation and the level of complexity of these different items of information for each screen page, as well as the order in which the pages are displayed, to take into account their own capacities, experience, and preferences.

Although this interface has the advantage of being better adapted to the capacities of each driver, it has the disadvantage, on the one hand, of not eliminating possible in-vehicle information overload and, on the other hand, of not taking into account the increase of the mental load affecting the driver due to the driving in difficult traffic conditions or in heavy traffic.

The Intelligent Interface in the Driving Compartment

This intelligent interface is an integrated and adaptative interface. It integrates the new informative and communicative functions on a single screen (eventually HUD and loudspeakers) installed on the automobile instrument panel (including a vocal transmission and recognition system). It is designed to automatically manage the ergonomic layout and design of the different information and commands, taking into account the driver's mental load, the type of information presented, and the type of traffic situation.

The Specific Automatic Commands

For a certain number of new command functions such as automatic speed control, obstacle detection and avoidance (AICC), the control of trajectory, and accidental leaving of the road, the reaction of the vehicle must be immediate and appropriate, especially because the situation is urgent and critical. In some cases, a warning transmitted in the driving compartment by an intelligent interface, or other device, may lead to a reaction from the driver that comes too late and will probably be inappropriate; thus, a specific automatic command, which does not require intervention from the driver, seems useful in order to take over automatically in emergency and critical situations.

The Electronic Copilot

This system integrates the intelligent interface and the specific automatic commands. A prototype of this type of system has undergone partial testing (cf. the presentation of the DRIVE-GIDS project by Michon, 1993). However, given the technological, economic, and ergonomic problems raised by the development of such a copilot, it seems unrealistic to envisage its full development in this stage.

DEVELOPMENTS OF INTELLIGENT INTERFACES

For realistic implementation in the driving compartment, it is first necessary to specify the functions of an intelligent interface and, second, to present its different architectures with respect to ergonomic requirements. It should be noted that this concept has already been the subject of significant developments in other areas, such as database interrogation, flying aircraft, and so forth (e.g., Chignell & Hancock, 1989; Sullivan & Tyler, 1991), and it appears interesting to extend it to new types of information transmitted in the driving compartment.

The Functions of the Intelligent Interface in the Driving Compartment

The intelligent interface is organized in three functional systems. The *integration system* receives, codes, and stores the different data from, on the one hand, the different information systems (navigation and route, traffic information, etc.) and, on the other, information from certain sensors (vehicle conditions, position on the route, distance from an obstacle or between vehicles, type of roadway, etc.), so that these data can be transmitted in a format that can be directly processed by the control system. The *control system* constitutes the "brain" of this intelligent interface with three main functions: dialog, strategy, and learning.

The *dialog function* controls the requests for information and driver-interface dialogs. Two scenarios may be encountered; the information is presented at the request of the driver, or else at the initiative of the intelligent interface. The exchange of information constitutes the dialog and a communication language must be developed in order to adapt it to the requirements of the whole driving population in all its diversity (learners, experienced drivers, etc.). This function is directly related to the strategy function.

The *strategy function* is involved in the management of all the data. This is done by performing different operations concerning the priority and the timing of the different items of information to be displayed (or queried), such as: the priority transmission of an item of information at the right time for the driver in a given situation, or to warn of a more or less serious incident (on the road or in the vehicle), or to prepare for or propose an action to the driver (within the framework of guidance: proposal to change lane in order to prepare for a forthcoming indication to change direction); delaying the display of the transmission of secondary information (a telephone call while the driver is in the process of overtaking another vehicle); the nondisplay (or nontransmission) of certain items of information (e.g., complicated navigation maps, lists of hotels) and the blocking of queries to the interface except under certain driving conditions or when stationary; the interruption of information transmission (following a telephone call, the automatic cutting-off of the radio, which can start again afterward).

The *learning function* records and processes certain reactions of the driver (reaction time, the use or nonuse of the information, etc.) linking it to information delivered by the system while driving, in order to better adapt or modify the strategy and information presentation functions to each individual driver.

The control system sends its data to the *presentation system.* On the one hand, the presentation system defines the presentation mode (visual or auditive) or combinations of modes (visual + auditive) and, on the other hand, it defines the ergonomy of the specific information presented or transmitted (complexity, quality, and organization). It is important to emphasize the usefulness of this presentation function: It is not simply a question of delivering (or not delivering) a given item of information in binary form (by displaying or not displaying it/sound presentation or visual display), but that of being able to transmit this information in a sensorially adapted manner and in a modular, specific ergonomic format, taking into account the type of information, the traffic situation (requiring different levels of mental load, such as the extremes of a clear expressway and a winding road with heavy traffic, etc.), and the driver's characteristics (sight and hearing problems, age, experience, perceptual-cognitive style, etc.).

Consider navigation. When drivers are confronted by a dense traffic situation, increasing their mental load, the guidance information can be simplified (an item of guidance information can be delivered consecutively in vocal message form to first advise the driver to change lanes, followed by the display of a very simple visual symbol on the screen), whereas the opposite can occur in a less demanding situation, as more detail can be provided to drivers (the same arrow symbol to which is added the next two or three changes of direction, and even the vehicle's position).

Intelligent Interface Architectures

Different artificial intelligence (AI) architectures have been developed and tested in relation to ergonomic requirements. More precisely, the advantages and disadvantages of models are analyzed using centralized and distributed architectures (see Labiale, Ouadou, & David, 1994).

The Models Required for the Adaptation. Four models seem indispensable to us to ensure the intelligent adaptive behavior of the interface. First, there is a *model of users*, describing their capacities, behavioral characteristics, preferences, and mental representation of the system. It comprises the rules of inference, enabling the control of these parameters, their modification, and the evaluation of the users. With an *operating model*, the system must be aware of its own operation in order to be able to determine the user's goals. A *utilization model* is responsible for taking into account the dynamic dimension of the interaction and convert the use of the system by a specific user. This particular model is a view of the operating model as seen through the prism of the user model. Finally, a *driving model* describes the different subtasks and the different driving situations, as well as their influence on drivers' mental load. The intelligent interface, with its adaptation mechanisms, must incorporate the consequences of driving conditions (traffic density, different road conditions, etc.) on the drivers' mental load and the risks of perturbing them while carrying out the driving task.

Centralized Architecture. A first version of centralized architecture has been developed, integrating the four models described, around an adapter (Labiale, David, & Ouadou, 1991). The adapter works on the data supplied by the four models (concerning the driving, the user, the operation, and the utilization) and its purpose is to manage (Fig. 9.1) the *dialog* (i.e., to determine what items should be displayed on the screens and the forms of dialog to be used with the driver), the *presentation* (i.e., to determine the type of screen-page display, the organization of items, and the presentation procedures), and the *strategy* (i.e., to manage the timing and the hierarchies of dialogs and presentations of data).

To perform the previous functions, the adapter synthesizes the data from these four models by using ergonomic rules and transmits them to the interface. Furthermore, the user's actions are transmitted to this adapter, which then updates the data of the models concerned. In reality, the adapter builds (according to the ergonomic rules contained in the adapt-

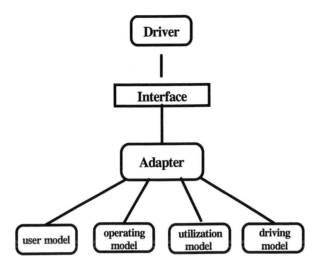

FIG. 9.1. Architecture of the centralized architecture.

er's knowledge base) a "status vector" that is transmitted to the interface control module.

By way of illustration, the following is an example of a rule used:

Rule 3 -Adaptation function-
If   Current Screen-Page = Menu + Navigation Map + Traffic Information
And   Driver-Mental-Capacity < Mental-Load-of-Driving + Associated-Mental-Load-Activity
Then   /*driving requires nearly all the driver's attention => reduce associated activities and only display priority data */
        /*Status-Vector [Current Screen- Page.display] = Delete (Menu + Traffic Information)*/

Although this centralized architecture has all the elements required to operate an intelligent interface, it proved to be poorly adapted for several reasons. First, the separation between the man–machine interface and its intelligence did not facilitate taking into account more accurate and specific knowledge concerning each element of the interface. Second, the user's capacities and preferences must be sufficiently detailed to take into account each element of the system. This increase in the user's knowledge base occurs to the detriment of the system's response time. In addition, the centralized approach requires that all the data transits via the adapter, which bottlenecks the data flow. Finally, the redundancies that coincide between the different models in the case of large knowledge bases lead to the system taking more time seeking in which part of the system it finds itself than in performing its work of adaptation.

Distributed Architecture

All these difficulties have oriented us toward a distributed architecture (for a detailed development, see Ouadou, Labiale, & David, 1993). In a distributed approach, the system is made up of intelligent interactive agents (Coutaz, 1990), each of which has specific responsibilities and is in control of its adaptive possibilities. The agents' behavior is the result of its observations, its knowledge of the user, and their interactions with the other agents. This agent therefore requires aptitudes for assessing the users' behavior, the capacity for memorizing this behavior, the possibilities of varying the presentation, the dialog as a function of context and the user, and so on. In this case, the intelligent interface comprises a group of agents organized hierarchically. Thus at the information systems level, there are agents responsible for managing these systems (e.g., automobile guidance management). For each system there is a service agent specialized in the different tasks (e.g., guidance agent, navigation agent). In this organization, the user and utilization models are no longer present as independent entities, but are grafted onto the different agents, each of which is responsible for the information it needs. When an agent instance is created, it automatically addresses the profile manager to request information on the user concerned by it (both on the instance and on the class). If no information is present in the base, the profile manager is responsible for producing the general characteristics of the user it knows by deduction.

When an agent dies, it transmits all the information on the user that it has deduced to the profile manager in order to store it in the database (to ensure their nonvolatility: The knowledge of the user in the agent must survive the instances of the agents.). The agents can have distinct architectures, either a PAC type (Coutaz, 1990) or an EPAC type (Extended PAC; David, 1991). This intelligent interface made use of EPAC-type agents. In this case, each agent is formed by two separate parts: a traditional part based on the dimensions of presentation, abstraction, control (PAC), and a new part comprising the dimensions of intelligent control, choice of presentation, goal evaluator, and help generator and profile (see Ouadou, Labiale, & David, 1993, which gives a detailed description of the model of the EPAC agents). In these conditions, the user profile manager, the incorporation of the context, and the environment enable the system to react dynamically and, above all, to explicitly take the user into account.

This distributed architecture of an intelligent interface should undergo development and testing because it ought to enable rapid and pertinent adaptation to modifications to the driving task, in-vehicle information, and driver behaviors.

THE PROBLEMS CONCERNING COGNITIVE ERGONOMICS

The taking into account of human factors involved by these new adaptable and intelligent systems in the driving compartment can be done using several distinct but complicated approaches: the analysis of the driver's requirements and expectations with respect to the information and intelligent systems in the driving compartment, the analysis of psycho-ergonomic problems concerning the presentation of information and dialog, the analysis of problems related to the use and acceptance of these systems, and the analysis of the influence of these systems on drivers' driving styles. This review deals mainly with the problems of cognitive ergonomics and with certain of the acceptance criteria linked with such systems (Labiale, 1992a).

The Ergonomic Criteria of In-Vehicle Interface

The presentation of information on the adaptable and intelligent interface must first satisfy a certain number of general ergonomic criteria (Labiale, 1992b, 1994a), such as:

1. Good information visibility and intelligibility, whatever the conditions of the road environment (background lighting and noise in the compartment) and the driver's sensorial capacities.
2. Rapid information legibility, in order not to perturb the driver's attention. Thus, the visual intake of information from an in-vehicle screen must not exceed a maximum duration of 1.2 sec per glance and not require more than two consecutive glances.
3. Immediate comprehension of the information, symbols, and messages presented, irrespective of the driver's cultural level.
4. The absence of competition between sources of information in the driving compartment (e.g., a visual warning on the instrument panel should not be transmitted at the same time as a navigation message on an in-vehicle screen, etc.) so as not to create a dangerous increase in the driver's mental load.

The first criteria are the same as for those already used for existing information systems (traditional instrument panel) and they are defined by what could be called *surface* ergonomy. The particularity of the intelligent interface, apart from these surface criteria (devoted to defining the quality of display visibility and legibility), is to call for *cognitive* ergonomics (Bastien, 1991), which take into account information comprehension and the perceptual-cognitive problems of drivers as regards the interface.

Furthermore, among the many ergonomic questions raised by the introduction of the intelligent interface in the driving compartment, this chapter investigates the problem of how mixed information (and dialog) display formats influence the interface. The ergonomic rules of man–computer screen interfaces recommend that each specific item of information be presented in the same sensorial mode (visual or auditory) and in the same format (for visual information, this implies that it is always displayed at the same place on the screen, in the same graphic form, etc.). However, it should be recalled that one of the possibilities of the intelligent interface enables the information presentation format to be adapted according to the driver's mental load in a real driving situation. In reality, the driver requesting a navigation aid can be presented with either a navigation map, when driving along a highway (completely free of traffic), or a direction arrow (linked or not with a vocal indication) when driving in heavy urban traffic. In these conditions, the intelligent interface might risk perturbing the driver through poor presentation and dialog format homogeneity. Ergonomic research should define whether the intelligent interface should only be used to manage the hierarchy of the information and the dialogs, or whether it is possible to adapt their presentation formats.

Influences on Mental Load

From a pragmatic point of view, ergonomists use the concept of mental load to evaluate the level of attention and effort required to carry out one or more tasks by an operator. Various theories of attention have been put forward, the first being the single channel theories (e.g., Broadbent, 1958), then the multichannel theories (e.g., Moray, 1967, 1979) and, lastly, Wicken's model (1984); in spite of their successive refinements, they appear to have a relatively limited power of explanation.

The intelligent interface consists in modifying its properties as a function of the driver's mental load, thus it is an adaptable interface that should increase the efficiency of man–machine–environment interaction. However, the problem of how to evaluate the automobile driver's mental load in a real driving situation remains to be determined. Several evaluation solutions can be formulated:

1. The measurement of the driver's mental load in real time. Because mental load is a multidimensional concept (Roscoe, 1987; Wickens, 1984), ideally, it must be possible to measure the respective mental loads due to

visual and auditory perception, to central cognitive processing, and to the driver's control movements. Evaluations can be carried out in the laboratory on certain interrelated aspects of mental load (estimation of the subjective load: Swat, Nasa-TLX, ... ; measurements of the cardiovascular component, of 0.1 Hz, of the evoked potential component P300, EEG, eye movements, EMG, performances of a secondary task, etc.). However, these evaluations cannot be used during normal driving, because they require fitting the driver with more or less cumbersome equipment. However, a certain number of indicators, such as the number of errors in carrying out the task have been proposed (Rouse et al., 1987), though it remains to be seen how the intelligent system can determine whether or not the driver's actions are appropriate. The pertinence of the other indicators, such as reaction time regarding the information presented by the interface, and so on, has still to be explored.

2. The evaluation of a model driver. This requires developing a model based, on the one hand, on taking into account the driver's relatively static variables (e.g., age, experience, driving style) and, on the other, on the real-time recording of dynamic driving variables, such as the vehicle's parameters (speeds, maneuvers, etc.) and road environment parameters (type of lanes, traffic density, weather conditions, etc.), to predict the driver's mental load.

Concretely, the method of estimation on the basis of a model is sufficient to implement a standardized intelligent interface, enabling the information and the dialog to be adapted to the driver using knowledge and databases concerning the road environment, vehicle performance, and certain relatively stable driver characteristics (age, years driving, etc.). Conversely, the real-time measurement of the driver's reactions will be necessary to implement a *personalized intelligent* interface, which, while including the previous knowledge bases, also integrates the driver's reactions to interface information (reaction types and times). Thus, the personalized intelligent interface requires a learning process so that it can adapt itself as well as possible in real time to each driver's dynamic particularities. Furthermore, this personalized intelligent interface must also take into account some of the driver's characteristics (sensorial and cognitive capacities, level of experience, motivations, etc.) so as to adapt to each user.

Concerning the specific automatic command, as Autonomous Intelligent Cruise Control (AICC) system, certain drivers feel slightly apprehensive during a situation in which they rapidly approach a vehicle traveling in front on the same lane. These drivers ask up to what point the AICC system will be able to set the distance between the vehicles in complete safety, possibly at what moment they should take over control of the vehicle and, afterward, if they will have enough time to brake or change lanes. In other words, everything happens in this situation as if the drivers, freed from regulating the speed and distance between the vehicles, switched their mental load to the supervision and monitoring of efficient system operation and reactions.

It seems that research must still be carried out regarding specific and application factors to determine the best methods of presenting information (in series and in parallel), which optimize attentional resources and in this approach it will be important to distinguish the peripheral processes (sensorial receivers and effectors) and central processes (cognitive), taking into account the information transmitted, the types of tasks carried out, and the driver's particularities.

The Problems of Shared Initiatives

As stated earlier, the adaptation of the interface can be carried out by a choice made by the driver or automatically by the intelligent system or by an intermediate process. According to Endsley (1987), it is possible to conceive of five progressive levels of system adaptation:

Level 1: The human operator performs the adaptation (this case does not call for the "intelligent" system).

Level 2: The system proposes several adaptation solutions to the operator but does not perform any of them, as it is the operator who decides.

Level 3: The system proposes a specific adaptation solution and waits for a possible validation by the driver in order to implement it.

Level 4: The system proposes an adaptation and carries it out if it does not receive a veto from the driver within a given period.

Level 5: The system automatically executes the adaptation without waiting for the driver to validate it.

The following examines the implications of these five levels, respectively, for the adaptable interface, the intelligent interface, and the specific command "intelligent" system.

Concerning the interface, Level 1 of adaptation by the driver has the advantage of enabling the driver to fine-tune the system and gain thorough knowledge of the interface's capacities of presenting information. This adaptation appears useful for a long-term adaptation taking into account, for example, an almost constant driving situation (on roads and expressways) and the driver's relatively static capacities (e.g., age, experience). On the other hand, in the case where short-term adaptation is required, particularly during rapidly changing driving situations (e.g., driving in town), this operation demands an additional mental effort from the driver

(Nowakowski, 1987) and may delay the adaptation of the interface, which risks losing touch with the traffic situation concerned. Lastly, in a certain number of difficult driving situations, it is possible that the driver will not be able to simultaneously handle the interface adaptation functions and those necessary for controlling the driving task. In brief, Levels 1 and 2 have implications requiring the driver to take several solutions into account and these cannot be interrogated unless the driver's mental load is very low (either at very low speed, or at a standstill). This removes any possibility of application from all the other driving situations.

Level 3 enables the driver to adapt an interface rapidly and does not necessarily require a response. Nevertheless, it may contribute to distracting the driver from the driving task during critical situations.

Levels 4 and 5 really define the adaptation of the interface by the intelligent system. On the one hand, they free drivers from this task and, on the other, they ensure an ergonomic presentation adapted to their mental load at the time. Their intervention appears to be crucial in driving situations requiring a considerable mental load. It should be noted that Level 4 permits the driver to retain a certain amount of initiative concerning the information transmitted in the driving compartment. However, Levels 4 and 5 lead to fears that there might be a fair number of cases where drivers are surprised by a change of presentation or of dialog with this interface, leading to possible perturbation of the driving task. Paradoxically, to solve this problem, the driver will have to learn through a process of familiarization with the different reactions and presentations of this intelligent interface.

Lastly, with regard to the specific command intelligent interface, Level 5 allows optimum control of the vehicle in a critical driving situation. However, the fact that the driver is excluded from this vehicle behavior control loop, even for a short time, risks reducing the psychological acceptance of such a system.

Interface Acceptance Problems and Perverse Effects

It is necessary to deal with the problems of driver acceptance of the intervention of an interface and intelligent system in the driving task. Besides the ergonomic criteria contributing to this acceptance, this question appears to depend on at least three main factors related to the use of the system or of the intelligent interface, that is, the relevance and reliability of the information transmitted on the one hand, and the scope of autonomy left to the driver on the other.

The relevance of the information transmitted concerns their content, their timing, and their mode of presentation so that they can be adapted respectively to each driving situation and each driver's expectations and requirements. Satisfying a criterion such as relevance requires that the designer of these systems integrates all these data and formulates different information and command management scenarios into different driving situations. It should be emphasized that any transmission of information in the driving compartment that occurs too early or too late, or is completely unrelated to the situation, risks leading to the driver deciding to do without the services of an intelligent interface of this type. However, to adapt this intelligent interface as well as possible, it would certainly be advisable to enable drivers to select (while the vehicle is stationary) the list of information they wish to interrogate or obtain during the journey.

Reliability is a vital factor on which to base the driver's confidence in the interface or the intelligent system. Unfortunately, it is difficult to obtain because it depends, respectively, on the technical reliability of the information systems and their sensors and on the intelligent software of the interface or system. This case, in particular, leads to fears of driving scenarios not taken into account from the beginning by the interface or the system; the system should then be able to learn these scenarios and correct its reactions. In this case, would the driver accept the inaccuracies and errors (possibly serious and which put his safety at risk) of a system of this type for long?

Furthermore, in the case where these automatic systems and intelligent interfaces fail, the question must be raised as to whether the drivers will be able to recover their capacities of reaction and control over the vehicle (especially in critical situations) prior to the installation of such systems. It is known, from a general point of view, that cockpit automation leads to a certain loss in the competency and performance of operators to carry out tasks in the case of failure.

The feeling of autonomy is in direct relation to the degree of initiative left to the driver concerning the control of communication with the interface and that of driving the vehicle. One might imagine that if an intelligent interface delays or even blocks the presentation of certain information (or commands) requested or, on the contrary, dictates the intrusion of messages in the compartment, drivers will risk feeling dispossessed of part of their autonomy in the management of information and commands in the driving compartment.

Furthermore, the introduction of an automatic specific command system may lead to drivers feeling dispossessed of their power of controlling the vehicle in all situations and generate a certain feeling of anxiety. A certain number of perverse short- and medium-term effects can be envisaged. The automatic specific command systems can contribute to modifying drivers' behavior and even their aptitudes. In particular, certain drivers, according to their driving styles, will be led to take more risks or else reduce their vigilance at the wheel, due to the confidence they place in the redeeming role of the system in critical driving situations. The possibility of letting the driver (if they so wish) select and validate the information to be presented and the commands to be activated (automatically or not), constitutes a first approach toward the recovery of autonomy and may contribute to reassuring drivers that they are fulfilling their role of "captain of the ship."

It is within this framework that the introduction of these new technologies into the driving compartment could be combined with an increased feeling of power and comfort at the wheel.

RESEARCH PERSPECTIVES AND CONCLUSIONS

It should be emphasized that the introduction of these new technologies of information, interfaces, and commands in the automobile driving compartment, which make use of AI, brings to light new fields of investigation into human factors.

In particular, research into cognitive ergonomy appears necessary, on the one hand, to build models of the driver that will enable the accurate evaluation and prediction of mental load and behavior in different driving situations and, on the other hand, to formulate the best scenarios concerning the methods of presenting the information and the driver–intelligent interface dialogs.

Lastly, a certain number of psychological implications can be foreseen regarding driving styles and drivers' attitudes toward safety. As can be ascertained from the analysis of the questions raised, the introduction of these intelligent systems in the driving compartment with the view to increasing comfort and safety appears uncertain, and it is advisable to give very serious consideration to the possibility of certain risks. In conclusion, the successful contribution of these new intelligent technologies in the automobile driving compartment can only take place on the basis of a systematic recognition of the essential involvement of cognitive ergonomy.

REFERENCES

Bastien, C. (1991). Modélisation de la tâche d'un opérateur [Operator task modeling]. In *Colloque Sciences et défense* (pp. 329–340). Paris: Dunob.

Broadbent, D. E. (1958). *Perception and communications*. London: Pergamon.

Coutaz, J. (1990). *Interfaces homme-Ordinateur: Conception et réalisation* [Man–computer interfaces: Design and development]. Paris: Dunob.

Chignell, M. H., & Hancock, P. A. (1989). *An introduction to intelligent interfaces*. North Holland: Intelligent interfaces.

David, B. T. C. (1991). *EPAC, extended PAC model for dialog design* (Rep. Research MIS-ECL). Ecole Centrale de Lyon.

Endsley, M. R. (1987). The application of human factors to the development of expert systems for advanced cockpits. *Proceedings of the Human Factor Society, 31,* 1388–1393.

Labiale, G. (1992a). Facteurs humains et systèmes intelligents dans le poste de conduite [Human factors and in-car intelligent systems]. *Revue Technologies Idéologies Pratiques, 10*(2–4), 161–176.

Labiale, G. (1992b). Visual displays. In B. Leiser & D. Carr (Eds.), *Analysis of input and output devices for in-car use, DRIVE-GIDS report,* 3.1–3.22.

Labiale, G. (1994). Ergonomie des systèmes d'informations dans la voiture: recherches et applications [Ergonomy of in-car information systems: Research and applications]. *Bulletin de psychologie, 48*(418), 90–98.

Labiale, G., David, B. T., & Ouadou, K. (1991). L'interface intelligente d'informations et de communication dans le poste de conduite [The intelligent information and communication interface in the automobile driving compartment]. In M. Brissaud et al. (Eds.), *Sciences humaines et Intelligence artificielle* (pp. 13–26). Paris: Hermès.

Labiale, G., Ouadou, K., & David, B. T. (1993). A software system for designing and evaluating in-car information system interfaces. In H. G. Stassen (Ed.), *Analysis, design, and evaluation of man–machine systems* (pp. 257–262). New York: Pergamon.

Labiale, G., Ouadou, K., & David, B. T. (1994). Ergonomics requirements of an in-car adaptable interface. In *Proceedings of 12th Triennial Congress of International Ergonomics Association, 4,* 166–168.

Michon, J. (Ed.). (1993). *Generic intelligent driver support.* London: Taylor & Francis.

Moray, N. (1967). Where is capacity limited? A survey and a model. *Acta Psychologica, 27,* 84–92.

Moray, N. (1979). *Mental workload: Its theory and measurement.* New York: Plenum.

Nowakowski, M. (1987). Personalized recognition of mental workload in human-computer systems. In G. Salvendy, S. L. Sauter, & J. J. Hurrell (Eds.), *Social, ergonomic and stress aspects of work with computers.* Amsterdam: Elsevier.

Noy, Y. I. (1990). Selective attention with auxiliary automobile displays. In *Proceedings of Human Factors Society 34th Annual Meeting,* 1533–1537.

Ouadou, K., Labiale, G., & David, B. T. (1993). Interface adaptable dans le poste de conduite et implications ergonomiques [Adaptable in-car interface and ergonomical implication]. In *Proceedings de "L'interface des mondes réels et virtuels."* Montpellier, France.

Roscoe, A. H. (1987). *The practical assessment of pilot workload.* Neuilly sur Seine: AGARD.

Rouse, W. B., Geddes, N. D., & Curry, R. E. (1987). An architecture for intelligent interfaces: Outline of an approach to supporting operators of complex systems. In T. P. Morand (Ed.), *Human–computer interaction.* Hillsdale, NJ: Lawrence Erlbaum Associates.

Sullivan, J. W., & Tyler, S. W. (1991). *Intelligent users interface.* ACM Press.

Wickens, C. D. (1984). Processing resources in attention. In R. Parasuraman & R. D. Davies (Eds.), *Varieties of attention* (pp. 63–102). London: Academic Press.

Chapter 10

An Evaluation of the Ability of Drivers to Assimilate and Retain In-Vehicle Traffic Messages

Robert Graham
Val A. Mitchell
HUSAT, Loughborough University, Leics., UK

The provision of weather, road, and traffic information to drivers via an in-vehicle system has potential benefits for both journey efficiency and traffic safety. Giving drivers advanced warnings of an event (where an event is defined as "any deviation from the normal traffic equilibrium state"; RDS-ALERT, 1990) can affect route choice and safety-related factors, such as driving speed. However, the success of such systems depends largely on the ability of drivers to assimilate, retain, and act on the information received. These processes rely on the application of ergonomics to the design of the system's man–machine interface (MMI). This chapter describes experiments evaluating various MMI aspects of a prototype in-vehicle system, and then makes recommendations accordingly.

The task of using a visually based in-vehicle information display can be broken down into a number of stages of information processing. Messages must first be detected, which involves the initial stimulation of vision, and recognized as familiar before being read and interpreted. The efficiency of this initial assimilation process (*assimilation* here refers to both visual and initial cognitive processing) may be affected by the layout, legibility, and complexity of the display. Some decision must then be made as to which elements of information are considered relevant. Finally, the relevant information is committed to memory until it must be acted on. The ability of drivers to recall traffic information depends on a number of factors, including the display complexity (Labiale, 1989), the modality of message presentation (Gatling, 1976), the length of messages (Labiale, 1990), the presence of other traffic (Luoma, 1993), and subject variables such as age and educational level (Labiale, 1992).

A road-based experiment was carried out to examine both the assimilation process and the retention of information over time. Measures of recall performance and eye glance behavior were used to assess three factors associated with the design of driver information systems: the *length* of messages (the display must be kept as simple as possible while maintaining adequate information for decision making), the *timing* of messages (information must be presented early enough to allow appropriate actions, but not so late that it is forgotten), and *driver age* (the system interface must be usable by the full range of drivers). With this last point in mind, the study compared the performances of two age groups of drivers using the system.

DRIVER AGE AND SYSTEM DESIGN

It is well known that the population of the "First World" is aging. Currently, 40% of British adults (16+) are over 50 years of age and by the year 2021 that figure is expected to reach about 48% of the adult population (Coleman, 1993). As car driving is already central to most people's social and business activities, the number of older drivers within the driving population can also be expected to increase significantly. Any information system intended to improve the safety and efficiency of the already congested road network must therefore be designed and evaluated with the older population in mind. This requires an understanding of the effects of aging on visual and cognitive processing.

A number of physiological changes in the eyes cause deterioration of the visual function with age. The lens begins to yellow and stiffen, leading to presbyopia (loss of accommodation), reduced transmission of light, and altered color rendering (Yanik, 1989). Atrophy of the eye muscles reduces the attainable angle of gaze and limits adjustments of the iris, causing problems with glare. Scarring of the cornea may cause scattering of the light entering the eye and further reduce the amount of light reaching the retina (Rabbitt & Collins, 1989). Overall, declining visual acuity means that older people may miss important information from road signs or the movements of other vehicles, and have difficulties processing in-vehicle displays.

Rabbitt and Collins (1989) outlined the cognitive changes that occur in old age. A general slowing of cognitive processing increases the reaction times of older people to external stimuli, and the "thinking time" required to interpret complex displays. In particular, older people have difficulties in situations requiring divided attention or simultaneous execution of two or more tasks. There is also a reduction in the ability to efficiently select from the world the information that is needed, at the moment in time when it is needed.

In terms of memory changes with age, Lovelace (1990) summarized the state of the research, as follows: "Only weak age effects are seen in primary memory, at least as far as the capacity of this memory component is concerned ... [but] ... when an active manipulation of the information in short-term memory is required, substantial age effects emerge" (p. 20). Any age effects may therefore reflect differences in the attentional resources available during encoding and storage of information, rather than a decline in capacity per se. Further, increasing the difficulty of the memory task, or requiring attention to be divided between more than one task, will affect the performance of older adults to a greater extent than young adults (Craik, 1977; Salthouse, 1982).

Changes in memory performance can lead to difficulties in using advanced in-vehicle equipment. Labiale (1992), for example, investigated the effects of age on the ability of drivers to recall routes from maps of differing complexity when displayed on an in-vehicle screen. The experiment took place while driving on quiet roads at a moderate speed and three groups of drivers were used (25–34 years, 35–44 years, and 45–63 years). It was found that the ability of drivers to recall routes decreased in proportion to age, particularly when the information presented was most complex.

Yanik (1989) noted that unlike young adults, older adults represent quite a heterogeneous population. It is therefore useful to classify drivers as young-old (age 55–64), middle-old (age 65–74), old-old (age 75–84), and the very old (age over 85). The present study focused on the young-old, as they make up the majority of the older driving population, and due to the scarcity of regular motorway drivers over 65 years of age.

THE LENGTH OF IN-VEHICLE MESSAGES

There is a difficult trade-off to be achieved in optimizing the length of messages for an in-vehicle information system. If visually displayed messages are too long, drivers will be required to take their eyes off the road ahead for long periods, thereby compromising safety. Moreover, drivers may have problems processing and filtering long messages, and retaining the necessary information until it is required. On the other hand, it is important to provide sufficient information concerning an event for drivers to make decisions regarding route efficiency and safety (e.g., whether to change routes, slow down, stop at services, etc.).

In much previous human factors research, eye glance analysis has been used to make recommendations about the optimal quantity of in-vehicle visual information. Zwahlen, Adams, and DeBald (1988) produced a tentative design guide for recommended glance durations and frequencies based on experiments into drivers' lane deviations while operating a CRT touch screen. They concluded that the number of consecutive glances to obtain a specific *chunk* of information must be limited to about 3, and greater than 4 was seen as unacceptable. Similarly, glance durations over 2 sec were rated unacceptable. The design guide also includes a grey area, between acceptable and unacceptable, for certain combinations of glance duration and frequency.

Memory retention performance can also be used to suggest acceptable message lengths. Gatling (1975) carried out a series of road-based experiments into the retention of route messages and concluded that drivers could handle up to four units of route information successfully (where a *unit* consisted of a distance or road number, e.g.). Labiale (1990) presented subjects with traffic messages on a simple text display and required them to repeat the message after a 30-sec delay. Messages of 4 *information units* (the definition of a unit is unclear, but Labiale gave the example "traffic jam on A40") were recalled with 100% accuracy. This fell to 96% for messages of 7–9 units, 75% for messages of 10–12 units, and 52% for the most complex messages of 14–18 units. These findings fit in well with the classic work by Miller (1956), who proposed that the maximum capacity of working memory was "seven, plus or minus two" chunks of information, with full attention deployed.

THE TIMING OF IN-VEHICLE MESSAGES

Little work has been carried out relating to the timing of messages in relation to events. In general, information must be given to drivers early enough for them to be able to act on it, but not so early that it is forgotten before it becomes relevant. Forgetting messages can be alleviated by repeating them, but this may annoy drivers, encourage them to take their eyes off the road, and interfere with the acquisition of subsequent messages. Also, if information is given too early, road conditions relating to an event may have changed by the time the driver reaches them. Studies into memory recall may be particularly relevant to the question of timing, as messages may relate to incidents a number of miles ahead.

A significant related issue is interference from the external traffic and road situation. A long retention period for messages will increase the probability that some external incident will disrupt memory during this interval. As traffic density increases and the road situation becomes more complex, the workload demands on the driver increase. This may affect the processing of traffic information; for example, Luoma (1993) found that the presence of other traffic systematically decreases the driver's recall performance for road signs. If a busy traffic environment adversely affects the assimilation and retention of in-vehicle information, then messages are likely to be missed or forgotten just when they are needed most critically.

Traditional road sign information on the motorway is given to the driver a mile or two before it must be acted on. Additional weather or traffic information can be provided by variable message signs (VMS) placed before junctions or known traffic trouble spots. Greater control over when messages are displayed in relation to events can be achieved by using short range communications (SRC) links to support the provision of in-vehicle driver information (Bhandal & Palmqvist, 1993). Strategically placed beacons, combined with message storage and dead-reckoning systems in the vehicle, will allow information to be provided to drivers at a specific point before an event. However, no guidelines currently exist to indicate this optimal point.

AIMS

The present study evaluated the MMI of a prototype in-vehicle information system, using measures of drivers' message recall performance and eye glance behavior. From these analyses, recommendations have been made concerning the amount of information that should be displayed on screen, the timing of messages in relation to events, and the presentation of message screens. The variable of driver age was also used to assess whether special attention needed to be paid to the older driver in designing the system MMI.

METHOD

Subjects

Two groups of subjects were used in separate experimental sessions. The first consisted of 11 younger subjects (6 males, 5 females; age range = 23–33 years, $M = 27.4$ years) and the second, 10 older subjects (5 males, 5 females; age range = 59–64 years, $M = 61.0$ years). All 21 subjects had at least 5 years driving experience, a clean license, and were regular highway drivers. Of the younger group, 3 wore glasses for driving and another had corrected vision through contact lenses. Eight out of the 10 older subjects wore glasses, and two of these specified needing reading glasses for the dashboard screen. They were paid £40 (US$64) for their participation.

Apparatus

An in-vehicle information system has been developed by Jaguar Cars Ltd. to demonstrate the potential for using SRC to support Road Transport Informatics applications (see Bhandal & Richardson, 1992, for a description of the initial prototype). The system is capable of receiving microwave transmissions from roadside beacons, and displaying messages to the driver via an LCD mounted in the vehicle's dashboard. However, for the purposes of the experiment in which no roadside infrastructure was available, the communication of messages was simulated by the experimenter.

Subjects drove a Jaguar Sovereign vehicle, equipped with a 4-in. diagonal color LCD screen (Sharp TFT-LCD model LQ4RA01) mounted in the dashboard, and an IBM-compatible PC in the boot to generate screens. The experimenter sat in the back seat of the vehicle and used a Macintosh Powerbook to control when the PC displayed messages. A small active loudspeaker was used to provide a beep from the Powerbook, whenever a new message appeared on the dashboard screen. An S-VHS video (Panasonic AG-7450A-E) recorded subjects' face and eye movements via a small camera (Panasonic WV-KSIS2E) mounted on the dashboard by the right side mirror.

Message Display Design

The messages were designed to provide the driver with three types of information, as presented in Table 10.1.

Road-sign graphics were used where appropriate, and the messages were partially color-coded with low saturation backgrounds whose color matched

TABLE 10.1
Design of Messages

Information Type	Message Content	Example of Elements	Color Code
Static information	Relating to junctions and services on the motorway	Junction no., road no./ destination of turn-offs, distance to junction	blue
Dynamic advisory information	Advanced warning of poor weather, road and traffic conditions, where the driver can still avoid the event	Event type, extent of event, traffic speed, distance to event	yellow
Dynamic hazard information	Urgent warnings of road and traffic events, where the driver can no longer avoid the event	Event type, speed limit, lane closures, distance to event	red

existing UK road signs. The messages in the study contained fictional information throughout. This ensured that subjects were forced to read and remember the information carried on each display, cancelling out any prior knowledge of the motorway used in the trials. However, it was recognized that the relevance of the messages to drivers, important in subsequent memory performance, could not be successfully manipulated in the experiment. The messages were displayed in the following pairs:

- A dynamic advisory screen, followed by a static junction screen (advisory/junction)
- A dynamic advisory screen, followed by a static services screen (advisory/services)
- A dynamic hazard screen followed by another dynamic hazard screen (hazard).

These are realistic pairings, in that drivers would first be presented with a possible problem, then with information to help change their journey if needed. The pairs differed in terms of the amounts of information contained in the messages. Advisory/junction message pairs carried a total of nine *elements*, advisory/services pairs had seven, and hazard pairs had four. For the purposes of the experiment, an element was defined as a single *chunk* of information (cf. Miller, 1956), such as a distance to an event, a road number, a sign graphic, or a speed limit. For example, a single screen including the text items "Roadworks," "speed limit 30 mph," and "3 miles ahead" was considered to have three message elements.

Procedure

The study was conducted in winter on the M1 motorway (the main arterial freeway in the UK). Both traffic and weather conditions throughout the study were demanding, providing a rigorous but realistic test environment. Each subject required two sessions to complete the experiment. The first was concerned with familiarizing the subject with the car, screens, and experimental procedure. The second session included the main experiment, followed by a brief postdrive questionnaire.

Subjects were presented with 36 messages via the in-vehicle display during a drive lasting approximately 75 min. Messages were presented in pairs (18 pairs in total), each screen being displayed for 8 sec and accompanied by a tone. Either 1, 2, or 3 min after the presentation of a pair, a simple question was asked about each screen (e.g., "How far ahead were the roadworks?"). The questions were chosen such that the different types of message element were included with equal frequency. There was a short pause of about 30 sec between the questions for one message pair and the presentation of the next pair.

Analysis of Eye Glance Behavior

Three measures of eye glance behavior were used in the study: visual allocation, glance frequency, and glance duration. *Visual allocation* considers the relative proportions of time spent looking at various areas of the visual scene and was used to assess the diversion of attention toward the display at the expense of other areas, such as the road ahead or rearview mirror. *Glance frequency* can reflect the complexity of a display, with poor design leading to frequent glances (Rockwell, 1988). Long dashboard *glance durations* may be caused by the display being illegible (Rockwell, 1988) and can lead to lack of attention to the road ahead and unacceptable lane deviations (Zwahlen et al., 1988).

The videos of subjects' faces allowed glance locations to be roughly determined, and glance durations to be measured to an accuracy of approximately $\frac{1}{25}$ sec. For analysis of visual allocation, the scene was divided into five areas: the left window and mirror, the right window and mirror, the rearview mirror, the road ahead, and the dashboard.

The dashboard area included the information display and the other dashboard instruments such as the speedometer. The assumption was made that the majority of glances to this area after message presentation were to extract traffic information from the display. Glance behavior was examined while each message pair was displayed (18 periods of about 16 sec for each subject) and during normal driving while no message was displayed or being remembered (2 periods of 2 min for each subject).

RESULTS

Memory Recall Performance

The number of incorrectly answered questions was converted to error rate data within the various categories. These data were then subjected to a repeated-measures ANOVA to assess the within-subjects factors of message pair type (advisory/junction, advisory/services, or hazard) and retention interval (1, 2, or 3 min), as well as the age factor.

There was no overall significant effect of age on the proportion of errors made, $F(1, 19) = .53$, $p = .48$; that is, older drivers were no worse at recalling information than younger drivers. The mean error rate for older subjects was 28.4% and for young subjects 24.5%. However, a significant interaction effect was observed between message type and age, $F(2, 38) = 4.97$, $p =$

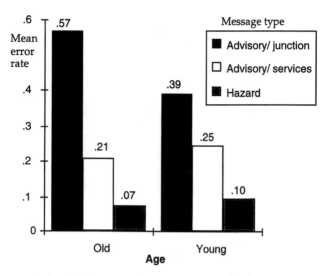

FIG. 10.1. Error rate by message type and driver age.

.012. It was found that older drivers had more problems than younger drivers recalling information from the more complex advisory/junction screens, but there was no age difference with the simpler advisory/services and hazard screens. This data is shown in Fig. 10.1.

The effect of message type was found to be highly significant, $F(2, 38)$ = 55.5, p = .0001. Information from advisory/junction messages (M = 47.6%) was recalled worse than that from advisory/services messages (M = 22.9%), which in turn was recalled worse than that from hazard pairs (M = 8.6%). Thus, as expected, subjects had problems retaining the longer, more complex messages in memory.

The time interval between presentation and recall of messages did not significantly affect the error rate, $F(2, 38)$ = 2.4, p = .10. However, there was a trend in the expected direction with more errors after a 3-min interval (M = 32.7%) than after a 1-min interval (M = 23.1%). No significant interaction was observed between retention interval and driver age with respect to the error rate. This data is shown in Fig. 10.2.

Eye Glance Behavior

During normal driving, the visual allocation to different areas of the visual field was very similar for the older and younger groups of drivers, with no significant effect of age on allocation to any of the five areas. When drivers' eye glance behavior was analyzed for the short periods while messages were displayed, however, large differences between the two age groups emerged. Older drivers spent a significantly greater proportion of time looking at the

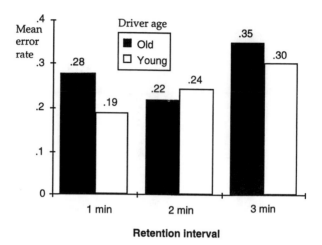

FIG. 10.2. Error rate by retention interval and driver age.

dashboard area than younger drivers, $F(1, 16) = 15.5$, $p = .0012$, and visual allocation was significantly smaller in older drivers to the road ahead, $F(1, 16) = 10.4$, $p = .005$. These results are illustrated in Fig. 10.3.

It was further found that older drivers' glances to the dashboard screen were of a significantly longer duration than younger drivers', $F(1, 318) = 155$, $p < .0001$. The mean duration of older subjects' glances was 1.31 sec, and for younger drivers, .95 sec. Perhaps more interesting was the fact that 7.6% of older drivers' glances were greater than 2 sec in duration, whereas only .4% of young drivers' glances exceeded this value. In terms of glance frequency, younger drivers required 6.5 glances on average to the dashboard to assimilate the displayed messages, whereas older drivers took 6.8 glances (this difference was not statistically significant).

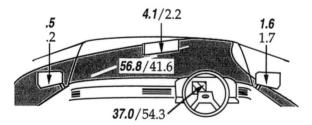

Percentage of total visual allocation
with messages displayed on screen
• *younger drivers*
• older drivers

FIG. 10.3. Total visual allocation with messages displayed on the dashboard screen.

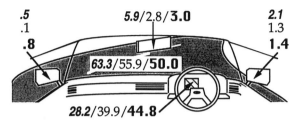

Percentage of total visual allocation
with messages displayed on screen
- **for hazard pair messages**
- for advisory/ services messages
- **for advisory/ junction messages**

FIG. 10.4. Total visual allocation for the different types of message.

While messages were displayed on screen, as messages became more complex, drivers allocated a greater proportion of their visual resources to the dash. This is clearly shown in Fig. 10.4; advisory/junction pairs required more visual attention to the dash than advisory/services pairs or hazard pairs.

The increased attention to the dashboard in turn reduced the attention paid to the road ahead and rearview mirrors. An ANOVA of the glance frequencies while messages were displayed revealed a highly significant effect of message type, $F(2, 318) = 18.9$, $p < .0001$; that is, more glances were made to advisory/junction message pairs ($M = 7.6$) than to advisory/services messages ($M = 6.7$) or hazard message pairs ($M = 5.7$). An ANOVA of the glance durations showed a similar effect, with message type significant, $F(2, 318) = 4.40$, $p = .013$; that is, the duration of glances made to advisory/junction message pairs ($M = 1.17$ sec) was longer than those to advisory/services message pairs ($M = 1.10$ sec) or hazard message pairs ($M = 1.07$ sec).

DISCUSSION

The study points not only to the way drivers assimilate and recall the information displayed on screen, but also to a number of differences in the way younger and older drivers are able to deal with this information. The ability of drivers to recall messages was affected by the length and type of message, and the length of the retention period. Older drivers required longer glances to read the information and experienced retention problems for the most complex messages.

Retention of Messages in Memory

The first item of interest is that there was no significant difference in overall memory recall performance between older and younger drivers. Only when the more complex advisory/junction message pairs were presented, did an age difference appear. Here older drivers had more difficulty recalling the information correctly; the error rate of 56.8% indicates a sizable problem. Young drivers also showed a high error rate for advisory/ junction screens (39.1%), suggesting that the complexity of those screens was too high.

The age results are consistent with previous research indicating that increasing the difficulty of a memory-related task will affect the performance of older subjects more than younger subjects (Salthouse, 1982). Also, the experiment required drivers to divide their attention between the memory task and the primary task of safe driving, and it is well known that such division of attention becomes increasingly problematic with old age (e.g., Craik, 1977). Given this divided attention and the difficulty of the task as a whole, it is perhaps surprising that older drivers performed just as well as younger ones for the simpler messages. Stokes (1992) pointed out that performance decrements for complex messages may be caused by the older person's reduced ability to assimilate information at speed, rather than the primary memory function (this idea is discussed later).

The advisory/junction screen pairs consisted of nine message elements, which represented the extreme of Miller's (1956) treatise that the maximum capacity of working memory was "seven, plus or minus two" chunks of information given full attention. It must be noted that this capacity will be reduced by the simultaneous task of driving. The better memory scores at lower complexities (e.g., messages containing four elements) suggest that the basic design of the messages is one that drivers can understand and remember with reasonable ease and accuracy. Therefore, if messages were to be simplified, all drivers would benefit, particularly the older population.

According to Gatling (1975), drivers have particular problems remembering route information containing numbers, and this is consistent with the results of the present experiment in which the more complex junction screens contained the most road numbers. In addition, the postdrive questionnaire revealed that subjects did not consider road numbers of destinations at junctions to be essential. The first step in reducing complexity may therefore be to pare down the road numbers displayed in each message.

The experiment also found that recall performance tended to be affected by the retention interval, with longer intervals leading to poorer performance. Previous driving studies have examined retention intervals of up to a minute and found such an effect (e.g., Luoma, 1993), but the periods of 1,

2, or 3 min tested in the present study may be more realistic for a system where drivers are required to make trip decisions based on the information received. The experiment showed that an error rate of 23% after a 1-min retention interval increased to a rate of 33% after a 3-min interval. This pattern of results must be considered when optimizing the timing of messages in relation to events. However, before making specific recommendations on timing, further research is required into situations where drivers have a realistic incentive to remember message elements.

The optimal timing of messages will vary according to their informational content. A message that is purely informative in nature (e.g., "Accident 5 miles ahead") will not need to be retained in full by drivers, but will allow them to expect traffic changes. However, when information is given requiring definite actions at a specific point (e.g., "30 mile speed limit—1 mile ahead"), then this must be provided close enough to the event for the driver to act accordingly.

It is likely that the retention interval effect was due, at least in part, to the increased chance of external traffic incidents occurring within longer retention intervals. Indeed, subjects commented that they were often distracted by other traffic during the retention period and forgot the message entirely. Thus, drivers may have problems remembering messages just when most messages are likely to be presented; that is, when traffic build-up is caused by an event ahead. One solution would be to build some "intelligence" into the interface of the system (see Michon, 1993, for a discussion). The traffic management center, or other information service provider, could make data about the local traffic density available to the system via the roadside beacon network. The system might then present only essential messages to the driver when traffic density is above a certain level.

Interpretation of the previous memory results, however, must take into account the fact that message realism had to be compromised in the experiment to avoid learning or experience effects. The information presented to drivers, although representative, was of no strategic value in terms of the current road environment, and lacked the contextual aspect to be expected in a real system. It is likely that the perceived importance of a message will determine the attentional resources allocated to the task of memory processing. Thus, drivers in the experiment might have used a strategy of shallow encoding of the information to memory and low attention to its retention, safe in the knowledge that it would not be needed later.

Assimilation of Message Content

The eye glance analysis revealed that, although both groups of drivers allocated the same attention to the various areas of the visual field during normal driving, older drivers spent a significantly greater proportion of time attending to the dashboard area while messages were displayed on screen. This, in turn, reduced the attention paid to the road ahead and rearview mirror. Although glance frequencies were similar across age groups, older drivers' glance durations to the screen were significantly longer than the younger group. In particular, 7.6% of older drivers' glances were in excess of 2 sec, a rate 19 times higher than those for younger drivers. If Zwahlen et al. (1988) are correct in their assertion that glance durations greater than 2 sec are generally unacceptable, in terms of dangerous lateral lane position deviations, then the older drivers in the study drove more dangerously than the young ones. It should be noted that traffic density and the speed of travel were high throughout the experimental sessions, so it is unlikely that the longest glances could have been made with full confidence of safety.

Wierwille, Hulse, Fischer, and Dingus (1988) cited evidence that both glance durations and glance frequencies increase with age due to the deterioration of vision and slowing of cognitive processes. Some of these changes were discussed earlier. They also proposed that inadequate character size or contrast ratios could lead to further decrements. Older drivers in the present experiment may have experienced specific problems with certain colors, such as blue (Yanik, 1989), and in accommodating to the brightness and contrast of the screen.

The older group suffered a further setback through correction for visual distance. Wearing reading glasses in the car was not a preferred strategy, and having to read the message while wearing distance glasses resulted in some drivers shifting their posture and peering over their glasses in order to accommodate to the close proximity of the display. Finally, a general slowing in the speed of processing with age may have led to problems assimilating the message content in the limited time available for each screen. This implies that if presentation of messages were driver paced rather than system paced, then the older population would benefit.

A novelty effect (that is, increased interest in the screen due to its unfamiliarity) is likely for both age groups, and may partly explain the large glance durations and frequencies. According to the guidelines proposed by Zwahlen et al. (1988), advisory/junction messages ($M = 3.8$ glances of 1.17 sec duration per screen) and advisory/services messages ($M = 3.4$ glances of 1.10 sec per screen) were found to lie in the gray area between acceptable and unacceptable. Zwahlen et al. further hypothesized that more than 3 glances to an in-vehicle system to obtain a specific chunk of information would lead to the task becoming uncomfortable to the driver. Given that the individual message screens required up to a mean of 3.8 glances, this implies that a two-screen message should be presented with a short intervening break to allow the driver to reorient themselves to the road ahead.

The analysis of eye glance behavior also supported the findings that some of the experimental messages were perhaps too complex. Rockwell (1988) commented that poor display design is usually reflected in more glances not longer glances, although illegible displays encourage longer glance durations. In the present study, both glance frequency and glance duration increased with screen complexity, suggesting that both the design and legibility of messages could be improved. Although Zwahlen et al.'s (1988) cut-off point of 2-sec durations is fairly arbitrary, effort must be spent in reducing glances as far as possible. In particular, the longest glances, of up to 3.8 sec in the present study, must be eradicated through improved message design. Some recommendations for improvements to all aspects of the system are presented next.

CONCLUSIONS

Although the experiment was carried out with a visual only system, the findings could be extended to auditory and auditory-plus-visual in-vehicle displays. Gatling (1976) found information recall performance in a real-road trial to be comparable across display modalities, but Labiale (1990) observed that longer messages were memorized better when presented visually. An optimal solution might therefore be a system using both modalities; the auditory for short summary messages, and the visual giving further details of these messages. In this case, drivers would only have to take their eyes off the road when particular message details were needed (e.g., for route diversion or for confirmation of the auditory message).

The results of the present experiment support the idea that only the simplest four element messages were satisfactory in terms of safety, usefulness, and driver acceptability. Messages with 9 elements were frequently forgotten and resulted in worrying eyes-off-the-road time. Even messages with seven elements seemed to be at the upper limits of drivers' capabilities. Further research is necessary to investigate whether certain types of message element can be more successfully processed than others, and which elements are not considered essential (see Graham, Mitchell, & Ashby, 1995). An initial recommendation to reduce complexity is that the road numbers of junction destinations be removed.

Increasing the display time of messages on screen or providing a "repeat last message" facility would alleviate some of the pacing that drivers experienced with the system. This would particularly benefit older drivers whose speed of cognitive processing is slower. In addition, a short break between screens in a message pair would allow drivers to reorient themselves to the road environment, reduce pacing and avoid excessive glance frequencies over short periods.

The study highlighted that longer retention intervals led to more errors (23% errors after 1 min, 33% after 3 min) and message providers should be aware of this pattern of forgetting over time. However, the ideal timing for messages requires further research, particularly into situations where the driver has a realistic incentive to remember the message content.

Finally, the study has shown that older drivers can remember travel and traffic information presented on screen as well as younger drivers, but have difficulties with the most complex messages. This suggests that although older drivers may be comfortable with the concept and implementation of an in-vehicle information system, they experience problems reading, encoding, and retaining complex messages. Visual difficulties found with the older eye could be overcome to a certain extent by providing controls to adjust the brightness, contrast, and saturation of the screen to drivers' personal preferences. Increasing the presentation duration or allowing the older driver to control pacing should improve the assimilation of messages. Overall, the simplification of message content, as described earlier, will improve both the assimilation and retention processes for the full range of drivers.

REFERENCES

Bhandal, A. S., & Palmqvist, U. (1993). Driver information and traffic control using short range vehicle roadside communications. In *Proceedings of the Autotech '93 Conference, Seminar 9: PROMETHEUS—Future Systems.* Birmingham, England: Institution of Mechanical Engineers.

Bhandal, A. S., & Richardson, M. J. (1992). Future applications for short-range vehicle/roadside communications links. In *IEE Colloquium on PROMETHEUS and DRIVE* (pp. 7.1–7.5). London: Institute of Electrical Engineers.

Coleman, R. (1993). A demographic overview of the ageing of First World populations. *Applied Ergonomics, 24*(1), 5–8.

Craik, F. I. M. (1977). Age differences in human memory. In J. E. Birren & K. W. Schaie (Eds.), *Handbook of the psychology of aging* (pp. 384–420). New York: Van Nostrand Reinhold.

Gatling, F. P. (1975). *Auditory message studies for route diversion* (Rep. No. FHWA-RD-75-73). Washington, DC: Federal Highway Administration, U.S. Dept. of Transportation.

Gatling, F. P. (1976). *The effect of auditory and visual presentation of navigational messages on message retention* (Rep. No. FHWA-RD-76-94). Washington, DC: Federal Highway Administration, U.S. Dept. of Transportation.

Graham, R., Mitchell, V. A., & Ashby, M. C. (1995). An analysis of driver requirements for motorway traffic information. In S. A. Robertson (Ed.), *Contemporary ergonomics 1995* (pp. 281–286). London: Taylor & Francis.

Labiale, G. (1989). Influence of in car navigation map displays on drivers performances. *Vehicle/highway automation: Technology and policy issues, SAE SP-791* (pp. 11–18). Warrendale, PA: Society of Automotive Engineers.

Labiale, G. (1990). In-car road information: comparison of auditory and visual presentations. In *Proceedings of the Human Factors Society 34th Annual Meeting* (pp. 623–627). Santa Monica, CA: Human Factors Society, Inc.

Labiale, G. (1992). Driver characteristics and in-car map display memory recall performance. In *Proceedings of the 3rd International Conference on Vehicle Navigation and Information Systems* (pp. 227–232). Piscataway, NJ: Institute of Electrical and Electronic Engineers.

Lovelace, E. A. (1990). Basic concepts in cognition and aging. In E. A. Lovelace (Ed.), *Aging and cognition: Mental processes, self-awareness and interventions* (pp. 1–28). Amsterdam: Elsevier Science.

Luoma, J. (1993). Effects of delay on recall of road signs: An evaluation of the validity of recall method. In A. G. Gale et al. (Eds.), *Vision in vehicles* (Vol. 4, pp. 169–175). Amsterdam: Elsevier Science.

Michon, J. A. (Ed.). (1993). *Generic intelligent driver support*. London: Taylor & Francis.

Miller, G. A. (1956). The magical number seven plus or minus two: Some limits on our capacity for processing information, *Psychological Review, 63*, 81–97.

Rabbitt, P., & Collins, S. (1989). *Age and Design*. Manchester, England: Age and Cognitive Performance Research Centre, University of Manchester.

RDS-ALERT (1990). *ALERT-C traffic message coding protocol proposed pre-standard* (DRIVE I Project V1029). Brussels, Belgium: DRIVE Central Office CEC.

Rockwell, T. H. (1988). Spare visual capacity in driving—revisited. In A. G. Gale et al. (Eds.), *Vision in vehicles* (Vol. 2, pp. 317–324). Amsterdam: Elsevier Science.

Salthouse, T. A. (1982). *Adult cognition: An experimental psychology of human aging*. New York: Springer-Verlag.

Stokes, G. (1992). *On being old*. London: Falmer Press.

Wierwille, W. W., Hulse, M. C., Fischer, T. J., & Dingus, T. A. (1988). Strategic use of visual resources by the driver while navigating with an in-car navigation display system. In *XXII FISITA Congress Technical Papers: Automotive Systems Technology: The Future* (Vol. 2, pp. 2.661–2.675). Warrendale, PA: Society of Automotive Engineers.

Yanik, A. J. (1989). Factors to consider when designing vehicles for older drivers. In *Proceedings of the Human Factors Society 33rd Annual Meeting* (pp. 164–168). Santa Monica, CA: Human Factors Society.

Zwahlen, H. T., Adams, C. C., Jr., & DeBald, D. P. (1988). Safety aspects of CRT touch panel controls in automobiles. In A. G. Gale et al. (Eds.), *Vision in vehicles* (Vol. 2, pp. 335–344). Amsterdam: Elsevier Science.

Chapter 11

The Format and Presentation of Collision Warnings

Stephen Hirst
Robert Graham
HUSAT, Loughborough University, Leics., UK

Many road collisions are caused by the driver's failure to detect objects or other road users at an early enough stage, or by misjudgment of the obstacle's movement. Recently the development of in-vehicle collision avoidance systems (CAS) has become possible due to advances in sensor technology. The purpose of these systems is to alert the driver to potentially hazardous situations, and thereby reduce the risk of accidents caused by detection failures. However, there are significant human factors problems to be overcome before the potential of anticollision technology can be realized. Assuming that the initial uptake of CAS will be for systems exerting a minimum level of intervention, the immediate problems concern the assignment of appropriate sensory modalities for warning presentation, and the criterion to be used for warning activation.

As driving is predominantly a visuospatial task, which places its greatest load on the visual attention system, warnings presented via a visual display may themselves go undetected. Alternatively, the presence of a visual display may place demands on visual attention that compete with those required for the immediate detection of a critical incident or those needed to perform any necessary evasive action. Theoretically, these problems should be less apparent for head-up display (HUD) technology, where the image is presented in the driver's field of view such that its content can be assimilated in conjunction with foveal viewing of the driving scene ahead. The HUD image may also be presented at near infinity. Such presentation seems most applicable to high load visual driving tasks (e.g., short headway monitoring) where drivers are reluctant to divert attention away from the scene ahead to a conventional dashboard display (Gantzer and Rockwell, 1968). Presentation of the image at optical infinity also reduces the amount of visual reaccommodation needed between the image and external scene (Stokes, Wickens, & Kite, 1990).

In contrast to the visual channel, the auditory mode of information presentation appears to offer a natural choice for designers of in-vehicle systems in that the output of both speech and nonspeech displays is eyes free: that is, the user can attend to an auditory display while engaged in a simultaneous task requiring both visual and physical performance, irrespective of where they are looking. Other advantages of sound are that it can be used to communicate the occurrence of discrete and continuous events, and it can be perceived in the dark or adverse visual conditions. Colavita (1974) showed that auditory warnings result in shorter reaction times than those presented visually. However, despite the potential flexibility of both speech and nonspeech displays for in-vehicle use, the range of individual differences in hearing acuity and the variability of noise levels in the vehicle environment preclude absolute reliability in the detection of auditory warnings. Furthermore, auditory displays also present a challenge in terms of user acceptability in that false alarms tend to be very irritating.

Given that visual and auditory warnings can pass undetected, it is likely that a successful CAS design will feature the provision of redundant information via the available sensory modalities (i.e., visual, auditory, and tactile). Tactile information can be used as a supplementary display channel when the primary visual or auditory channel is degraded or overburdened, or as a substitute display channel if the primary channel is closed (Sorkin, 1987). Janssen and Nilsson (1991) compared several experimental CASs in a simulator study where subjects were required to engage in a vehicle-following task. Collision warnings were provided by a continuous "bar" display, a warning light, a warning buzzer, or by smart gas pedal. In the latter case, the pedal was activated by a $25N$ increase in pedal force whenever the warning criterion was met. The results showed that the effect of the various CASs was to reduce the frequency of short headways, and increase the overall driving speed, acceleration and deceleration levels, and time spent in the left lane (i.e., that used by oncoming vehicles), where subjects could avoid irritation caused by the warnings. However, the smart gas pedal produced the desired reduction in short headways and suffered least from the counterproductive effects of increased overall speed, speed irregularity, and driving in the left lane.

With respect to the selection of a warning activation criterion, Horst (1984) analyzed videotaped footage of naturally occurring vehicle conflict situations using various time-related measures and presented evidence to suggest that a time-to-collision (TTC) measure of 4 sec could be used to discriminate between cases where drivers unintentionally found themselves in a dangerous situations from those where they remained in control. The TTC measure, as defined by Hayward (1971), is the time taken for two vehicles to collide if they maintain their present speed and heading, and is obtained by the following equation:

$$TTC \text{ (sec)} = \text{intervehicle distance (ft)/relative speed (ft/sec)}$$

Färber (1991) advocated that in the design of CASs, high reliability should be sacrificed in favor of high validity: That is, the system should provide warnings only in critical situations but without any guarantee that it will do so in all situations that are potentially dangerous. In practice, this could be achieved by adopting a selective collision warning criterion, and for example, Färber suggested that an algorithm based on a TTC of 4 sec may be suitable. The advantages claimed for this arrangement are that the system would not give rise to a high number of false alarms, and drivers would be unable to compensate for the gains in safety by corresponding changes in behavior. However, Nilsson, Alm, and Janssen (1991) conducted an experiment in which they compared three experimental CASs based on a TTC activation criterion in a car-following task. The subjective ratings indicated that a TTC value of 4 sec may be "too short" an activation criterion. Also, Maretzke and Jacob (1992) reported that drivers anticipating problems tend to start braking at around TTC = 4 sec, and given that a certain amount of reaction time must be built into in a collision warning, they proposed a warning activation criterion of TTC = 5 sec.

Although the calculation of the TTC criterion is speed dependent (i.e., greater stopping distances are allowed at higher relative speeds), it does not take into account the fact that, for a similar braking force, it takes longer to brake to a stop from a higher speed than from a slower speed. This brings into question whether a fixed TTC measure would be appropriate for CAS warning activation across the range of driver–vehicle–highway situations: There appears to be evidence to suggest that it would not. Malaterre, Peytavin, Jaumier, and Kleinmann (1987) instructed subjects to indicate the last moment at which they considered that they would be able to stop successfully while approaching a stationary line. They found that the mean deceleration rates necessary to avoid overshoot increased with speed. Similarly, Horst (1991) carried out a field study in which subjects approached a stationary object with instructions to start braking at the latest point they thought they could stop in front of the object. He found that TTC at the onset of braking increased with speed. However, results did not fit a constant deceleration model (i.e., at higher speeds, subjects had problems avoiding collision due to the excessive deceleration rates that were required).

Two laboratory studies are reported here. In the first study, the aim is to examine interface issues relating to the presentation of headway information to drivers. Given that existing research has compared the efficacy and acceptability of warnings presented in different modalities in isolation (Janssen & Nilsson, 1991; Panik, 1984); the study will investigate combined modality warning presentations (i.e., integrated visual and auditory warnings). Haptic displays are not considered because the available experimental facilities do not to support this mode of information delivery. The purpose of the second study is to examine two particular factors that may affect drivers' braking responses in their use of an experimental CAS. First, any effect of timing of TTC collision warnings on drivers' braking will indicate the utility of the CAS. If drivers make full use of the collision warning information, one would expect earlier warnings to cause earlier braking. Furthermore, a timing criterion that causes drivers to brake earlier will be preferable to one that has no effect. Second, given that video-based collision situations involving either a moving or a stationary target vehicle are to be employed, the study presents a unique opportunity to examine the effects of both the absolute speeds of the target and following vehicle and the relative speed differences between the two vehicles. Other published studies have to date only considered stationary obstacles. If speed variables (relative or absolute) are found to affect the braking decisions, then it may be appropriate to incorporate a speed-dependent variable in the algorithm for the activation of warnings.

STUDY 1

Method

A number of approaches to a stationary vehicle, at constant speeds of 30 and 40 mph, were recorded on video. The video sequences were edited to start at initial TTC values of either 8 or 12 sec, and to finish at TTC 2 sec. The edited recordings were used to produce six blocks (display conditions) of 16 sequences recorded in random order. Various combinations of visual and auditory displays were overlayed on five of the six blocks. *Abstract* and *pictorial* visual displays (Figs. 11.1 and 11.2, respectively) combined with speech and nonspeech auditory displays yielded four *warning* display conditions; and a *fill bar* type visual display (Fig. 11.3) comprised an *informative* display condition. The remaining block served as a nondisplay mediated control condition. A warning activation criterion of TTC = 4 sec was employed. The abstract display consisted of five horizontal, colored bars that illuminated in a traffic light sequence: for example, green, amber, and red, according to the TTC from the target vehicle. The pictorial

safe   close   too close

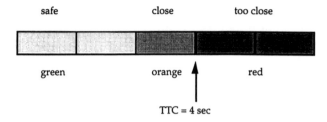

green   orange   red

TTC = 4 sec

FIG. 11.1. Abstract visual display.

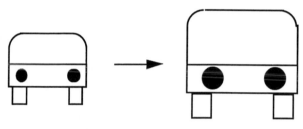

FIG. 11.2. Pictorial visual display.

FIG. 11.3. Informative visual display.

display comprised a vehicle icon that appeared at warning activation, and was replaced .2 sec after by a similar but larger icon. The informative display consisted of a horizontal bar whose size changed to correspond with the intervehicle headway. The visual displays were presented above and to the right of the target vehicle to simulate head-up presentation. The auditory warnings consisted of a single warning tone (500 Hz), or the phrase "Danger Ahead": Both were of .86 sec duration. The informative display was accompanied by a series of auditory tones whose temporal separation was reduced as headway decreased.

Twenty-four subjects (12 male and 12 female; $M = 31.2$ years) were recruited from the general public and were paid £15.00 ($23 U.S.) for participation in the experiment. A projection facility comprising an overhead multiscan projector (Sony FX-VPH–1271QM), a Super VHS video player (Panasonic AG 7350), and large projection screen was used in conjunction with a stationary experimental vehicle. The positioning of the vehicle in front of the screen facilitated a 25° vertical by 40° horizontal viewing angle consistent with the field of view of the lens used to record the video image. Subjects seated in the experimental vehicle were instructed to brake at the last moment they judged collision with the vehicle shown on video could be avoided. Braking responses were captured via activation of the brake pedal, but no subject feedback concerning the outcome of braking responses was provided. On completion, a questionnaire was presented to subjects to elicit preferences for the various displays.

Results

Objective Data. An analysis of variance (ANOVA) was carried out on the subjects' braking data. The ANOVA contained three within-subjects factors: Display, Approach Speed (i.e., 30 & 40 mph), and Duration of Approach Sequence.

Display Type and Braking Decision Points. The results of the ANOVA suggested that there was a significant difference between display conditions, $F = 5.1$, $p < .01$. A plot of the mean braking scores (i.e., TTC value at braking point) for each display condition can be seen in Fig. 11.4, which indicates that the previous difference is largely due to the influence of the abstract/speech condition (C4), which produced higher braking scores than all other conditions.

A subsequent Fishers Protected Least Significant Difference (PLSD) test indicated that there were several differences, significant at the 95% level, between the conditions. The only difference between the free braking control condition (C1) and the experimental conditions was for the abstract/speech condition (C4), where subjects braked earlier with the abstract/speech display. The abstract/speech condition (C4) also produced significantly higher braking scores than all other experimental conditions

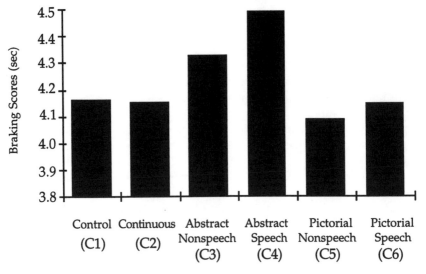

FIG. 11.4.   Mean braking point for each condition.

other than the abstract/nonspeech condition (C3). The braking scores produced for the abstract/nonspeech condition (C3) were also significantly higher than those in the pictorial/nonspeech (C5). This suggests that the abstract-type display, particularly when paired with speech, leads to an earlier braking response. There appears to be no difference between the results produced by the other conditions and the nondisplay mediated response.

Approach Speed and Duration. The speed and duration of approach were varied during the trial in order to find out if these variables affected the subjects' judgment of time to collision. The ANOVA showed a significant difference, $F = 32.1$, $p < .01$, to exist between braking scores pertaining to the two approach durations (i.e., initial TTC values of 8 and 12 sec). Inspection of the subjects' braking scores for these two conditions revealed that all subjects braked earlier in the longer approach condition (initial TTC = 12 sec).

Minimum Braking. In order to determine the number of braking decision points, which in a realistic situation would have resulted in collision, minimum braking times based on a deceleration rate of -23 ft/sec^2 were subtracted from the subject braking response times. For example, at 30 mph, any braking response later than a TTC value of 1.92 sec (i.e., momentarily after the video sequence ended at TTC 2 sec) would have resulted in a collision. As shown in Fig. 11.5, most collisions occurred in the informative display condition (C2), and less than half the number of collisions obtained in the free braking condition (C1) were produced in both the abstract/nonspeech and the abstract/speech conditions (C3 and C4, respectively).

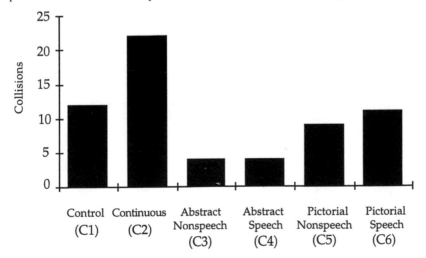

FIG. 11.5. Number of collisions in each of the display conditions.

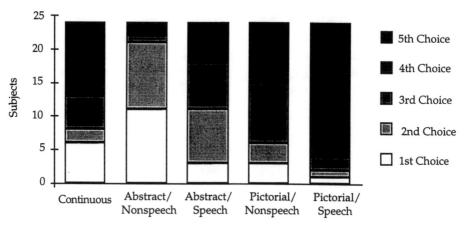

FIG. 11.6.   Subjects' ranking of display preference.

Questionnaire Data.   After completion of the trials, subjects were given a final questionnaire where they were asked to rank their order of preference for displays. Figure 11.6 shows the rankings of the individual conditions where the display attributes are combined. Here most subjects ranked the abstract/nonspeech display (C3) as their first choice. The informative display (C2) received more first choices than the other abstract display (C4), which had produced the best objective responses (i.e., earlier braking). A Wilcoxon Signed Rank Test confirmed the abstract/nonspeech display (C3) as the subjects' first choice, but there was no significant difference between the rankings given to the abstract/speech and informative displays (C4 and C2, respectively).

STUDY 2

Method

Video recordings were obtained in a similar fashion to that described for the first study, except that three target vehicle speeds of 0, 10, and 20 mph were employed. The abstract visual display and a warning tone was overlayed on all video sequences. Three different warning activation timings were employed such that the warning tone and first red bar of the abstract display appeared at a TTC of 3, 4, or 5 sec. The edited video sequences were used to produce six blocks, each containing 18 sequences recorded in random order: that is, resulting in 6 repetitions for each following vehicle speed/target vehicle speed/warning activation time combinations. All video sequences started at TTC 12 sec and ended at TTC 2 sec. Twenty-four members of the general public (12 male and 12 female; $M = 33.7$ yrs) participated in the study and were paid £15.00. The procedure followed was similar to that described previously, except that the video sequences continued until either the brake was pressed or the sequence ended (i.e., at TTC 2 sec).

Results

Timing of Collision Warnings and Vehicle Speed. The braking data was subjected to a repeated measures ANOVA containing three factors: Following Vehicle Speed (i.e., 30 and 40 mph), Target Vehicle Speed (i.e., 0, 10, and 20 mph), and Warning Activation Time (i.e., 3, 4, and 5 sec TTC). A wide variation in subjects' braking criteria was observed, where the mean braking point varied from approximately TTC 2 to 7 sec across the subjects. The main effect of warning time was found to be highly significant, $F = 42.0$, $p < .01$: That is, subjects tended to brake earlier for the earlier warnings, as can be seen in Fig. 11.7. There was also an interaction effect between the timing of warnings and vehicle speeds, $F = 4.19$, $p < .01$. Figure 11.8 shows the interaction, where it can be seen that subjects tended to brake earlier at slower relative speeds. However, it was apparent that earlier warnings produced earlier braking at all speed combinations.

The main effects of following and target vehicle speed (FS and TS, respectively) were highly significant ($F = 8.15$, $p < .01$ and $F = 31.7$, $p < .01$, respectively), and there was also a significant interaction between the two, $F = 7.9$, $p < .01$. The direction of the effect was counterintuitive in that at faster relative approach speeds (e.g., approaches to a stationary target), subjects braked later than at slow relative speeds (e.g., approaches to 20 mph targets). The effect is illustrated in Fig. 11.9, which also shows that, for similar moving targets speeds, the slower (30 mph) rather than the faster following speed (40 mph) produced earlier braking. For stationary targets, similar braking scores were obtained for both approach speeds.

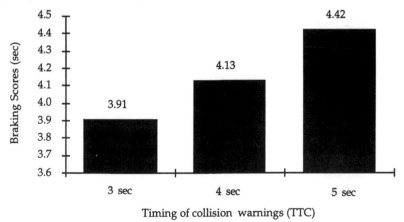

FIG. 11.7.   Mean braking point for the different warning presentation times.

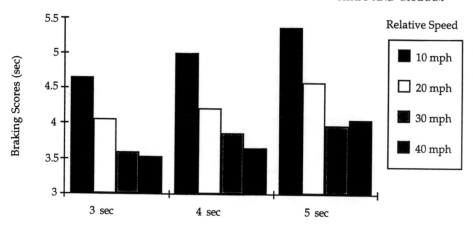

FIG. 11.8. Interaction between timing of warnings and relative speed of approach.

Subjects' braking scores across the different relative speed conditions were also expressed in terms of distance to target and are shown in Fig. 11.10. Theoretical plots for constant braking at TTC 3, 4, and 5 sec are also shown for comparison. Interestingly, the line produced by the braking scores is approximately equal to the line produced by plotting TTC 3.0 sec plus 1 ft for every mile per hour of following speed. This is expressed in the following equation:

$$D = 3.0 \text{ sec} \cdot (FS - TS) + 0.7 \text{ sec} \cdot FS$$

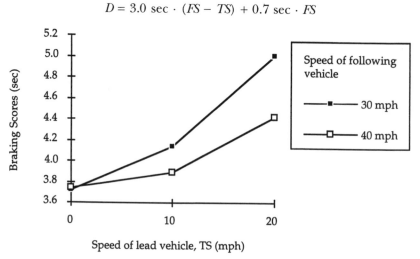

FIG. 11.9. Mean braking point for the different absolute vehicle speed conditions.

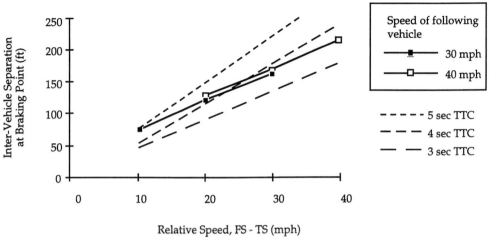

FIG. 11.10. Mean braking points for different relative speed conditions.

where D = intervehicle distance, FS = following speed, and TS = target speed. However, it can be seen across the speed conditions that the subjects' braking was more conservative than the 3 sec TTC criterion, and less so than 5 sec TTC. In contrast, braking tended to be more cautious than 4 sec TTC at low relative speed differences, but less so at high relative speeds.

Finally, the consistency of subjects' braking (i.e., the spread of braking points within each condition) was analyzed, and revealed a highly significant effect of relative speed, $F = 9.28$, $p < .01$. At lower relative speeds, there was more variability in the braking scores than at higher relative speeds.

Discussion

It is apparent from the objective and subjective data obtained in the first study that the abstract visual warnings were superior to the pictorial variety. The abstract-type display led to earlier braking responses, produced the least number of collisions, and was ranked first by most subjects. This finding tends to complement the conclusion of Gantzer and Rockwell (1968) that whereas improved performance in vehicle control can be obtained by the provision of supplementary visual headway information, drivers engaged in highly time critical situations are unable to benefit from such information because they are reluctant to divert attention away from the primary visual source. In the case of the abstract warnings, the visual display was available during the entire approach to the stationary vehicle, and so subjects were able to use the supplementary information to augment their braking judgments. The fact that the abstract/speech warnings gave rise to earlier braking responses than the nondisplay-mediated control condition tends to support this rationale. In contrast, the pictorial warnings were only displayed when collision was imminent: at precisely the moment when the driver least needed a source of visual distraction.

With regard to the auditory displays, the results of the objective and subjective data analyses are less clear cut. Overall, no significant difference was found between the braking scores for the speech and nonspeech displays, but subjects clearly preferred the nonspeech warnings. Although a few subjects specifically mentioned that they disliked the voice used, it may be that speech warnings are generally more irritating than discrete auditory tones. Whatever the case, it appears that the speech warnings caused the greater annoyance, and several factors may have exacerbated the effect. First, subjects were unable to escape the presentation of the auditory warnings irrespective of when they activated the brake, and therefore were potentially exposed to a large number of false alarms. Second, the fact that only one type of warning was displayed meant that in the case of the speech conditions, the same speech warning was presented over and over again. Third, given that there were no response alternatives to braking, the actual informational content of the speech warning was largely redundant.

In contrast to the subjects' preference for nonspeech auditory warnings, the comparison of the individual conditions showed that the abstract display, especially when combined with speech warnings, led to earlier braking behavior. One possible explanation is that the speech warnings caused annoyance, and subjects used feedback from the visual display to preempt these warnings and activate the brake before their occurrence. As stated previously, only the abstract warnings were available during the entire approach sequence. Whereas this strategy, employed in the abstract/speech condition, would not enable subjects to avoid the speech warnings altogether, it may have served to reduce the level of annoyance. Admittedly, this interpretation of the results would be stronger had subjects been able to avoid the auditory warnings altogether. An alternative explanation is that subjects were concerned that the speech warnings were too slow to give them enough time to react in time to avoid collision, and compensated by braking earlier. As previously, they were able to use the feedback from the visual display to preempt the speech warnings.

The main finding of the second study is that subjects' braking behavior can be influenced by the presence of collision avoidance warnings. Subjects instructed to brake at the last moment to avoid collision with the vehicle shown on video, tended to brake earlier when presented with earlier warnings. This pattern of results was apparent in most subjects, although there were a few who braked consistently, irrespective of the timing of the warnings. The main effect may have been partly due to the repetitive nature of the experiment: that is, subjects stereotypically pressing the brake pedal in response to the auditory warning rather than to the approaching target in the visual scene. However, if that were the case, then one would not expect to find, as in the previous study, differences between patterns of braking points for different types of CAS display employing a uniform warning criterion (i.e., at 4 sec TTC). Caution must be exercised when attempting to extrapolate the findings of a laboratory-based experiment to the real world. For example, in typical collision situations, drivers do not knowingly approach a relatively slow moving vehicle and brake at the last possible moment to avoid collision. On the other hand, given that drivers are likely to be more safety conscious than subjects participating in a laboratory experiment, it is possible that the effect of the timing of warnings could be stronger in the use of a CAS on the road.

It is not possible to conclude from the above finding alone whether earlier warnings (e.g., TTC = 5 sec) are more appropriate than later ones (e.g., TTC = 3 sec). Whereas earlier warnings gave rise to earlier braking, the trade-off between the timing of warnings and the frequency of false alarms must also be considered. False alarms occur when the system alerts the driver to a situation they have under control, and are obviously undesirable in terms of user acceptability. Unfortunately, the earlier the warnings, the greater the number of false alarms. Inspection of the braking point data, where the different speed components are separated (e.g., Fig. 11.10), reveals a number of points that are relevant to this matter. Whereas subjects appear to have employed a linear distance versus relative speed rule to judge their braking point, it was not one based on a constant TTC rule. In fact, none of the TTC timings employed seem to offer a satisfactory solution. Generally speaking, the 5-sec TTC criterion gave rise to a high number of false alarms (i.e., where warnings were given after braking commenced) in all but the lowest relative speed condition. The 4-sec rule gave rise to alarms in fewer situations: that is, those involving high relative speed differences (30 and 40 mph). But even so, in the context of an actual CAS, it would be likely to produce an unacceptably high number of false alarms. Finally, the 3-sec TTC criterion produced the least number of alarms, but at high relative speeds, taking reaction time into account, may allow insufficient time to brake to successfully avoid collision.

It appears that one solution may be to adjust the timing of the TTC criterion to somewhere between 3 and 4 sec, so that it is more discriminative: That is, so that it only produces warnings before the mean subject braking at the highest relative speed. However, it is possible that a more appropriate warning criterion may obtained by modeling the braking data obtained in the present experiment—that is, where warnings would be activated at a distance from the target equivalent to TTC 3 sec plus 1 ft for every mile per hour of following car speed. Such a criterion would have certain advantages over a system based solely on TTC. First, false alarms would be minimized across a range of speed situations. Second, in the case of an algorithm based solely on TTC, it is possible to achieve a very short headway to the vehicle in front by maintaining a low relative speed: making a collision unavoidable if the lead vehicle were to brake suddenly. A criterion with an added distance factor (e.g., dependent on following car speed) solves this problem by issuing a proximity warning even when the relative speed is zero. However, given that the present findings must be validated to ensure that they are representative of actual drivers braking behavior, it would be imprudent to recommend a specific algorithm from this study alone.

The other main finding of the experiment was an effect of speed on drivers' braking points. Faster following speeds and slower target speeds led to later braking. Greater relative speeds of approach (the combination of the above two factors) also produced later braking. The reasons for these speed effects are not intuitively obvious. Generally speaking, the laws of dynamics indicate that drivers should tend to brake earlier at fast rather than slow speeds to bring their vehicle to rest at a particular point. Therefore, one would expect drivers' to have learned this by experience and it be reflected in their braking behavior. Indeed, field studies by both Horst (1990, 1991) and Malaterre et al. (1987) found that drivers began braking earlier at faster approaches to stationary targets. However, differences between laboratory- and field-based experiments may account for the discrepancy between the results of the present study and those already cited. First, subjects participating in simulator trials are under less stress than those in an actual driving scenario. Braking judgments are clearly more crucial when an actual collision is imminent, and drivers may employ a different basis for their braking decision in such situations. Second, it has been argued (Cavallo & Laurent, 1988) that speed and distance information cannot be reliably judged from two-dimensional film sequences. Third, the importance of subject instructions has been confirmed by Horst (1991) and these also varied between the studies. Finally, the absence of sound cues and feedback as to whether or not a collision would have occurred may both affect drivers' decisions; Groeger and Brown (1988) and Groeger, Grande, and Brown (1991), respectively, found that these factors improved subjects' judgments of TTC.

None of the aforementioned factors fully explain the direction of the speed effect. However, there is one other main difference between the present study and previous ones: The present experiment included moving targets as well as stationary ones. The results indicated that moving targets led to earlier, more conservative braking than stationary ones, and faster moving targets gave rise to earlier braking than the slower ones. Therefore, one possible explanation is that drivers were using different mechanisms to judge their point of braking for a moving target compared to a stationary one. Indeed, the experimental data shows that for the stationary target conditions, there was no significant speed effect. Judging the relative motion for a moving target is a more complicated task than the same judgment for a stationary target. Using the Gibsonian model for TTC judgment (e.g., Lee, 1976), in the case of a moving target, the optic flow originating from the target is inconsistent with that originating from the background scene. The estimation of relative motion and the subsequent calculation of required braking point based on this information will therefore be more difficult. It was noticeable that the variability in subjects' brake points was significantly higher for slow relative speeds (i.e., faster targets) than for fast speeds of approach (slower or stationary targets). This inconsistency can be taken to support the previous hypothesis. If moving target situations are more complex, the results could be due to the calculation of relative motion simply taking longer, leading to later braking decisions.

CONCLUSIONS

There are several implications of the findings of the first study to the design of collision avoidance headway displays. First, given that only the abstract visual warning achieved better performance results than the non-display-mediated control, and that in general it was liked by subjects, it is recommended that this type of visual warning display should be employed. The display should also provide continuously available feedback as the one used in this study. It should be noted that the visual displays used in the experiment were presented in head-up mode. The assumption was that if one display was more effective than the others, then it was certain to be apparent in this superior presentation mode. Of course, there is no guarantee that the previous findings can be extrapolated to other modes of presentation (i.e., midhead and head-down displays).

Second, in collision avoidance systems dedicated to a particular type of hazard (e.g., a headway warning system), the use of a discrete nonspeech auditory warning combined with the visual display is recommended. This conclusion is based on the grounds that subjects clearly preferred the nonspeech warnings. Also, whereas this type of display was not shown to result in braking responses that were significantly earlier than the nondisplay-mediated control, it incurred fewer collisions. Third, where an integrated collision avoidance system is to be designed the use of speech warnings is advisable. Bertone (1982) noted that speech warnings are more informative than simple auditory tones, and they not only alert the user to the problem but also provide more cues as to its nature. The fact that language is highly overlearned means that speech is likely to be more effective in conditions of high workload or stress, where the meaning of coded auditory tones may be forgotten (Edman, 1982). As stated previously, a number of experimental factors contributed to the unpopularity of the speech warnings. However, it is apparent that speech can cause irritation, and therefore it would be prudent to provide speech warnings as a default option and make nonspeech auditory warnings available as a user option. Of course, as in all cases where auditory displays are used, system designers must ensure that it is possible to turn the displays off and that false alarms are kept to a minimum.

Regarding future work on CAS displays, based on the assertion that "the research platform that can deliver a realistic and ecologically valid traffic negotiation environment is a vehicle operating in real traffic" (Zaidel, 1991), the previous recommendations require validation in actual road trials. As previously stated, the optimum CAS design will feature a combined visual, auditory, and tactile display. Given that work on the "smart" gas pedal has indicated particular advantages of the haptic channel, there is a need for further research on integrated modality displays.

The results of the second study suggest that a CAS warning criterion should not be based solely on a TTC measure but one augmented by an additional distance factor. The distance/speed plot produced by braking scores obtained in the study was approximately equal to the line produced by plotting TTC 3 sec plus 1 ft for every mile per hour of following speed, but it is not possible to recommend a specific algorithm. Further research is needed, in actual driving scenarios, to examine drivers braking judgments for moving targets over a wide range of speed conditions—including those situations where the target vehicle is decelerating. An observed speed effect was in the opposite direction to that hypothesized: That is, subjects tended to brake later at faster rather than slow approach speeds. The faster relative speeds involved approaches to stationary target vehicles, and moving targets in the slower approaches. It was argued that subjects were using different mechanisms to judge braking in the two situations, and that calculation of a moving target is more complex, takes longer to process, and leads to later braking decisions.

REFERENCES

Bertone, C. M. (1982). Human factors considerations in the development of a voice warning system for helicopters. In *Behavioral Objectives in Aviation Automated Systems Symposium* (pp. 133–142). Warrendale, PA: Society of Automotive Engineers.

Colavita, B. A. (1974). Human sensory dominance. *Perception and Psychophysics, 16*, 409–412.

Cavallo, V., & Laurent, M. (1988). Visual Information and skill level in time-to-collision estimation, *Perception, 17*, 623–632.

Edman, T. R. (1982). Human factors guidelines for the use of synthetic speech devices. *Human Factors Society 26th Annual Meeting* (pp. 212–216). Santa Monica, CA: Human Factors Society.

Färber, B. (1991). *Designing a distance warning system from the users point of view*, APSIS report. Glonn-Haslach: Institut fur Arbeitspsychologie und Interdisziplinare Systemforschung.

Gantzer, D., & Rockwell, T. H. (1968). The effects of discrete headway and relative velocity information in car-following performance. *Ergonomics, 11*(1), 1–12.

Groeger, J. A., & Brown, I. D. (1988). Motion perception is not direct with indirect viewing systems. In A. G. Gale et al. (Eds.), *Vision in vehicles* (Vol. 2, pp. 35–43). Amsterdam: Elsevier Science.

Groeger, J. A., Grande, G., & Brown, I. D. (1991). Accuracy and safety: Effects of different training procedures on a time-to-coincidence task. In A. G. Gale et al. (Eds.), *Vision in vehicles* (Vol. 3, pp. 27–34). Amsterdam: Elsevier Science.

Hayward, J. C. (1971). *Near misses as a measure of safety at urban intersections.* Doctoral dissertation, Pennsylvania State University.

Horst, A. R. A., van der (1984). *The ICTCT calibration study at Malmo: a quantitative analysis of video recordings* (Rep. No. IZF-37). Soesterberg, Netherlands: TNO Institute for Perception.

Horst, A. R. A., van der (1990). *A time based analysis of road user behaviour in normal and critical encounters.* Soesterberg, Netherlands: TNO Institute for Perception.

Horst, A. R. A., van der (1991). Time-to-collision as a cue for decision-making in braking. In A. G. Gale et al. (Eds.), *Vision in vehicles* (Vol. 3, pp. 19–26). Amsterdam: Elsevier Science.

Janssen, W. H., & Nilsson, L. (1991). An experimental evaluation of in-vehicle collision avoidance systems. *24th ISATA International Symposium on Automotive Technology and Automation,* 209–214.

Lee, D. N. (1976). A theory of visual control of braking based on information about time-to-collision. *Perception, 5,* 437–459.

Malaterre, G., Peytavin, J. F., Jaumier, F., & Kleinmann, A. (1987). *L'estimation des manoeuvres realisables en situation d'urgence au volant d'une automobile* (Rep. INRETS No. 46). Arcueil-Cedex, France: Institut de Recherche sur les Transport et leur Securite.

Maretzke, J., & Jacob, U. (1992). Distance warning and control as a means of increasing road safety and ease of operation. In *FISITA '92: Safety, the Vehicle and the Road, XXIV FISITA Congress* (pp. 105–114). London: Institute of Mechanical Engineers.

Nilsson, L., Alm, H., & Janssen, W. (1991). *Collision avoidance systems: Effects of different levels of task allocation on driver behaviour,* DRIVE Project V1041. Haren, The Netherlands: Generic Intelligent Driver Support Systems, Deliverable GIDS/ MAN3, Traffic Research Centre.

Panik, F. (1984). Fahrzeugkybernetik. In *Papers XX FISITA Congress,* "Das Automobil in der Zukunft" (SAE-P143, Paper No. 845101). Warrendale, PA: Society of Automotive Engineers.

Sorkin, R. D. (1987). Design of auditory and tactile displays. In G. Salvendy (Ed.), *Handbook of human factors* (pp. 294–309). New York: Wiley.

Stokes, A., Wickens, C., & Kite, K. (1990). *Display technology: Human factors concepts.* Warrendale, PA: Society of Automotive Engineers.

Zaidel, D. M. (1991). *Specification of a methodology for investigating the human factors of advanced driver information systems* (Rep. No. TP-11199E). Ergonomics Division, Transport Canada.

Chapter 12

In-Vehicle Collision Avoidance Support Under Adverse Visibility Conditions

Wiel Janssen
TNO Human Factors Research Institute, Soesterberg, The Netherlands

Hugh Thomas
British Aerospace, Bristol, UK

Advanced sensor technology for road traffic promises a reduction of accident risk, which is to be achieved by means of so-called in-vehicle collision avoidance systems (CAS). The basis lies in the conviction that it should be possible to support human drivers by technical means in avoiding an impending collision that is unnoticed or perhaps not noticeable at all.

A CAS recognizes critical configurations just before they actually happen, after which it initiates some form of corrective action. The action may be a warning to the driver or (the beginning of) a specific evasive action itself. This chapter deals with so-called longitudinal collision avoidance systems, where the target is a preceding vehicle in the same lane. The sensor technology for this configuration, and therefore the potential for future application, is relatively well developed. Before considering CAS design issues in more detail, it seems wise to reflect on what it actually hopes to achieve.

The average Western automobile driver experiences a rear-end collision—the case to which the current generation of sensors has most to offer—about once every 25 to 30 years. If narrow escapes are also counted as events to be avoided, then support would, of course, be needed somewhat more frequently. Nevertheless, it is clear that the event a CAS should detect is so rare that serious doubts should be entertained regarding the possibility that detection of critical configurations could ever be performed flawlessly, let alone that it could be achieved without false alarms. Given that this is the case, it is probably better to take a modest stance on what to expect from a CAS, and to ask how different CAS designs would affect overall driver behavior rather than to focus directly on expected, or estimated, accident reductions.

This chapter examines what is, in behavioral terms, the best way of giving CAS support in conditions that confront drivers with problems of reduced visibility while car following. It elaborates on earlier work on CAS support by asking whether the design of a CAS should be indifferent to varying external conditions, or whether it should be adapted to those conditions by introducing, for example, new parameter settings. The logic of the experiment rests on some assumptions to be presented before describing the experiment proper.

WHAT SHOULD A WELL-DESIGNED COLLISION AVOIDANCE SYSTEM ACHIEVE?

What, then, are the behavioral effects to look for in a well-behaved CAS? The following are the parameters of driving behavior on which there is at least some consensus in the literature as to their relevance:

1. The distribution of (time) headways between preceding vehicles and the CAS vehicle, in particular, the proportion of short headways (below .5 or 1.0 sec).
2. The speed distribution of the CAS vehicle.
3. The distribution of its momentary accelerations/decelerations.

The primary behavioral aim of a CAS is to shift the distribution of headways away from the tail that contains short values. Evans and Wasielewski (1983) reported empirical evidence corroborating the intuitive assumption that there is a relation between time headway and accident involvement. Presumably, when the time headway is below the following driver's reaction time, the probability of a rear-end collision increases in a stepwise fashion. On the basis of this result, the proportion of short headways occurring is now routinely used as a quality index for any system that exerts control over longitudinal elements of the driving task (e.g., Zhang, 1991). This use of the proportion of short headways is not in conflict with Hitchcock's (cited in Shladover, 1993) observation that following of a preceding vehicle at short headways may actually be less risky than less close following, given that a sudden deceleration of the preceding vehicle will lead to a rear-end collision with a smaller Δv in the former case. There is no contradiction because Hitchcock's observation pertains to the *severity* of a rear-end collision once it happens, whereas the Evans and Wasielewski result pertains to the increased accident *probability* at short headways. Insofar as the proportion of short headways is related to accident probability, it remains, therefore, an acceptable proxy for what a CAS should achieve.

If all a CAS would do is to reduce the amount of short headways, then matters, in terms of assessing what the best CAS is, would be simple. However, it can be expected that drivers will show more general changes in their behavior when armed with a CAS, and some of these may be counterproductive to safety. This may occur even in the distribution of time headways themselves: When there is a CAS that warns at a certain criterion headway an overall compression of the headway distribution may result because drivers get to know when they will be warned. Time headways would then begin to pile up against the criterion. Insofar as this entails a more frequent occurrence of headways that are above the criterion, but still below the driver's reaction time, this will be counterproductive to the net safety effect. Figure 12.1 illustrates the reasoning.

Apart from affecting the time headway distribution, the availability of a CAS might induce changes in other parameters, notably speed and speed variability. It should not come as a surprise if it were found that the availability of a CAS induces faster and more irregular driving, both of which are counterproductive to safety. This is why driving speed and its variability should be measured as part of the overall assessment of CAS-supported driving behavior.

Finally, there is the possibility that drivers may just sit back and relax (i.e., let their attentiveness drop) when being fortunate enough to possess a CAS. Again, this would have a degrading effect on net safety. Granted

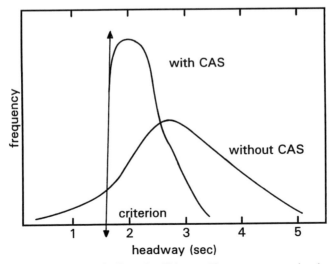

FIG. 12.1. Hypothetical effect of collision avoidance support on headway distribution. Criterion is arbitrarily set at a 1.5 sec headway.

that counterproductive changes induced by a CAS will probably be mixed with primary beneficial effects on short headways (see, e.g., Janssen & Nilsson, 1990, for illustrations that this does indeed happen), what would then be the best CAS?

Unfortunately, traffic safety science as it presently stands is incapable of estimating a net effect from different, and opposing, sources, if only because the relations between the individual behavioral parameters and aggregate safety are poorly known. All that can be done, therefore, is to define the best CAS as the one that has beneficial effects on headways while minimally suffering from counterproductive effects on other relevant parameters, and hope that this ideal exists and that it will be found.

Having more or less defined how the ideal CAS looks, the next issue is the choices to make in CAS design so that the ideal will be approached as closely as possible. Basic design issues include the following: (a) What is the definition of a critical car-following configuration (i.e., what is the criterion that should trigger system action)? There is a range of options here in terms of critical times and/or distances, of which time headway and time to collision (the momentary distance between two vehicles divided by their speed difference) are natural candidates. Janssen (1989) discussed their respective a priori merits and demerits. (b) What should the system action be once a critical configuration has been detected? As in the case of the criterion, there is a range of choices, from alerting the driver to starting an overrulable action to fully automatic braking. (c) Are the criterion and the action always to be considered as independent factors, or can there be some synergy in considering them as combinations? The idea is that some system actions may be more naturally coupled to particular criteria, so that some extra gain will be achieved from their combination (Janssen, 1989). (d) Are criteria and actions, or combinations thereof, to be considered as fixed, or would they have to depend on other conditions, like traffic conditions, visibility conditions, or even driver characteristics?

Considerable research efforts, notably within the PROMETHEUS and DRIVE programs, have been directed toward the first three of the previous questions. For example, within the DRIVE–I program, Janssen and colleagues (Janssen, 1989; Janssen & Nilsson, 1990; Nilsson, Alm, & Janssen, 1991) identified a promising candidate CAS that uses a 4 sec time-to-collision criterion and that acts through the CAS vehicle's accelerator by increasing the pedal's counterforce. As it turned out, this was the one among a set of plausible systems studied that did not suffer from counterproductive behavioral changes, while considerably reducing the occurrence of short headways. One particular feature of this result, however, is that it was obtained in conditions in which there was no problem with visibility. Therefore, if subjects in these studies had a problem, it was probably not with detecting the presence of the preceding vehicle per se, but rather with assessing its movement relative to the own vehicle. Yet, it remains to be seen whether CAS criteria or actions should not in some way depend on external conditions. For example, when a preceding vehicle is hardly visible because of fog or darkness, there may be reason to warn the following driver when that vehicle is within a "worst case" distance, that is, within a range that would lead to a collision if this vehicle were to brake suddenly. In clear weather, on the other hand, it could be assumed that a driver does not need to be warned for the presence of a vehicle per se, but only for those momentary configurations of relative speeds and distances from which a collision would ensue. That is, a time-to-collision criterion might be more appropriate to, but possibly also restricted to, that case.

In order to investigate the possibility that CAS system features may have to be adapted to prevailing visibility conditions, the present study compared the performance of three candidate CAS systems under normal and adverse visibility conditions. Two of these were active supports that used an added—but overrulable—counterforce on the accelerator when triggered. The third was a purely informative, HUD-like system.

A SIMULATOR EXPERIMENT ON DIFFERENT FORMS OF COLLISION AVOIDANCE SUPPORT

Design and Procedure

Twenty-four male subjects drove the TNO simulator on a standard two-lane road in a number of car-following episodes in which they received different forms of CAS support. The simulator consisted of three subsystems: The *supervisor computer* (PC, 80486 microprocessor) took care of communications within the experiment. The *vehicle model computer* (PC, 80486 microprocessor) was used for calculating the position of the simulated vehicle; it has the dynamic characteristics of a Volvo 240. And the *computer-generated image system* (CGI; Evans & Sutherland ESIG 2000) generated real-time images. The subject in the simulator was seated in a fixed-base mock-up of a Volvo 240 with all normal controls. The CGI system provided the visual scene in the form of an image that was projected in front of the mock-up by means of a high resolution BARCOGRAPHICS 800 projector (visual angles: 50 degrees horizontally, 35 degrees vertically).

Several types of CAS support were compared. First, there is a CAS that continuously displayed the subject's momentary braking distance on the vehicle's windscreen in the form of a red horizontal line projected onto the road surface. This was a purely informative CAS, which gave drivers an indication of their own braking distance relative to the position of the preceding vehicle. This CAS is a conservative one because it does not take into account that the preceding vehicle will have a braking distance of its own that should in fact be added to the CAS vehicle's braking distance. Its inclusion in the present comparison allows the assessment of what could be achieved by a relatively simple system that does not require the technology for making speed and distance measurements relative to preceding vehicles. Second is the earlier "4 sec TTC + Added Counterforce" CAS. Then there is a CAS that used either the 4 sec TTC or the 1 sec simple headway criterion, whichever came first, followed by an increased pedal counterforce (TTC + 1 sec). The rationale of this criterion was that a "pure" TTC criterion permits following drivers to approach a leading vehicle as closely as they want, as long as they do so by staying under the relative speed threshold that, at a given distance, would trigger the TTC criterion. In the extreme, this would permit the CAS driver to follow at a very short headway at about the leading vehicle's speed. Adding the 1 sec simple headway criterion would prevent this from occurring.

There was also a control condition in which subjects drove without CAS support. Because of the possibility of carryover effects, type of CAS support was a between-subjects variable in the experiment. Thus, there were four groups of six subjects each. The other main variable of the experiment, visibility, was treated as a blocked within-subjects variable. There were three visibility conditions: *normal* (daytime) visibility, *degraded visibility because of fog* (meteorological visibility range = 40 m), and *degraded visibility because of darkness.* The simulator's ESIG 2000 CGI system provides a standard fog function, which allows the visibility distance to be set at any desired value. The closer an object is to the eye, the less fog is applied and the more visible the object. Therefore, the visibility range of the CGI system is closely related to the standard meteorological visibility range.

Car-following episodes (48 per subject per visibility condition) were induced by presenting preceding vehicles on the road that drove 10, 20, 30, or 40 kph slower than the subject. These were preset relative speeds, to be produced by taking real-time measurement of the subject's speed at the instant the preceding vehicle appeared on the road (which was timed to occur at a 7 sec initial time headway relative to the subject's vehicle).

In a quarter of all episodes, the preceding vehicle performed a braking maneuver (3 m/sec^2, which was maintained for 2 sec) at exactly the instant the subject's vehicle came within a following distance of 40 m. Preceding vehicles in the degraded conditions always had their (rear) running lights on.

A final aspect of the experimental procedure was that oncoming vehicles were programmed to appear in the opposite lane (of the two-lane road) frequently, though at irregular intervals, so that overtaking by the subject was never a feat to be accomplished lightly.

Dependent Variables

Driving behavior was evaluated by four variables: average driving speed, the standard deviation of the driving speed, percentage of time headway to preceding vehicle was below 1.0 sec, and percentage of time subject was in left lane. This variable was added because previous studies have observed that subjects may perform the evasive action of moving into the left lane in order to avoid being bothered by the CAS (Janssen & Nilsson, 1990).

Results

Figure 12.2 shows the results obtained for the four support conditions (three different CAS supports + control) and the three visibility conditions. The effects shown are only for those variables reaching significance in an ANOVA, which were the effects of type of CAS support on average speed (p = .03) and on the percentage of time headway to the preceding vehicle less than 1.0 sec (p = .05).

There was also a significant interaction between the CAS and the visibility conditions with respect to the proportion of short headways (p = .04). The interaction was due to a different ordering of the support conditions for the normal (daytime) visibility condition. In this condition, the HUD-like CAS showed no apparent increase, relative to control, in the proportion of short headways. Apart from this, there was no evidence that CAS effects were in any way moderated by visibility conditions.

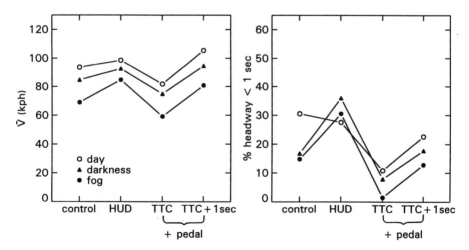

FIG. 12.2. Average speed (left) and percentage of time headway was less than 1 sec (right) for different CAS supports under different visibility conditions.

The effects of visibility conditions per se are also apparent from Fig. 12.2. They were considerable, though hardly surprising.

The overall pattern in the results is that there appears to be a single CAS that brings down the amount of close following without generating an increase in average driving speed. This is the earlier "4 sec TTC + Added Counterforce" CAS. Somewhat surprisingly, the addition to this CAS of an extra constraint in the form of a 1-sec simple headway criterion worsened performance, both in terms of average driving speed and of the proportion of short headways that occurred.

The braking distance HUD showed no demonstrable beneficial effect on behavior, not even a reduction in short headways.

CONCLUSIONS

This study compared the behavioral effects of a number of reasonable CAS systems, that is, systems that constituted plausible ways of reducing the probability of rear-end collisions. There may, of course, be other plausible solutions and other reasonable parameter settings for the systems studied here. However, the present systems are themselves the more promising candidates from a larger set of systems (e.g., Janssen & Nilsson, 1990). For this reason, the results of the present experiment can be considered to have general implications.

The CAS systems evaluated here were found to differ substantially in their effects on driver behavior. Beneficial effects were consistently obtained only for the CAS that had already proven its worth in earlier work, the 4 sec TTC + Added Counterforce CAS. Compared to a control condition, this CAS reduced the amount of close following while not increasing overall driving speed. Thus, this CAS met the requirement of showing a beneficial effect on the primary car-following parameter that was not offset by counterproductive changes in another essential parameter. Also, the favorable effects associated with this CAS did not depend on the visibility conditions. The suspicion that the design of a well-functioning CAS should be differentiated according to external conditions therefore proved to be unwarranted.

It is interesting to speculate on the reasons, first, why there were significant differences between different forms of CAS support and, second, why there was no differential effect on performance of the supports in degraded visibility conditions. With regard to the first question, the failure to achieve favorable behavioral effects of both the HUD-like CAS and the CAS that used the combined 4 sec TTC + 1 sec headway remains to be explained. Post hoc, the HUD-like device may have failed exactly because it gives information in a form unrelated to the relative movement of the preceding vehicle. That is, displaying the own momentary braking distance with little or no reference to the position of the preceding vehicle may simply not assist in staying out of a critical zone. This explanation is supported by the finding that the HUD-like CAS showed no increase in short headways when the preceding vehicle was actually visible, that is, in the normal visibility condition.

That the CAS with the extra 1 sec headway criterion performed worse than its counterpart—without the extra—may reflect basic complexities that arise from combining an absolute criterion with a relative one. A time-to-collision criterion takes relative speed into account, whereas a headway criterion depends only on the following vehicle's speed. Combining the two may make it hard to understand for a driver when the CAS is going to act. What drivers notice is that the distance to the leading vehicle at which the accelerator pedal starts to produce an extra counterforce varies in a complex way as a function of their own speed as well as of the relative speed with which they approach the leading vehicle. The net result of this may be that the driver chooses to disregard what the CAS is suggesting, relying instead on their own judgment.

The absence of differential CAS effects in different visibility conditions shows that a CAS may help irrespective of what perceptual problem (detecting a vehicle versus recognizing its movement) the driver has. It is, in fact, a fundamental characteristic of a time-to-collision criterion that it captures both the detection and the recognition aspect. That is, if a time-to-collision criterion is met this indicates at the same time that there is a preceding vehicle and that its relative speed is such that it will be reached within, say, 4 sec. Other criteria, including the simple headway criterion, do not have this characteristic. It may be for this reason that a time-to-collision criterion is capable of providing support to drivers even under conditions of adverse visibility.

REFERENCES

Evans, L., & Wasielewski, P. (1983). Risky driving related to driver and vehicle characteristics. *Accident Analysis and Prevention, 15*, 121–136.

Janssen, W. H. (1989). The impact of collision avoidance systems on driver behavior and traffic safety. *DRIVE–I: Deliverable GIDS/MAN 1.*

Janssen, W. H., & Nilsson, L. (1990). An experimental evaluation of in-vehicle collision avoidance systems. *DRIVE–I: Deliverable GIDS/MAN 2.*

Nilsson, L., Alm, H., & Janssen, W. H. (1991). Collision avoidance systems—Effects of different levels of task allocation on driver behavior. *DRIVE–I: Deliverable GIDS/MAN 3.*

Shladover, S. (1993). California PATH research on AVCS: Recent accomplishments and future plans. *Proceedings IVHS America*, 58–65.

Zhang, X. (1991). Intelligent driving—PROMETHEUS approaches to longitudinal traffic flow control. *Vehicle Navigation & Information Systems, Conference Proceedings SAE*, 999–1010.

Chapter 13

Cognitive ITS: On Cognitive Integration of ITS Functions Around the Driver's Task

Håkan Alm
Swedish Road and Transport Research, Linköping, Sweden

Ove Svidén
Arise, Linköping, Sweden

Yvonne Waern
Linköping University, Sweden

CAR DRIVING AND INTELLIGENT HELPSYSTEMS

The first step in any cognitive ergonomics endeavor is to analyze the task to be supported. Car driving can be analyzed in a number of different ways. McKnight and Adams (1970) suggested 43 separate main tasks, and it is possible to break these down into about 1,700 subtasks. A less detailed, but for some purposes more useful classification, was suggested by Rumar (1986) and included the following categories: to plan the trip, to navigate during the trip, to follow the road, to interact with other road users, to decide the speed, to follow rules and regulations, to handle the car, and to control other tasks in the car.

The next step consists in analyzing the opportunities and consequences of the prospective technical support. Intelligent helpsystems in the future car can, in principle, give the driver help in each and every task. A driver can be helped with the task of planning a trip, finding the way to the destination, avoiding accidents on the way to the destination, and so forth. One danger in this possible development is that drivers may have a number of different intelligent helpsystems in the future car. The subtask "to control (and interact with) other tasks in the car" may increase its proportion of the driver's different subtasks. Failure to allocate attentional resources in an optimal way may increase the risk of distraction. Distraction from the inside of the car (internal distraction) has been reported as one important precrash factor (Treat, 1980). To avoid the risks of information overload and distraction, it is necessary to take a perspective where intelligent helpsystems are designed so that they are adapted to the drivers' cognitive abilities and limitations. The ideal goal is to give the drivers the information they need, at the right moment, in the right situation, and in the right way.

INHERENT LIMITATIONS AND LEARNING OPPORTUNITIES

It may be proposed that the limitations of human beings can be overcome by training or education. It has been shown that learning of skills proceeds from declarative processing (slow and rational) toward procedural (automated) skill (Anderson, 1982). Thus, it could be suggested that people may learn to meet the demands of handling both the traffic situation and the intelligent helpsystems after some period of learning.

There are, however, definite limits on the skilled processes, a limit that varies with situational factors as well as individual ones. These limits are defined by the sensorimotor characteristics of human beings (cf. Card, Moran, & Newell, 1983). Consider as an example the time taken to make a simple decision under a simple condition. This time can, at best, be as short as 200 msec, an estimation arrived at by taking young students into the laboratory and giving them high preparation, good light conditions, and no disturbing circumstances. For decisions in slightly more complex situations, like those of braking when seeing the braking light of the car in front, the reaction time increases by at least five times, up to 1 sec or more in average.

Another example concerns the number of independent factors that a human being can attend to simultaneously. If these factors are visual, only one single object can be focused, whereas other objects in the neighborhood of this focused one may be caught by indirect sight. If the factors are auditive, two factors may be attended to only if the processing does not require any details. Simple beeps may thus be distinguished from other auditory signals (like listening to the radio or road instructions). Of course, several factors may be paid attention by time sharing, but then the time for moving attention from one factor (or object) to another has to be considered in the real-time driving situation.

It is further known that people's information processing is hierarchically controlled. People have the possibility to make crude plans on a high (rational) level and to refine these plans on a lower (skilled) level according to the requirements of the particular situation. However, if the situation changes to the extent that the higher levels have to be involved, people will have to use the slow, rational processes again.

These limitations of the human being should be considered in designing any support in the driving situation. In the following section, some requirements on the support are suggested, taking the human limitations into account.

COGNITIVE REQUIREMENTS ON SUPPORT FOR THE DRIVING TASK

It is self-evident that a support in the driving situation should not disturb the driver. However, how it is possible to assess the support versus the disturbance before actually having the support? General analyses of the relations between human error and automation may be found in, for instance, Reason (1992). Also, particular analyzes of cockpit automation have been performed, from which other lessons can be learned (see, e.g., Hughes, 1995).

However, in order to get into the particularities of a driving situation, the driving task has to be analyzed in detail, as well as its possible support. There are at least three kinds of support: automatization of some of the driver's tasks, informing the driver about the road, and alerting the driver to critical information in the current driving situation (Micron, 1993; Svidén, 1993).

The cruise control devices are one example of automatization of actions that otherwise demand the driver's attention. Drivers do not always notice changes in speed limits, neither do they always remember the actual speed limit. Slowing down the vehicle according to changes in the traffic pace may seem to be a nonintrusive and effective means of supporting the driver in keeping a safe distance to the car in front. Ongoing research indicates, however, that people prefer having control over the car to using this automated device.

Another automatization that seems feasible concerns lateral control. Keeping the car within the lane is a task that, in the long run, gets tiring. The resulting decreased vigilance may prevent the driver from reacting adequately and quickly enough in unexpected situations, such as meeting animals or skidding. Automated clutch handling, braking, and signaling to cars behind might be useful in such situations. However, human drivers' reactions to these proposals have to be carefully investigated.

Alerting the driver to critical information is useful both when the drivers' attention is overloaded by the complexity of the situation and when the drivers' vigilance is low, due to a long, boring driving time. This kind of alert should be sparse, however, in order to be effective, carefully timed, and given in an adequate modality. A possible solution would be to make the presentation of information from helpsystems dependent on the static

parts of the traffic environment (Verwey, 1991). For instance, do not present any message when the drivers are driving in a situation where their
workload is high (e.g., a roundabout).

Usually, the traffic situation is overloaded by visual cues. That is why
beeps are so effective. Following this line of thinking, it is possible that
mainly auditory messages can be used for navigation systems (Davies &
Schmandt, 1989).

Support may also be tactile or kinesthetic. For vigilance problems, auditory, tactile, or kinesthetic signals seem to be helpful. For instance, a tactile
message from the gas pedal was used as a signal from the anticollision
system (Janssen & Nilsson, 1990). This led to more positive results compared to any other system tested.

The last kind of support, road and traffic information, is most relevant
to deliberate, rational decisions on a high level, and should be offered
only when the situation demands or when the driver requests it. In the
first case, the information should be as unobtrusive as possible, optimally
visible (or audible) without interference with the lane keeping and traffic
checking tasks. In the second case, it is possible to envisage a more involved
information, for instance, given for trip planning before the trip or during
the trip at some parking lot at the side of the road. This information then
serves the higher level, rational plan.

Only a thorough analysis of the whole driving situation can suggest the
conditions for choosing a relevant kind of support. Further, the integration
of these different kinds of support requires an idea of different traffic
situations in order to show when the supports will interact smoothly with
each other and the driver and when they may interfere with each other.

LESSONS FROM DRIVE AND PROMETHEUS

The European research projects PROMETHEUS (Program for European
Traffic with Highest Efficiency and Unprecedented Safety) and DRIVE
(Dedicated Road Infrastructure for Vehicle Safety in Europe) have taught
a number of important lessons for the future work with Road Transport
Informatics (RTI) systems. The size of these projects makes it impossible
to review all of the important lessons that have been learned. Instead, this
chapter concentrates on a small sample of knowledge gained in what
loosely can be called the *behavioral domain.*

It may be argued that many of the most important problems lie in this
behavioral domain. Whether the support potential mentioned previously
will be reached or not depends, to a large degree, not only on cognitive
characteristics but to a large extent on the acceptance that different RTI
systems get from the public. Thus, if the intention is to maximize the

potential of different RTI systems, then it is important to know how they should be designed to be generally accepted. In order to do so, it is necessary to know more about people's knowledge and beliefs about computerized systems, what they see as positive and negative aspects, how they believe the use of computerized systems in the car of tomorrow will change their task of driving, and so on.

In the PROMETHEUS project PRO-GEN, questions about people's knowledge and beliefs about different RTI systems have been studied to some extent. One consistent result in these studies is that RTI systems that interfere with important aspects of the drivers' direct control of the car are not approved by future users. For instance, given the choice between a system that informs the driver about a correct distance to a leading vehicle and a system that actually keeps the correct distance, the subjects studied preferred the system that only informs the driver, and does not take over the control of the headway keeping task. There is also a fear having to do with the possibility that the authorities may use RTI systems to control and monitor individual drivers. These, and other results, give some important general guidelines for the design of future RTI systems.

The introduction of RTI systems into the car of tomorrow is supposed to inform, assist, or help the driver in many different ways. But, there is also a risk that the introduction of RTI systems can lead to unwanted changes in drivers' behavior, changes that are hard, or even impossible to predict in advance. The purpose with the DRIVE project BERTIE (V 1017, Changes in Driver Behavior due to the Introduction of RTI Systems) was to look at changes in driver behavior when different RTI systems were introduced. A number of field and simulator studies were made in the project in order to study changes in driver behavior when mobile telephones and navigation systems were used during driving. During an investigation of the use of mobile telephones during driving, a number of effects that may affect the drivers' safety in a negative way were detected.

One important implication of these studies is that the car of tomorrow, equipped with a number of RTI systems, needs some kind of internal intelligence that can prevent the driver from being overloaded and/or distracted by RTI systems in critical driving situations. The aim of the DRIVE project GIDS (V 1041 Generic Intelligent Driver Support Systems) was to determine how a class of intelligent co-driver systems should be designed to be maximally consistent with the drivers' information needs and performance capabilities (see Micron, 1993, for an excellent summary of the project). The aim of the GIDS project was to protect the driver from the risk of being overloaded and/or distracted in critical driving situations, which very well may be a risk if RTI systems are developed from the perspective of what is possible from a technological point of view. The GIDS project, in which a large number of cognitive psychologists took part, started with the perspective of the driver, and stressed the importance of adapting the system to what is possible from the driver's point of view. To adapt different RTI systems, and combinations of different RTI systems, to drivers' possibilities and limitations, the project developed the notion of an "information refinery" (Micron, 1993) that could prioritize and delay information, and present it in such a way that the driver understands it quickly without being distracted or overloaded.

To summarize, a psychological (cognitive as well as attitudinal) perspective is important in the work with RTI systems for a number of reasons. It is needed in order to understand what kind of RTI functions that have a potential to be accepted by the public, and also to get general guidelines for how these functions should work. It is needed in order to adapt the different RTI systems, and the totality of all RTI systems, to the drivers' possibilities and limitations. Finally, it is important in order to assess how the implementation of different RTI systems, and groups of RTI systems, will affect the drivers' behavior.

CONCLUSIONS

A logical conclusion from what has been said so far is that more knowledge is needed to answer a large number of questions. On the level of single intelligent helpsystems, more knowledge of cognitive characteristics and learning is needed in order to optimize the information presentation, that is, to improve the drivers' possibilities to perceive, interpret, and understand the messages from the systems. On the level of several interacting intelligent helpsystems, more knowledge is needed concerning the relation between drivers' workload (in terms of static as well as dynamic aspects of different traffic situations) and the helpsystems. This can help identify situations where presentation of extra information should either be avoided, or made in some special way. Such a program needs to test different ways of envisaging the new technology in a realistic context without endangering the traffic situation. For cognitive ergonomic studies, simulations afford a rich source of information.

For attitudinal studies, tossing ideas around and testing their approval among people, video films seem promising. Here, alternative display techniques can be shown in a dynamic traffic setting (cf. Borgström & Svidén, 1990). A video can act as a screening of different display ideas, later to be simulated and tested in dynamic vehicle simulators.

An early testing of the man–machine interaction of drivers and potential IVHS functions and display units is crucial to decide the usability before the costly prototype and field trials stage. If IVHS is to become a blessing to the driver and traffic, a lot of effort is needed to specify the dynamic man–machine interaction between drivers and cognitive IVHS support.

REFERENCES

Anderson, J. R. (1982). Acquisition of cognitive skill. *Psychological Review, 89*, 369–406.

Borgström, R., & Svidén, O. (1990). *An Invitation to DRIVE.* Video of simulated RTI symbols presented by Head-Up Display in a car. Swedish National Road Administration.

Card, S. K., Moran, T. P., & Newell, H. A. (1983). *The psychology of human–computer interaction.* Hillsdale, NJ: Lawrence Erlbaum Associates.

Davies, J. R., & Schmandt, C. M. (1989). The back seat driver: Real time spoken instructions. In D. H. M. Reekie, E. R. Case, & J. Tsai (Eds.), *Proceedings of the First Vehicle Navigation and Information Systems Conference (VNIS '89)* (pp. 146–150). Toronto: IEEE.

Hughes, D. (1995). Studies highlight automation "surprises." *Aviation Week & Space Technology,* February, 48–49.

Janssen, W. H., & Nilsson, L. (1990). *An experimental evaluation of in-vehicle collision avoidance systems* (Deliverable Rep. No. DRIVE V 1041 GIDS/MAN2). Haren, The Netherlands: Traffic Research Centre, University of Groningen.

McKnight, A. J., & Adams, B. B. (1970). *Driver education task analysis: Vol. 1. Task descriptions* (Final Rep., Contract No. FH 11-7336). Alexandria, VA: Human Resources Research Organization.

Micron, J. A. (Ed.). (1993). *Generic intelligent driver support.* London: Taylor & Francis.

Reason, J. (1992). *Human error.* Cambridge, England: Cambridge University Press.

Rumar, K. (1986). *Age and road user behavior.* Paper presented at the Fourth Nordic Congress of Traffic Medicine, Esbo, Finland.

Svidén, O. (1993). MMI scenarios for the future road service informatics. In A. Parkes (Ed.), *Driving future vehicles.* London: Taylor & Francis.

Treat, J. R. (1980). *A study of precrash factors involved in traffic accidents* (Highway Safety Research Institute Review, HSRI 10/11, 6/1). Ann Arbor, MI: Highway Safety Research Institute.

Verwey, W. B. (1991). *Towards guidelines for in-car information management: Driver workload in specific driving situations* (Rep. No. IZF 1992 C-4). Soesterberg, The Netherlands: TNO Institute for Perception.

Chapter 14

The Effect of Vision Enhancement Systems on Driver Peripheral Visual Performance

Linda L. Bossi[1]
Loughborough University, Leics., UK

Nicholas J. Ward
Andrew M. Parkes
HUSAT, Loughborough University, Leics., UK

Peter A. Howarth
Loughborough University, Leics., UK

THE NEED FOR VISION ENHANCEMENT

Few would dispute the importance of vision to the driving task. The vast majority of information required for driving is obtained through the visual system (Mourant & Rockwell, 1972; Olson, 1993), and it can be severely degraded at night, in fog, or with inclement weather, not only in terms of the lack of visual information available, but also because what information is available may be misleading (Rheinhardt-Rutland, 1986). Night driving is generally accepted to be two to three times more hazardous than daylight driving per kilometer driven (Rumar, 1990b; Vanstrum & Landen, 1984). There is a particularly high rate of accidents involving pedestrians at night (Brown, 1980; Evans, 1991; Hall, 1983).[1]

Visual perception problems arise at night for a number of reasons: Major visual functions such as acuity, contrast sensitivity, and depth perception are reduced substantially at lower illumination levels (Bullimore, Fulton, & Howarth, 1990; Leibowitz & Owens, 1975a, 1975b; Olson, 1993); the glare of opposing headlights can reduce the visibility of low contrast objects such as pedestrians (Rumar, 1990a); and pedestrians grossly overestimate how visible they are to motorists who are facing opposing head-

---

[1]Now at DCIEM, Downsview, Ontario, Canada.

lights (Allen, Hazlett, Tacker, & Graham, 1970). These problems may be even worse for older drivers, a growing proportion of the driving population (Olson, 1988, 1993), because decreased retinal illumination and increased light scattering in the eye (Weale, 1963) decrease visual performance and could increase the time and distance required for older drivers to see and react to potential obstacles (Olson, 1988; Olson & Sivak, 1983). Despite these impairments, motorists tend to drive at speeds for which the stopping distance is greater than the visibility distance to objects such as pedestrians (Olson & Sivak, 1983).

TECHNOLOGICAL SOLUTIONS

Much effort has been directed at roadway lighting improvements to overcome some of these problems (Sivak & Flannagan, 1993). While clearly reducing the incidence of accidents (Baldrey, 1988), provision of roadway lighting, except at the most critical sites, has been considered too costly to be a viable solution to visual performance impairment at night (Schwab & Hemion, 1971). Direct illumination of the road scene through the use of main beam headlights is the most common and straightforward alternative. However, well-known problems associated with their use include glare for oncoming motorists, and light reflection in rain and fog.

Other technologies offer potential solutions to the problems experienced by drivers under conditions of limited visibility. Developments specifically in illumination, sensor, and display technology may be able to expand the range of human visual capabilities under adverse driving conditions (Parkes, Ward, & Bossi, 1995). It is technologically possible to use sensors in a vehicle that are capable of detecting radiation outside the visible spectrum. Infrared sensor technology, for example, has the capability to "see through" darkness, fog, or rain, even in the absence of any illumination within the visible spectrum (Brickner & Staveland, 1989; Stark, 1987). Whereas the human eye is sensitive within a narrow range of electromagnetic energy (roughly .38 to .7 μ, or 380 to 700 nm), infrared imaging systems are sensitive to heat radiation in the infrared bands (3–5 μ for "near infrared" or 8–14 μ for "far infrared" systems) and are capable of transforming the energy from outside the visible spectrum (i.e., distribution of relative temperatures) into a picture within the visible spectrum viewed on a TV screen (Brickner & Staveland, 1989), thus dramatically extending the range of human visual capacities. Such technology has been successfully employed in many military applications, especially for targeting and weapon guidance.

Until recently, thermal imagery was typically displayed on small screens within military aircraft or weapon platforms. This type of display would be of limited utility for vehicle navigation. However, it is now possible to display thermal images in the operators line-of-sight, superimposed on the real-world scene through the use of helmet-mounted or head-up displays (Nordwall, 1991; Pickering, 1987; Scott, 1988). These sophisticated FLIR-like systems (Forward Looking Infrared) are now in widespread use in aviation, extending operational capabilities to include low altitude navigation and flight control under conditions of limited visibility (Stark, 1987). The success of these technologies has led to their consideration as potential vision enhancement systems (VES) for automotive applications. Head-up displays (HUDs) are the most promising of these alternatives because drivers are unlikely to accept the restrictions imposed by helmet-mounted displays. It has become possible only recently to combine infrared sensor and HUD technology as a demonstration VES in road-going vehicles.

Head-up displays have been used in the aviation environment for many years, and more recently in commercial vehicles, to display critical information in the operator or driver's line of sight, superimposed on the real-world scene. The vast majority of HUDs display qualitative, quantitative, or representational information in digital, symbol, or text format (e.g., flight dynamics and aircraft status information in aviation, vehicle speed, confirmation of indicators, alerting messages in automobiles). The HUD images are collimated, or presented at a distance that is optically equivalent to the objects being viewed outside the vehicle (Stokes, Wickens, & Kite, 1990). HUD images for aviation applications are collimated for viewing at infinity, whereas those used in most automotive applications typically display information at a closer optical distance, because much of what a driver needs to see is closer than optical infinity (Weintraub & Ensing, 1992).

Figure 14.1 shows a schematic of a driver VES that uses both infrared sensor and HUD technology. Essentially, the infrared sensor image is proc-

Infrared Sensor + Image Processor + Head-up Display Technology

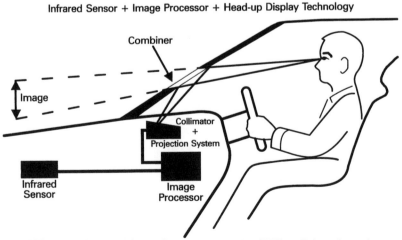

FIG. 14.1. Driver vision enhancement system (VES) utilizing thermal imagery and HUD technology.

essed to visible wavelength, and then presented to the driver with a head-up display so that the enhanced image is overlaid on the actual driving scene ahead. The collimator, through refraction (lenses), reflection (mirrors), or diffraction (holograms), bends the light rays from the image source so that they are delivered at a prescribed focal distance (i.e., parallel rays for optical infinity). Once collimated, the display image is then combined with the real-world scene viewed by the driver using a combiner, which may be positioned between the driver and the windscreen or may be integrated into the windscreen itself (Lee, 1983).

Conventional HUDs and those currently affordable for driving applications use refractive collimation and provide the operator with a total field of view of only 15 degrees horizontal by 10 degrees vertical (Swift & Freeman, 1986). Thus, use of a VES will provide drivers with only a small window of enhanced vision. HUD images are monochromatic, usually green. Future developments in HUD technology may allow for larger fields of view (Coonrod, 1983), although these will not likely be affordable for driving applications in the near future.

CONCERN WITH NEW TECHNOLOGIES

The human factors implications of introducing VES must be thoroughly investigated to ensure that the use of these new technologies will accommodate the diverse needs and capabilities of the driving population, the wide range of situational and environmental conditions under which it will be used, and the potential for misuse or system failure. Any new type of driver information system has the potential to distract driver attention, overload sensory capabilities, and create dangerous compensatory driver reactions (Rumar, 1990b; Zwahlen, 1985; Zwahlen & DeBald, 1986). For example, VES may allow more drivers to go faster under worse conditions rather than actually enhance safety (Evans, 1991).

The integration of multiple new technologies and their application to the nonaviation nonmilitary driving environment may create even more problems. All of the technologies involved in driver VES were developed for military and aviation applications. There are huge differences (i.e., age, selection, training, visual function, range in abilities) between the general driving population and those for whom the technology was designed. There are also significant differences between the aviation and driving tasks, the major difference being the higher level of visual attention to the outside world that is required in driving (Schlegel, 1993). Also consider the differences in technological system capabilities and reliability between those procured and maintained for the military and those intended for the average automotive consumer.

Each of the VES technologies, on its own, has characteristics that may limit its effectiveness in the driving situation. Infrared thermal images, for example, are phenomenologically quite distinct from those to which humans are regularly exposed and their use may carry a cost of increased workload, even for those who are highly trained in their interpretation (Brickner, 1989; Brickner & Staveland, 1989; Stark, 1987). A number of problems have also been associated with the use of HUDs that could impact on their effectiveness in automotive VES applications. These include interaction of HUD images with other light sources to produce unwanted reflections (Wilson, 1983); masking by high luminance HUD imagery of otherwise visible background scene items (Weintraub & Ensing, 1992); imperfect alignment of HUD image with scene over which it is superimposed, causing apparent HUD "jitter"; inappropriate collimation of HUD image causing binocular disparity, diplopia (double images), and discomfort (Genco, 1983; Gibson, 1980); and eye accommodation that is inappropriate for the degree of collimation (J. Iavecchia, H. Iavecchia, & Roscoe, 1988; Norman & Ehrlich, 1986).

VES ATTENTIONAL ISSUES

Most significant among the problems associated with the use of HUDs may be their potential for distracting the driver from paying adequate attention to the real scene. Although HUDs were designed to facilitate rapid switching of attention between instrument information and the outside world (Dudfield, 1988; Sojourner & Antin, 1990; Weintraub & Ensing, 1992; Weintraub, Haines, & Randle, 1984), the proximity of information from both sources may actually interfere with that process. Dashboard or panel-mounted displays offer powerful cues to switch attention that are not present with HUDs: the need to look up, change focus, and change convergence (Weintraub & Ensing, 1992). Furthermore, humans may not be accustomed to dividing attention between information that is superimposed in visual space, and the display of information in the visual field may draw attention away from the outside scene without the vehicle operator even being aware that this is happening (Weintraub, 1987). This attention-grabbing effect of HUD images has been termed *cognitive capture* in the literature.

Anecdotal evidence of cognitive capture exists: Military pilots have described informational HUDs as "compelling," for example, and fixation on the display is claimed to decrease the frequency of pilot's visual scanning of the outside scene (Dopping-Hepenstal, 1981). A number of carefully controlled laboratory studies have also demonstrated that such cognitive capture can occur when images are superimposed in visual space (Fischer, Haines, & Price, 1980; Neisser & Becklan, 1975), although there is evidence to suggest that if the two images share a common frame of reference and are at the same optical distance, parallel processing without cost may be possible (Foyle, Sanford, & McCann, 1991). This may be the case when thermal images are superimposed on the driver's forward field of view, as in an ideal driver VES. However, it will be difficult to achieve ideal collimation of the thermal image given the depth of information in the driving environment. The unique characteristics of thermal images may also preclude their simultaneous assimilation with the real driving scene.

The previous evidence and the majority of HUD-related literature is concerned with the problems of dividing attention between the HUD image and the immediate real-world background over which it is superimposed. However, this background represents only a small fraction of the environment and available information to which the driver must attend; attention must be paid to events occurring peripheral to the driver's direct line of sight as well. Given that a driver VES will be capable of enhancing only a small portion of the driver's forward field of view for the foreseeable future, together with the evidence of HUD cognitive capture, it is possible that the VES may negatively affect driver visual scanning or peripheral visual capabilities, which are vital attributes of safe driving practice.

Studies outside of the HUD literature also suggest that a VES might impair peripheral object detection. Increased arousal or increasing workload in even a simple central primary task can decrease attention available to peripheral cues (Baddeley, 1972; Leibowitz & Appelle, 1969; Mackworth, 1965; Webster & Haslerud, 1964; Zahn & Haines, 1971); the effect has been demonstrated over a wide range of eccentricities. Such peripheral visual performance degradation has been interpreted as a narrowing or focusing of attention wherein the "functional" or "effective" field of view shrinks and expands depending on the perceptual demands of a task (Eriksen & Yeh, 1985; Sanders, 1970). In the literature, this effect of focused attention has been called "tunnel vision" (Mackworth, 1965; Williams, 1985), "cognitive tunneling" (Dirkin, 1983), or "perceptual narrowing" (Lee & Triggs, 1976).

Perceptual tunneling has even been convincingly demonstrated in a real driving environment. Lee and Triggs (1976) conducted a series of experiments to test driver peripheral vision detections in a moving vehicle, under varying levels of driving task demand and in different task situations. Drivers were required to respond to lights presented semi-randomly on the dashboard (at 30 and 70 degrees from the direct line of sight) while driving through areas differing in attentional demand (highway through suburban to shopping center). Targets were only presented when the driver was directly viewing the forward scene. Significantly fewer peripheral detections occurred in the high attentional demand situations (shopping center and suburban environments), indicating a narrowing of the functional field of view about the point of fixation. A similar study (Miura, 1986) showed that eye movement patterns must adjust to high driving task demands, indicating that perceptual tunneling also affects the dynamic functional field of view.

The results of these driver peripheral vision performance studies may have direct bearing on the use of vision enhancement systems for driving under conditions of limited visibility. Conventional HUDs have been demonstrated to grab attention (Dopping-Hepenstal, 1981; Fischer et al., 1980; Weintraub & Ensing, 1992), either because of their novelty, or the increase in information displayed. Thermal imagery is known to be difficult to interpret (Brickner, 1989; Brickner & Staveland, 1989) and may be so even when superimposed on the actual driving scene. Unless otherwise demonstrated, it is not unreasonable to suggest that the presentation of thermal images on vehicle HUDs may be demanding or attention grabbing for the general driving population and may result in perceptual tunneling. The small size of conventional HUDs may also contribute to a narrowing of the focus of attention. It is possible that the normal eye scanning behavior of the driver will be affected by the introduction of a new enhanced portion of the forward scene and that the benefits accrued from the VES in terms of object recognition in the enhanced area, may be offset to an extent by a reduced rate of detection of events occurring in the periphery (Fischer et al., 1980).

This study presents part of the results from a comprehensive examination of the effects of the presence of an idealized VES on recognition and discrimination of brief, irregularly presented targets at various eccentricities outside the enhanced central area (Bossi, 1993). It was hypothesized that if perceptual tunneling exists, it would vary as a function of the difference in information available in the VES and the periphery. It was therefore decided to simulate a VES operating in both dusk and night conditions. Because of the need to maintain close control over the variables of the visual scene a laboratory environment was required. Previous studies have shown that target detection will degrade as a function of eccentricity (e.g., Dirkin, 1983; LeGrand, 1967). If perceptual tunneling is present, then it would be expected that target detection and identification performance would be worse in the periphery when a compelling image is presented in the central area. It would further be expected that a greater fall off in performance would be experienced in the darker ambient illumination condition, due to the greater disparity in quality of visual information available between the HUD image and surrounding real-world scene.

The VES in this experiment was idealized in order to control for any performance decrement that may be caused by problems with current technology: The VES "thermal image" was simulated using monochromatic daytime images, phenomenologically easier to interpret than real thermal imagery; and the VES images were matched perfectly in time and space with the background driving scene, eliminating performance problems that could occur due to binocular disparity, diplopia (double images), image jitter, or inappropriate collimation. If perceptual tunneling is demonstrated even under these idealized conditions, it may be that the use of a real VES, with all its current technological limitations, could exacerbate the problem further.

METHOD

This repeated-measures study was laboratory based, using a simple part-task driving simulator. Control and VES driving conditions were simulated under two conditions of illumination (NIGHT and DUSK) by editing daytime driving footage and projecting it onto a large screen in a darkened room. Subjects were required to perform peripheral target detection and identification tasks while tracking a vehicle on the screen, thus simulating the critical driving tasks of potential obstacle detection while steering the vehicle.

Thirteen subjects—6 male and 7 female, ages from 24 to 39 ($M =$ 31.5)—were recruited and paid for their participation in the study. All subjects had vision, with their normal optical correction (for those applicable), which was well within the requirements for driving in the United Kingdom (all met or exceeded the experimental minimum metric static acuity of 6/9, normal color vision, no field loss, acceptable phoria as measured using Keystone VS-II Telebinocular Vision Screener). All were experienced drivers with an average annual driving mileage of 18,000 km per year.

Actual daytime driving scenes were recorded onto videotape to show a field of view of approximately 50 degrees horizontal and 33 degrees vertical. The scenes depicted a route along rural and semi-rural single and dual carriageway roads in England. The scenes were recorded at a stable speed between 50–80 kph and at 15–20 m distance behind a lead vehicle during moderate traffic flow. Filming was intentionally conducted on a dull overcast day to avoid large differences in luminance (i.e., between sunshine and shadows), which might reduce the phenomenological realism of the night and dusk simulations. The same 30-min segment of driving footage was used for the preparation of all experimental condition videotapes to control for the possible effects of road geometry and scene complexity, because these factors have been shown to affect eye movement behavior and target detection performance (Cohen & Studach, 1977; Cole & Hughes, 1984; Cole & Jenkins, 1984; Shinar, McDowell, & Rockwell, 1977).

The driving scenes were then edited. First, artificial targets were superimposed onto the periphery of the daytime driving scenes and then the films were degraded to simulate the two illumination conditions. This order was necessary to ensure that target contrasts would be consistent between the illumination conditions.

Targets in the form of relatively constant contrast, $M = .85$, $SD = .26$, using formula $(L_{task} - L_{background})/L_{background}$, Landolt Cs were superimposed onto the video images at various eccentricities, as shown in Fig. 14.2. Targets were randomly presented in one of four orientations: gap up, down, right, or left. Each target was presented for 200 msec and subtended .92 degrees of arc at the eye at the simulator viewing distance (11 min of arc for the gap in the Landolt C). A total of 80 targets, 10 at each location, were randomly presented in each condition, with intervals between targets ranging from 10 to 90 sec (random target intervals were modified only to maintain consistent target conspicuity; targets were not superimposed unless the lead vehicle was within the central $10 \times 15°$ portion of the scene and the target background was uniform in complexity and color).

Target locations were selected to occur at 5-degree increments outside of where the VES image would appear (10, 15, 20, and 25 degrees left and right of center). These areas were considered to be the most relevant to the task of obstacle detection in real driving. The target locations were intentionally staggered along two bisecting meridians to minimize the possibility that subjects would develop successful target search patterns.

The size, duration, and contrast of targets were selected after numerous pilot trials to avoid ceiling and floor effects and to minimize search or eye movement and re-fixation after target detection. Subjects would only be able to identify the target orientation if their eyes were fixated within approximately 10 degrees of the targets being presented. They would be able to detect the targets only if they were attending to the periphery. In

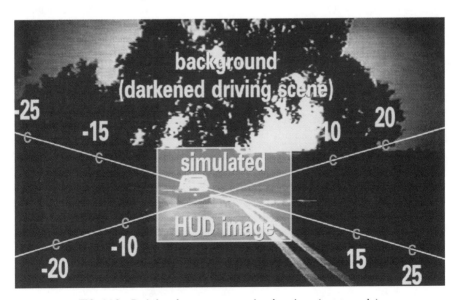

FIG. 14.2. Peripheral target presentation locations (not to scale).

this way, if eye movement patterns were altered (they were not actually measured in this experiment), then one would expect this change in pattern to affect the results. And, more importantly, the results would give some indication of the effect of the VES on attention to the periphery (Eriksen & Yeh, 1985; Shepherd, Findlay, & Hockey, 1986).

The same driving footage was used for each condition, thus different sets of target presentations were prepared for each condition so that subjects would not learn where to search for targets across conditions. Their presentation was counterbalanced across subjects and conditions to minimize any confounding effects due to possible differences in target set difficulty.

The films were then degraded to replicate night or dusk driving control conditions (estimated mean road surface luminances of .27 and .90 cd/m^2). Further duplicate films were produced that had similar peripheries, but that were edited to include a simulated idealized VES central area, representing a 15-degree horizontal by 10-degree vertical field of view (for the simulator viewing distance of 4.75 m). The central portion of the video image was manipulated to replicate a monochromatic green high quality HUD image (estimated mean luminance contrasts—between central road surface with and without the VES image manipulation—of 10.7 for VES-NIGHT, 2.6 for VES-DUSK condition). Essentially, the idealized VES image was achieved by superimposing the central 15 × 10 degree portion of original daytime driving footage with monochromatic green shading. This created a simulated HUD image with perfect collimation and alignment with the background scene (i.e., a single image) and avoided the image interpretation problems that real thermal imagery might have imposed. Figure 14.3 shows the same sample video frame from each of the four experimental conditions.

Subjects were required to perform a tracking task while seated in a vehicle steering mock-up. A hoop was secured to a metal frame in front of the subject to allow radial and vertical travel. Attached to the hoop was a laser pointer that provided a tracking light. The primary task for the subjects was to keep the spot of light projected by the laser on the rear number plate of the lead vehicle shown on the video screen, thus providing a visual task analogous to the important driving task of steering. The secondary task was to respond to peripheral target presentation by pressing a "brake" pedal, and identifying the target orientation verbally. Pedal position was adjustable for subject comfort.

Subjects were given 5 min training on the tracking task, and 15 min to practice both tracking and target tasks under daylight conditions. The experimental conditions (NIGHT, DUSK, NIGHT plus VES, DUSK plus VES), each of 25 min duration, were presented over two test sessions on different days. Order was counterbalanced. Subjects were instructed to

FIG. 14.3. Sample video frame from each of the four experimental conditions (DUSK conditions on the left, NIGHT conditions on the right).

track the lead vehicle using the laser pointer, to press the brake pedal when they saw a target, and to say "left," "right," "up," "down," or "don't know" in respect of the orientation of the Landolt C. Subject reaction time was recorded directly from the video time code signal corresponding to target onset, and a signal from the brake pedal on depression. A log was kept of verbal responses. A monitor manually recorded the cumulative time that the laser tracking light strayed from the lead vehicle, using a stopwatch; this provided a measure of primary task performance and motivated subjects to attend to the primary tracking task rather than actively search for targets. Subjects were not given any feedback as to their tracking or target task performance.

RESULTS

To assess the effects on visual performance of the presence of the VES, each dependent measure was analyzed using a 2(VES, no VES) × 2(NIGHT, DUSK) × 8(ECCENTRICITY of target location) within-subjects ANOVA. Statistical results are presented in Table 14.1. Figure 14.4 presents the results graphically, showing the mean number of targets detected and identified, by target eccentricity, in each of the four experimental conditions.

TABLE 14.1
Significance of Main Effects

Measure	Effect	F Ratio and Significance Level
Detection	VES	$F(1, 12) = 28.87$, $p < .001$
	DARKNESS (D)	$F(1, 12) = 60.30$, $p < .001$
	ECCENTRICITY (E)	$F(4, 40) = 33.29$, $p < .001*$
	DARKNESS × VES	$F(3, 33) = 6.47$, $p < .01*$
	VES × E	not significant
	D × VES × E	not significant
Identification	VES	$F(1, 12) = 8.44$, $p < .05$
	DARKNESS (D)	$F(1, 12) = 110.05$, $p < .001$
	ECCENTRICITY (E)	$F(7, 84) = 57.15$, $p < .001$
	DARKNESS X VES	$F(1, 12) = 26.82$, $p < .001$
	VES × E	not significant
	D × VES × E	not significant

*F ratios adjusted for nonhomogeneity of variance.

At NIGHT, it can be seen that target detection and identification performance was consistently poorer with the VES across all eccentricities. Although not statistically significant, this decrement in performance appears to be greatest for the more central targets, and an unusual dip in mean detection performance is also evident for targets presented 20 degrees right of center for both the control and VES conditions. At DUSK, there appears

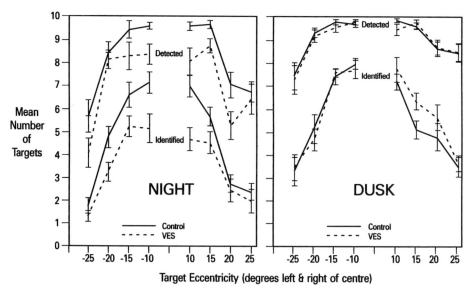

FIG. 14.4. Mean target detection and identification performance at night and dusk with and without a simulated driver VES (standard error indicated).

to be little or no difference between VES conditions, although slightly more targets were identified on the right when the VES image was present. A ceiling effect is apparent for target detection at dusk, although the consistency of results suggests that this did not mask any VES effect.

Comparison of the two graphs in Fig. 14.4 reveals that ambient illumination affected target detection and identification regardless of VES condition. For the majority of target eccentricities, fewer targets were detected or identified at night. It can also be seen that target task performance decreased with increasing target eccentricity.

No significant effect of VES presence on reaction times was found. However, scene darkness did have a consistent effect, with reaction times being longer in the NIGHT condition than at DUSK, $F(1, 11) = 26.30$, $p < .001$. Also reaction times increased as a function of target eccentricity, $F(7, 77) = 13.73$, $p < .001$. Tracking performance was consistent across all conditions, minimizing the possibility that subjects compensated for the target tasks by taking their attention away from the primary task of tracking.

DISCUSSION

Effect of Eccentricity and Background Illumination Level

Target task performance decreased with increasing target eccentricity regardless of VES or illumination condition, an effect that is also well-documented in the literature (Bartz, 1962; LeGrand, 1967; Leibowitz & Appelle, 1969). This finding is not surprising considering that the subjects' primary task was to track the lead vehicle; their eyes were likely directed to the most central area of the driving scene predominantly, consistent with normal driver visual behavior in a car-following situation (Sanders & McCormick, 1993), and the more extreme the target presentation location, the longer it would take to move the eyes and re-fixate to identify the target detected.

Target task performance was also better at DUSK than at NIGHT, regardless of VES condition. This is likely due to the increased conspicuity of those targets presented at DUSK. Although target contrast was held relatively constant between conditions, the luminance of targets in the DUSK conditions was higher, and target luminance is known to affect conspicuity and detection probability (Engel, 1971) as well as performance with Landolt C targets (Mandelbaum & Sloan, 1947). The difference in target task performance between DUSK and NIGHT is particularly evident at the more extreme target presentation locations (refer to Fig. 14.4), a finding that may be explained by the increased visiblity of peripheral events and objects at DUSK to direct driver attention and eye movements to the periphery.

Regardless of VES condition, the noticeable peak in NIGHT detection performance for targets presented 15 degrees right of center, and the apparent drop in performance for targets located at either 20 degrees right or 25 degrees left of center may relate to the intentional staggering of target presentation locations and the relation of these positions to normal eye scanning behavior. Target conspicuity may also play a role. At 15 degrees right, targets were typically presented on the road surface, and this relatively large, unchanging, and simple background might have made the targets at that site more conspicuous than those seen against a more complex background (Cole & Jenkins, 1984; Ward, Parkes, & Crone, 1994). The drop in performance at 20 degrees right and 25 degrees left of center at NIGHT can be similarly explained. Reference to Fig. 14.2 indicates that these targets were presented at a higher location than all others. Perhaps the poorer target detection performance at these sites reflects the lack of driving-relevant information in these locations. Target conspicuity is known to be affected by expectancy and information needs (Cole & Hughes, 1990; Hills, 1980; Hughes & Cole, 1986) and, at times, targets were actually superimposed on the sky, an unlikely source of traffic-relevant information. The fact that these target detection performance peaks and drops were evident only at NIGHT (with and without the VES) may be a reflection of possible ceiling effects for detection performance at DUSK. In addition, detection performance may not have dropped as significantly at DUSK because attention and eye movements were more frequently directed to the periphery by the more visible objects and events occurring there.

All of these findings suggest that the subjects might indeed have adopted realistic driver eye movement patterns, and lend further indirect support to the contention that eye movement patterns in simple laboratory-based experiments such as this one can represent those in the real driving situation (Hughes & Cole, 1986).

Effect of VES

The results clearly show that the presence of the VES degraded peripheral target detection and identification performance at NIGHT. Although not statistically significant, the most noticeable performance degradation occurred for those targets presented closest to the VES image. There was no VES-induced degradation in performance at DUSK.

The degradation observed in peripheral visual performance at night may simply be due to the relative brightness of the central image and its contrast with the background scene. Just as our eyes are automatically drawn to oncoming vehicle headlights, it is possible that the VES or HUD images grab attention simply because they stand out from a dark background. Visual attention is known to be drawn to items in a display that are large and bright (Wickens, 1992). This explanation is also consistent with the lack of any observed VES-induced performance degradation at dusk, because the dusk condition central VES image had a relatively lower contrast with the peripheral background scene.

It may also be that the VES image affected retinal adaptation, thereby inducing disability glare effects (Howarth, 1990) and decreasing sensitivity to the lower contrast targets in the peripheral visual field under the night conditions (Bhise, Farber, & McMahan, 1976). The targets may have become invisible in the night condition because the luminance difference between the target and background was small compared with the luminance of the bright HUD image. This explanation is also consistent with the lack of any VES-induced degradation at dusk, because the VES image had a lower contrast with the background scene at dusk and the targets presented at dusk were higher in luminance than those presented at night.

Alternatively, the VES-induced degradation in target performance at night may have been due to the relative richness of the information provided within the VES, enabling subjects to perform well on the tracking task without the need to scan the periphery. The lack of performance degradation at dusk may also be associated with the relative ease of vehicular guidance at dusk when compared with night; spare attentional resources may be available to cope with the demands of peripheral event detection and recognition while using a VES.

Regardless of the reason for the observed decrement in performance at night, the results confirm those found by other researchers in a driving situation with high central task demand (Lee & Triggs, 1976; Miura, 1986). The fact that detection performance was degraded with the VES at night may indicate that the functional field of view became more focused or narrow. Degradation in identification performance at night with the VES suggests that subjects might have been spending a higher proportion of time fixating the central tracking task rather than scanning the scene as compared with the no-VES control condition. This can only be confirmed by actual measurement of eye scanning behavior. However, if subjects tend to concentrate on the HUD image at the expense of the periphery, and in particular the near periphery, then there is the potential for safety-relevant information in the scene to be missed.

These results provide strong indirect evidence that the presence of a VES in night conditions induces perceptual tunneling. Results are also consistent with the premise that the VES might have altered normal eye scanning behavior at night. It would appear that when the VES produces an image markedly different from the nonenhanced periphery, attention to the periphery may be reduced. It is particularly noteworthy and of concern that the greatest decrement in target detection and identification appeared to occur at eccentricities closer to the VES, because sudden events occurring at these locations might require the most immediate attention of the driver.

Application of Results

As in any simulation of a complex task, there were many aspects of the real driving task that were not represented in this experiment. However, the simulated situation in this study might be regarded as the optimum for the use of a driver VES. If perceptual tunneling can occur in the simulator while performing very simple tracking tasks, then degradation of peripheral attention may be even more severe in actual traffic. The added stresses associated with speed control, obstacle avoidance, monitoring of a much larger field of view, rearview mirrors, and other in-vehicle displays and the consequence of error are likely to exacerbate VES-induced perceptual tunneling (Baddeley, 1972). Potential obstacles in the real driving environment are likely to be even less conspicuous and less frequent than the artificial abrupt-onset targets used in this study (Yantis & Jonides, 1984). Furthermore, subjects in the study were alerted to the task of target detection, whereas drivers using the system may not be conscious of the need to scan or consciously attend to the periphery. The subjects were also ideal in terms of their ability to detect and react to peripheral events while using a driver VES, being relatively young with visual capabilities above those required for most driver licenses.

Limitations of illumination and VES simulation in this study raise even more concerns about the potential for perceptual tunneling when real vision enhancement systems are eventually used in real driving situations. Night and dusk conditions were achieved by simply darkening daylight driving scenes. Real night and dusk conditions are likely to have fewer visible peripheral cues to direct attention and eye movements outside the area of the VES display.

The VES was also idealized in that it did not present real thermal imagery. As discussed previously, thermal imagery may be very difficult to interpret, being phenomenologically quite distinct from normal visible images (Brickner, 1989; Brickner & Staveland, 1989; Stark, 1987). Essentially, the images represent the distribution of temperatures and thermal contrasts in a scene; because they are created by emitted rather than reflected radiation, they lack the type of shading (brightness and contrast) that is characteristic with daylight images. Thermal images also change over time in a rather unpredictable manner, being affected by such factors as time of day, adjacent sources, moving air masses, changes in humidity, and many other influences of which the operator is unlikely to be aware. Finally, many man-made or natural sources of heat create unique "thermal signatures," which differ significantly from the regular representation of the same object. Any or all of these characteristics could create image interpretation problems for the general driving population, increasing the central driving task demand, with potential to further degrade peripheral event detection. The thermal image may also negatively interact with other light sources such as headlights. A real VES that superimposes actual thermal images could therefore distract the driver from attending to peripheral events even more than was demonstrated using the idealized VES in this study.

Finally, the simulated VES used in this study also idealized the HUD technology achievable. Current limitations with HUD technology will preclude perfect alignment of the VES image with the real-world scene over which it is superimposed. Any time lag in the presentation of the HUD image or misalignment of the image over the real-world scene image could prevent the driver from simultaneously assimilating both the HUD and background images, thereby enhancing potential for cognitive capture by the HUD image. Imperfect HUD image alignment might also result in apparent HUD jitter, which may further distract the driver from attending to outside peripheral events.

Excessive binocular disparity, diplopia (double images), or discomfort can also result if the HUD image is not correctly collimated with the plane of the outside world (Genco, 1983; Gibson, 1980); any of these might impair attention allocation to the real world. And, given the highly variable distances of objects to which drivers must attend in the real driving environment, it will be difficult to optimize the VES collimation distance so that these HUD-associated problems do not occur.

Another potentially serious problem with the use of current HUD technology in VES may be resultant eye accommodation that is inappropriate for the degree of collimation (J. Iavecchia et al., 1988; Norman & Ehrlich, 1986), with the possibility that this could affect distance judgment. The extent to which pilot distance judgment is affected by HUD misaccommodation is hotly debated (R. A. Benel, 1980a, 1980b; R. A. Benel & D. C. R. Benel, 1981; Newman, 1987; Roscoe, 1979, 1982, 1984; Weintraub, 1987; Weintraub & Ensing, 1992). However, if HUD misaccommodation is also a problem in the driving environment, VES may actually worsen already existing driver problems with distance judgment at night and in fog (Brown, 1970; Ross, 1975), the very conditions under which the VES was designed to operate.

Other previously mentioned problems with the use of HUDs may also further impair driver ability to attend to peripheral events in the real world. Interaction of HUD images with other sources of illumination (e.g, moonlight, headlights) can produce unwanted reflections (Wilson, 1983), possibly affecting driver contrast sensitivity and peripheral object detection. The luminance of thermal HUD images may mask the visibility of items in the scene over which they are presented, by reducing the luminance contrast of any visible light that may be seen through the HUD with the naked eye (Weintraub & Ensing, 1992). The development trend to increase thermal image luminance (Coonrod, 1983) would only serve to exacerbate this problem. A central high luminance source can also increase the adaptation level of the eye (Howarth, 1990), decreasing visual sensitivity to lower luminance low contrast objects in the periphery, the very type of object (e.g., pedestrians) critical for a driver to detect.

Any or all of the aforementioned realistic VES technological limitations could increase the demands on driver attention in the real traffic situation or reduce the detectability of low contrast objects in the peripheral field of view, the practical effects of which may be to inhibit cognitive switching of attention between the VES image and scene even further than that demonstrated by this study.

Recommendations for Further Research

Given the importance of peripheral vision in driving, there is a clear need for further research to assess the safety implications of VES-induced perceptual tunneling. Further research is recommended to determine if the performance decrement revealed in this study is affected by driver age, experience, visual function, training, or experience, each of which has been shown to have an effect on normal driver visual behavior (Mourant & Rockwell, 1972; Shinar, McDowell, Rackoff, & Rockwell, 1978). The effects on peripheral visual performance under varying environmental and visibility conditions should also be examined because this study was limited to examination of ambient illumination only.

A number of VES variables also need to be evaluated in terms of their impact on attentional resources. These may include HUD image position, size, shape, collimation and luminance, thermal image-processing options, degrees of image disparity, and interactions with typical sources of illumination in the driving environment, each of which alone may affect driver performance with VES technologies.

Finally, more realistic simulation of the driving task, HUD attributes, environmental conditions, and presentation of potential obstacles is needed to permit more accurate generalization to the real driving environment.

ACKNOWLEDGMENTS

We would like to thank B. Laffoley of Pilkington p.l.c., P. Marsh, and K. Cooper of HUSAT for their contribution in producing stimulus material, editing, and analysis software.

REFERENCES

Allen, M. J., Hazlett, R. D., Tacker, H. L., & Graham, B. V. (1970). Actual pedestrian visibility and the pedestrian's estimate of his own visibility. *American Journal of Optometry and Archives of American Society of Optometry, 47*(1), 44–49.

Baddeley, A. D. (1972). Selective attention and performance in dangerous environments. *British Journal of Psychology, 63*(4), 537–546.

Baldrey, P. E. (1988). Road lighting. In A. G. Gale, M. H. Freeman, C. M. Haslegrave, P. Smith, & S. P. Taylor (Eds.), *Vision in vehicles—II* (pp. 163–165). Amsterdam: Elsevier Science.

Bartz, A. E. (1962). Eye-movement latency, duration and response time as a function of angular displacement. *Journal of Experimental Psychology, 64*(3), 318–324.

Benel, R. A. (1980a). Vision through interposed surfaces: Implications for vehicle control. In D. J. Osborne & J. A. Levis (Eds.), *Human factors in transportation research* (pp. 328–336). London: Academic Press.

Benel, R. A. (1980b). Eyes and glass curtains: Visual accommodation, the Mandelbaum effect, and apparent size. In G. E. Corrick, E. C. Haseltine, & R. T. Durst, Jr. (Eds.), *Proceedings of the Human Factors Society 24th Annual Meeting* (pp. 616–620). Santa Monica, CA: Human Factors Society.

Benel, R. A., & Benel, D. C. R. (1981). Background influence on visual accommodation: Implications for target acquisition. In R. C. Sugarman (Ed.), *Proceedings of the Human Factors Society 25th Annual Meeting* (pp. 277–281). Santa Monica, CA: Human Factors Society.

Bhise, V. D., Farber, E. I., & McMahan, P. B. (1976). Predicting target-detection distance with headlights. *Transportation Research Record, 611,* 1–16.

Bossi, L. L. M. (1993). *The effect of enhanced image head-up displays on driver peripheral visual performance.* Unpublished master's thesis, Loughborough, Leics, UK: Loughborough University of Technology.

Brickner, M. S. (1989). Apparent limitations of head-up displays and thermal imaging systems. In R. S. Jensen (Ed.), *Proceedings of the Fifth International Symposium on Aviation Psychology* (Vol. 2, pp. 703–707). Columbus, OH: Ohio State University.

Brickner, M. S., & Staveland, L. E. (1989). Comparison of thermal (FLIR) and television images. In R. S. Jensen (Ed.), *Proceedings of the Fifth International Symposium on Aviation Psychology* (Vol. 1, pp. 276–281). Columbus, OH: Ohio State University.

Brown, I. D. (1970). Motorway crashes in fog—who's to blame? *New Scientist, 48* (24 December 1970), 543–545.

Brown, I. D. (1980). Are pedestrians and drivers really compatible? In D. J. Oborne & J. A. Lewis (Eds.), *Human factors in transport research* (Vol. 2, pp. 371–379). London: Academic Press.

Bullimore, M. A., Fulton, E. J., & Howarth, P. A. (1990). Assessment of visual performance. In J. R. Wilson & E. N. Corlett (Eds.), *Evaluation of human work* (pp. 648–681). London: Taylor & Francis.

Cohen, A. S., & Studach, H. (1977). Eye movements while driving cars around curves. *Perceptual and Motor Skills, 44,* 683–689.

Cole, B. L., & Hughes, P. K. (1984). A field trial of attention and search conspicuity. *Human Factors, 26*(3), 299–313.

Cole, B. L., & Hughes, P. K. (1990). Drivers don't search: They just notice. In D. Brogan (Ed.), *Visual Search—Proceedings of the First International Conference on Visual Search* (pp. 407–417). London: Taylor & Francis.

Cole, B. L., & Jenkins, S. E. (1984). The effect of variability of background elements on the conspicuity of objects. *Vision Research, 24,* 261–270.

Coonrod, J. F. (1983). Future development trends for head-up displays. In W. L. Martin (Ed.), *Optical and human performance evaluation of HUD systems design* (Rep. No. AFAMRL-TR-83-095 [AD-P003 164], pp. 74–82). Wright-Patterson Air Force Base, OH: Air Force Aerospace Medical Research Laboratory, Aerospace Medical Division.

Dirkin, G. R. (1983). Cognitive tunneling—use of visual information under stress. *Perceptual and Motor Skills, 56*(1), 191–198.

Dopping-Hepenstal, L. L. (1981). Head-up displays—the integrity of flight information. *IEE Proceedings—F Communications Radar and Signal Processing, 128*(Pt. F, No. 7), 440–442.

Dudfield, H. (1988). Optimising the pilot-display interface using part-task simulation: A case description. In M. A. Life, C. S. Narborough-Hall, & W. I. Hamilton (Eds.), *Simulation and the user interface* (pp. 213–225). London: Taylor & Francis.

Engel, F. L. (1971). Visual conspicuity, directed attention and retinal locus. *Vision Research, 11*, 563–576.

Eriksen, C. W., & Yeh, Y. (1985). Allocation of attention in the visual field. *Journal of Experimental Psychology: Human Perception and Performance, 11*(5), 583–597.

Evans, L. (1991). *Traffic safety and the driver.* New York: Van Nostrand Reinhold.

Fischer, E., Haines, R., & Price, T. (1980). *Cognitive issues in head-up-displays* (Tech. Rep. No. 1711). Washington, DC: NASA Ames Research Center.

Foyle, D. C., Sanford, B. D., & McCann, R. S. (1991). Attentional issues in superimposed flight symbology. In R. S. Jensen (Ed.), *Proceedings of the Sixth International Symposium on Aviation Psychology* (pp. 577–582). Columbus, OH: Ohio State University.

Genco, L. V. (1983). Optical interactions of aircraft windscreens and HUDs producing diplopia. In W. L. Martin (Ed.), *Optical and human performance evaluation of HUD systems design* (Rep. No. AFAMRL-TR-83-095 [AD-P003 164], pp. 20–27). Wright-Patterson Air Force Base, OH: Air Force Aerospace Medical Research Laboratory, Aerospace Medical Division.

Gibson, C. P. (1980). Binocular disparity and head-up displays. *Human Factors, 22*(4), 435–444.

Hall, J. W. (1983). Pedestrian accidents on rural highways. *Transportation Research Record, 904*, 46–50.

Hills, B. L. (1980). Vision, visibility and perception in driving. *Perception, 9*, 183–216.

Howarth, P. A. (1990). Assessment of the visual environment. In J. R. Wilson & E. N. Corlett (Eds.), *Evaluation of human work* (pp. 351–386). London: Taylor & Francis.

Hughes, P. K., & Cole, B. L. (1986). What attracts attention when driving? *Ergonomics, 29*(3), 377–391.

Iavecchia, J., Iavecchia, H., & Roscoe, S. (1988). Eye accommodation to head-up virtual images. *Human Factors, 30*(6), 689–702.

Lee, P. N. J., & Triggs, T. J. (1976). The effect of driving demand and roadway environment on peripheral visual detections. In *Proceedings of the 8th Conference of the Australian Road Research Board* (pp. 7–12). Melbourne: Australian Road Research Board.

Lee, R. (1983). Overview of HUD optical design. In W. L. Martin (Ed.), *Optical and human performance evaluation of HUD systems design* (Rep. No. AFAMRL-TR-83-095 [AD-P003 164], pp. 4–10). Wright-Patterson Air Force Base, OH: Air Force Aerospace Medical Research Laboratory, Aerospace Medical Division.

LeGrand, Y. (1967). *Form and space vision* (rev. ed., trans. M. Millodot & G. Heath). Bloomington, IN: Indiana University Press.

Leibowitz, H. W., & Appelle, S. (1969). The effect of a central task on luminance thresholds for peripherally presented stimuli. *Human Factors, 11*(4), 387–392.

Leibowitz, H. W., & Owens, D. A. (1975a). Anomalous myopias and the intermediate dark focus of accommodation. *Science, 189*, 646–648.

Leibowitz, H. W., & Owens, D. A. (1975b). Night myopia and the intermediate dark focus of accommodation. *Journal of the Optical Society of America, 65*(10), 1121–1128.

Mackworth, N. H. (1965). Visual noise causes tunnel vision. *Psychonomic Science, 3*, 67–68.

Mandelbaum, J., & Sloan, L. L. (1947). Peripheral visual acuity. *American Journal of Opthalmology, 30,* 581–588.

Miura, T. (1986). Coping with situational demands: a study of eye movements and peripheral vision performance. In A. G. Gale, M. H. Freeman, C. M. Haslegrave, P. Smith, & S. P. Taylor (Eds.), *Vision in vehicles* (pp. 205–216). Amsterdam: Elsevier Science.

Mourant, R. R. M., & Rockwell, T. H. (1972). Strategies of visual search by novice and experienced drivers. *Human Factors, 14*(4), 325–335.

Neisser, V., & Becklan, R. (1975). Selective looking: Attention to visually specified events. *Cognitive Psychology, 7,* 480–494.

Newman, R. L. (1987). Response to Roscoe, "The Trouble with HUDs and HMDs." *Forum—Human Factors Society Bulletin, 30*(10), 3–5.

Nordwall, B. D. (1991). Radar, targeting infrared give Navy Marines precision weapons capability. *Aviation Week and Space Technology, 134*(4 February 1991), 54–65.

Norman, J., & Ehrlich, S. (1986). Visual accommodation and virtual image displays: Target detection and recognition. *Human Factors, 28*(2), 135–151.

Olson, P. L. (1988). Problems of nighttime visibility and glare for older drivers. *Effects of aging on driver performance* (SAE SP-762, Paper No. 881756), 53–60.

Olson, P. L. (1993). Vision and perception. In B. Peacock & W. Karwowski (Eds.), *Automotive ergonomics* (pp. 161–183). London: Taylor & Francis.

Olson, P. L., & Sivak, M. (1983). Comparison of headlamp visibility distance and stopping distance. *Perceptual and Motor Skills, 57,* 1177–1178.

Parkes, A. M., Ward, N. J., & Bossi, L. L. M. (1995). The potential of vision enhancement systems to improve driver safety. *Le Travail Humain, 58*(2), 151–169.

Pickering, S. (1987). Turning night into day. *Electronics and Power, 33*(7), 447–450.

Rheinhardt-Rutland, A. H. (1986). Misleading perception and vehicle guidance under poor conditions of visibility. In A. G. Gale, M. H. Freeman, C. M. Haslegrave, P. Smith, & S. P. Taylor (Eds.), *Vision in vehicles* (pp. 413–416). Amsterdam: Elsevier Science.

Roscoe, S. N. (1979). When day is done and shadows fall, we miss the airport most of all. *Human Factors, 21*(6), 721–731.

Roscoe, S. N. (1982). Landing airplanes, detecting traffic, and the dark focus. *Aviation, Space and Environmental Medicine, 53,* 970–976.

Roscoe, S. N. (1984). Judgements of size and distance with imaging displays. *Human Factors, 26*(6), 617–629.

Ross, H. (1975). Mist, murk and visual perception. *New Scientist, 1975,* 658–660.

Rumar, K. (1990a). The basic driver error: Late detection. *Ergonomics, 33*(10/11), 1281–1290.

Rumar, K. (1990b). Driver requirements and road traffic informatics. *Transportation, 17,* 215–229.

Sanders, A. F. (1970). Some aspects of the selective process in the functional visual field. *Ergonomics, 13*(1), 101–117.

Sanders, M. S., & McCormick, E. J. (1993). *Human factors in engineering and design* (7th ed.). New York: McGraw-Hill.

Schlegel, R. E. (1993). Driver mental workload. In B. Peacock & W. Karwowski (Eds.), *Automotive ergonomics* (pp. 359–381). London: Taylor & Francis.

Schwab, R. N., & Hemion, R. H. (1971). Improvements of visibility for night driving. *Highway Research Record, 377,* 1–23.

Scott, W. B. (1988). LANTIRN provides breakthrough in night-fighting capabilities (low-altitude navigation and targeting infrared for night). *Aviation Week and Space Technology, 128*(25 April 1988), 34–38.

Shepherd, M., Findlay, J. M., & Hockey, R. J. (1986). The relationship between eye movements and spatial attention. *Quarterly Journal of Experimental Psychology, 38*(A), 475–491.

Shinar, D., McDowell, E. D., & Rockwell, T. H. (1977). Eye movements in curve negotiation. *Human Factors, 19*(1), 63–71.

Shinar, D., McDowell, E. D., Rackoff, N. J., & Rockwell, T. H. (1978). Field dependence and driver visual search behaviour. *Human Factors, 20*(5), 553–559.

Sivak, M., & Flannagan, M. (1993). Human factors considerations in the design of vehicle headlamps and signal lamps. In B. Peacock & W. Karwowski (Eds.), *Automotive ergonomics* (pp. 185–204). London: Taylor & Francis.

Sojourner, R. J., & Antin, J. F. (1990). The effects of a simulated head-up display speedometer on perceptual task-performance. *Human Factors, 32*(3), 329–339.

Stark, E. A. (1987). FLIR: What you don't see is what you get. In R. S. Jensen (Ed.), *Proceedings of the Fourth Symposium on Aviation Psychology* (pp. 554–563). Columbus, Ohio: OSU Aviation Psychology Laboratory.

Stokes, A., Wickens, C., & Kite, K. (1990). *Display technology: Human factors concepts.* Warrendale, PA: Society of Automotive Engineers.

Swift, D. W., & Freeman, M. H. (1986). Application of head-up displays to cars. *Displays,* July 1986, 107–110.

Vanstrum, R. C., & Landen, J. C. (1984). The dark side of driving. *Transportation Quarterly, 38*(4), 491–505.

Ward, N. J., Parkes, A. M., & Crone, P. (1994, August–September). *The legibility of head-up displays within the driving environment: The effect of background scene complexity.* Presentation to the Vehicle and Navigation and Information Systems International Conference, Yokohama, Japan.

Weale, R. A. (1963). *The ageing eye.* London: H. K. Lewis.

Webster, R. G., & Haslerud, G. M. (1964). Influence on extreme peripheral vision of attention to a visual or auditory task. *Journal of Experimental Psychology, 68*(3), 269–272.

Weintraub, D. J. (1987). HUDs, HMDs and common sense: Polishing virtual images. *Human Factors Society Bulletin, 30*(10), 1–3.

Weintraub, D. J., Haines, R. J., & Randle, R. J. (1984). The utility of head-up displays: Eye focus versus decision times. In M. J. Alluisi, S. de Groot, & E. A. Alluisi (Eds.), *Proceedings of the Human Factors Society 28th Annual Meeting* (pp. 529–533). Santa Monica, CA: Human Factors Society.

Weintraub, D. J., & Ensing, M. (1992). *Human factors in head-up display design: The book of HUD.* Dayton, OH: CSERIAC, Wright Patterson Air Force Base.

Wickens, C. D. (1992). *Engineering psychology and human performance* (2nd ed.). New York: HarperCollins.

Williams, L. J. (1985). Tunnel vision induced by foveal load manipulation. *Human Factors, 27*(2), 221–227.

Wilson, W. (1983). Sun/moon capture evaluation. In W. L. Martin (Ed.), *Optical and human performance evaluation of HUD systems design* (Rep. No. AFAMRL-TR-83-095 [AD-P003 164], pp. 28–30). Wright-Patterson Air Force Base, OH: Air Force Aerospace Medical Research Laboratory, Aerospace Medical Division.

Yantis, S., & Jonides, J. (1984). Abrupt onsets and selective attention: Evidence from visual search. *Journal of Experimental Psychology: Human Perception and Performance, 10*(5), 601–621.

Zahn, J. R., & Haines, R. F. (1971). The influence of central search task luminance upon peripheral visual detection time. *Psychonomic Science, 24*(6), 271–273.

Zwahlen, H. T. (1985). Driver eye scanning, the information acquisition process and sophisticated in-vehicle information displays and controls. In I. D. Brown (Ed.), *Ergonomics International 85 Proceedings* (pp. 508–510). London: Taylor & Francis.

Zwahlen, H. T., & DeBald, D. P. (1986). Safety aspects of sophisticated in-vehicle information displays and controls. In E. L. Wiener (Ed.), *Proceedings of the Human Factors Society 30th Meeting* (pp. 256–260). Santa Monica, CA: Human Factors Society.

Chapter 15

Perceptual Attention to Contact Analogue Head-Up Displays

Leonard Stapleton
HUSAT, Loughborough University, Leics., UK

VISION ENHANCEMENT USING CONTACT ANALOGUE HEAD-UP DISPLAYS

Driving under adverse viewing conditions (e.g., at night, during dusk, in fog, etc.) is commonly recognized as a contributing factor in accident causation (e.g., Hills, 1980; Olson, 1988). Recent developments address providing drivers with additional visual information under such conditions (see, e.g., Mulvanny, 1993). Such technologies are generically referred to as vision enhancement systems (VESs). These systems acquire an image of the forward portion of the road scene ahead, which is then presented in an enhanced format to the driver. The system of interest in the present study uses a head-up display (HUD) format to provide the driver with the enhanced information.

With a HUD system, real-time video footage of the central portion of the road scene ahead is obtained, using, for example, an infrared camera. The obtained image is processed to convert it into a visible format. This image is then reflected off the inside of the front windshield or an intermediate virtually transparent screen. The optics that project the image simultaneously collimate it so that it appears on a focal plane at some distance in front of the driver (see Fig. 15.1).

The fields of view of the camera and the HUD image are matched so that the image overlays the portion of the forward road scene that is viewed by the camera. Thus, elements within the image overlay the objects and

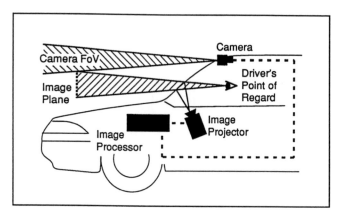

FIG. 15.1. Schematic of a head-up display presenting a contact analogue image of the forward road scene (Key: FoV; Field of View).

features in the driving scene that give rise to them. The image is simultaneously filmed and presented so that the movement of objects in the road scene, and movement of the vehicle through the road environment give rise to corresponding movements in the image. This matching of the image and environment is termed *contact analogue.*

PROBLEMS WITH ACHIEVING CONTACT ANALOGUE

A basic premise of the concept of contact analogue is that the image is perceptually integrated with the underlying and surrounding external scene. There is an assumption that the reaction of the perceptual system to the upgraded elements within the image will parallel its reaction to the real external elements giving rise to the images. Several factors conspire against the integration of the enhanced image with the visual background (i.e., the achievement of contact analogue). These include the following:

The HUD image is presented at a fixed focal distance from the driver. Thus, small head movements by the driver will induce small movement parallax effects between the background and image (e.g., Hold a pen at arm's length and align it with a distant object. Then move your head from side to side. The pen appears to move relative to the object.).

When both eyes converge on the image, further away objects appear double-imaged or blurred (e.g., Swap attention back and forth between a pen at arm's length and an aligned background object.).

The camera generating the HUD image is not on the same line of sight as the driver's eyes. This produces a slight mismatch between the perspective within the image and the perspective within the environment as seen from the point of view of the driver.

If time delay between filming and displaying the image creeps in, then a temporal mismatch could result. That is movement in the image might lag behind movement in (or through) the background.

For a comprehensive review of advantages and drawbacks associated with HUD usage, see Weintraub and Ensing (1992).

CONTACT ANALOGUE AND PERCEPTUAL ATTENTION

The present study addressed the issue of integration of the HUD image with the background scene from the perspective of attentional allocation. In general, when attending to an object within the visual field, the object is fixated on the acute central region of the retina, namely the fovea. Stabilizing images of such fixated objects involves tracking movements of the eyes that match the object's motion (or, if the observer is moving, counterbalance the observer's eye, head, or body movements relative to the fixated object under attention). This is achieved principally through smooth pursuit eye movements, with an occasional corrective saccadic (i.e., ballistic) eye movement (e.g., Sperling, 1990). The control of eye movements and the allocation of attention are thus intimately related (see Kowler, 1990; Sperling, 1990).

Given that the human observer is in almost constant movement, the images of nonfixated objects generate continuous optic flow across the retina. Thus, attention and the oculomotor processes controlling eye movements operate in tandem by isolating the moving object from its background.

This tendency is demonstrated in an experiment by Kowler, van der Steen, Tamminga, and Collewijn (1984) in which subjects observed superimposed visual fields made up of patterns of dots. One field was held stationary while the other moved. The subjects attended to either the stationary or moving field. Measurement of the subjects' eye movements revealed a consistent fixation of the field under attention, with a negligible influence on tracking from the unattended field.

Similarly, a selective attention experiment by Khurana and Kowler (1987) further supports the relation between attentional allocation and the oculomotor control of eye movements. They presented subjects with two moving arrays of letters. The subjects were instructed to attend to, and track, one array (tracking accuracy being monitored by measuring eye movements). Targets (one of the letters being replaced by a numeral) were presented in either array. The results showed that target detection in the attended array was superior to target detection in the unattended array. Subjects also demonstrated a relative inability to attend to (i.e., detect targets in) the array that was not being tracked.

These findings imply that the oculomotor system isolates a moving section of the optic array for attention, to the exclusion of other moving sections. This suppression of competing moving elements in the visual array has repercussions for the perceptual integration of a HUD image and background scene in a contact analogue application. If motion parallax or temporal lag create discrepancies between elements in the HUD image and overlaid background scene, attention to one may be expected to impair attention to the other. The present study addresses this hypothesis.

EXPERIMENTAL METHOD AND PROCEDURE

Design

The present experiment manipulated subjects' allocation of attention by having them monitor a primary task presented in either of two media: a free-standing HUD unit (a holographic combiner projecting video images) or a background projection screen. The effects of this manipulation were assessed by measuring subjects' reactions to secondary targets presented through either medium. It was hypothesized that attending to one medium (e.g., the HUD) impairs event detection in the alternative medium (the screen), and vice versa.

The relation between eye movements and attentional allocation is central to this hypothesis. A critical design concern was that the events presented in either medium were integral components of that medium, that is, they should exhibit movements that paralleled the movements of adjacent display elements. Simply overlaying graphics symbols (which would be stationary relative to global movements within the display medium) as stimuli was insufficient.

Furthermore, a discrepancy between the focal distance of the HUD image and the projection screen would be expected to increase movement parallax effects due to small head and body movements of the subject. As attentional allocation is associated with the perceptual isolation of visual elements moving within either medium, it was hypothesized that target perception in the unattended medium would be worse when the HUD image was at an intermediate focal distance, that is, not focused on the same plane as the screen.

Stimulus Materials and Equipment

Contact analogue was mimicked by generating two videos, one to be played on the projection screen and the other through the HUD unit. The HUD footage covered the central area of the screen footage and was appropriately matched in size and content so that it appeared that both videos were taken simultaneously. Both videos were obtained using a moving belt driving simulator and a model lead vehicle at constant headway (30 m to scale) and constant speed (45 kph to scale).

The lead vehicle had three "Chroma Key" blue discs mounted on the back. Using image processing, this hue allowed frame-by-frame graphics overlay that made the discs adopt block colors chosen by the experimenter. Using this technique, second generation tapes were generated containing the stimulus sequences used in the trials. This technique also insured that the stimuli shared common motion (e.g., from vibrations) with adjacent image elements. The discs were approximately 1.2 degrees in diameter. This relatively large size was necessary to provide sufficient Chroma Key blue for overlaying in the wide-angle image.

Two shades of gray were used. The central disc alternatively swapped between the light and dark shades of gray (duration of presentation of each was .5 sec). After random intervals, the lighter shade persisted for a 1 sec period. This constituted a primary task event and subjects were required to register this event by depressing a foot pedal. Each side disc remained light gray, and after a random interval either disc briefly changed to dark gray (duration equalling approximately .25 sec). This shade change represented a target presentation that the subject responded to by pressing a left or right hand push button.

Prior to generating the overlaid tapes, pilot tests were undertaken using four experimental staff as subjects to determine the optimum shades of gray for use in overlaying. As there was a limited range of shades available, matching stimulus perceptibility in the screen and the HUD was imperfect. HUD targets were slightly more readily perceived during these trial runs than screen targets.

Each video was time coded so that reaction times for subject responses could be determined by comparing the video frame at which an event occurred with the frame at which the subject reacted. At 25 frames per second, one frame represented 40 msec.

Two Super VHS video players (Panasonic Model AG-7350) were coupled with a time code reader, which in turn was connected to the push buttons and foot pedal. One video was projected (via a Sony Multiscan Projector— Model VPH/1271QN Super Data EX) on a large projection screen (approximately 3 × 4 m presenting a horizontal viewing angle of approximately 53 degrees, for a viewing distance of 4 m). The other was projected through the free-standing HUD unit and filled approximately 12 degrees (horizontal) of the forward view. The entire arrangement was situated in a lightproof laboratory area, with the subject seated in a steering buck (an enclosing frame containing car front seats and the HUD unit).

Subjects

Twenty subjects, 13 male and 7 female, with a mean age of 28.4 (range 17–37) participated in the study. Each subject viewed 1 hr (including rest breaks) of video tape.

Procedure

Prior to experimental trials, subjects were given instruction on the use of the pedal and buttons, and then were shown two pairs of practice tapes, each 2 min long, during which they made practice responses. One pair contained primary task events and targets (4 primary task events and 15 targets) in the HUD and no events in the screen. The second pair contained the same number of primary task and target events presented in the screen, with no events in the HUD.

Each block of experimental trials lasted 6 min (consisting of two 3 min sessions, with a brief rest period between, to avoid eye fatigue). Each block contained 6 primary task events and 12 targets. The 12 targets were divided equally between the primary and secondary media, with an equal number presented to either the left or right. Half the subjects received two blocks with the primary task in the HUD, followed by two blocks with the primary task in the screen. For the remaining subjects, the presentation order was reversed.

For each pair of HUD or screen primary task blocks, one of the pair was presented with the HUD image on a focal plane that was coplanar with the projection screen. For the other pair, the HUD was focused at an intermediate distance of 2.5 m from the subject.

In total, the 48 target presentations for each subject consisted of 6 Targets × 2 Focal Plane Distances × 2 Target Media × 2 Attended Media.

Between blocks (and after practice) subjects were given breaks of several minutes. Before starting trials, subjects were told that if at any stage they experienced signs of simulator nausea or eye strain, then they could stop the trials. No subject opted to do so.

RESULTS

For each subject response a printout of the time codes on both HUD and screen videos, and a symbol indicating the subject's response (left or right button, or pedal press), was produced. Comparison of the actual time codes at which event onset took place allowed reaction times to be computed. These were in multiples of video frames (40 msec). As the duration and contrast of targets were chosen to make detection uncertain, the numbers of events detected in each condition were also used for comparison.

The two data types (subjects' reaction times to targets and numbers of targets detected) were treated separately. Within each condition, two HUD focal planes were used, namely: coplanar with the screen (C), or at an intermediate distance of 2.5 m (I). Median reaction times for target detection and numbers of detected targets for each subject and within each subcondition were calculated (based on 6 targets for each plane [× 2] for each target medium [× 2] for each attended medium [× 2]). For two subjects, the order of presentation of focal distance was not recorded. Their data was omitted from analysis.

Each data type was analyzed separately within a 2(Plane) × 2(Attended Medium) × 2(Target Medium) repeated measures ANOVA.

No statistically significant interaction for focal plane was observed. To utilize the data from the two subjects whose order of presentation of co-planar and intermediate distance images was not recorded, all co-planar and intermediate results were respectively pooled and medians recalculated (based on 12 targets for each target medium [× 2] for each attended medium [× 2]). These median responses were then analyzed in a 2(Attended Medium) × 2(Target Medium) repeated measures ANOVA. Only significant results are reported.

Reaction Time Data

There was a significant main effect for attended medium, $F(1, 18) = 6.73$, $p < .05$, reaction time being significantly faster when subjects attended to the screen ($M = 16.184$) rather than the HUD ($M = 17.368$). There was also a significant main effect for target medium, $F(1, 18) = 12.63$, $p < .005$, reaction time being significantly faster for HUD targets ($M = 15.987$) rather than for screen targets ($M = 17.655$). (Note: A corrupted data file reduced reaction time data to 19 subjects.)

Numbers Detected

There was a significant main effect for attended medium, $F(1, 19) = 40.576$, $p < .0001$, numbers detected being significantly higher when subjects attended to the screen ($M = 9.65$) rather than the HUD ($M = 7.55$). Also, there was a significant main effect for target medium, $F(1, 19) = 46.654$, $p < .0001$, numbers detected being significantly higher for HUD targets ($M = 10.25$) rather than for screen targets ($M = 6.95$).

There was a significant interaction between attended medium and target medium, $F(1, 19) = 34.342$, $p < .0001$. This interaction is illustrated in Fig. 15.2. Based on a post-hoc comparison of means using Tukey's HSD test ($d_{crit} = .883$), all HUD targets for both attention conditions showed superior detection to screen targets. Also, screen targets were significantly more readily detected when attending to the screen than attending to the HUD.

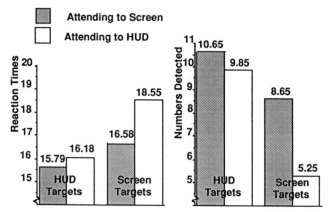

FIG. 15.2. Reaction time data (in multiples of 40 msec video frames) and numbers detected.

Summary

The experimental hypothesis suggests that the medium attended to facilitates target perception in that medium at the expense of target perception in the alternative medium. Thus, reaction to HUD targets should be superior when attending to the HUD (i.e., when the primary task is presented in the HUD) than when attending to the screen. Similarly, screen targets should be more readily perceived when attending to the screen than when attending to the HUD.

Reaction times for targets compared across both primary task conditions failed to show this hypothesized difference, though the observed direction of difference for the screen targets was in the hypothesized direction (viz., screen detection) was superior when attending to the screen. However, HUD targets appear more readily perceived when the primary task is in the screen rather than in the HUD.

The detection data did demonstrate a significant effect in the hypothesized direction; namely, more screen targets were detected when attending to the screen than when attending to the HUD. However, contrary to prediction, HUD detection is also superior when the primary task is in the screen, though this effect is considerably weaker than that for screen detection. Also, for both the reaction time data and numbers detected HUD targets were more readily perceived than screen targets.

DISCUSSION

One of the design considerations in generating the stimulus materials was that targets were sufficiently close to threshold levels of brightness and contrast for detectability. This was to ensure that differences between conditions would be evident without needing to run large numbers of subjects over extended periods of time. Unfortunately, characteristics of video presentation biased target detectability in favor of the HUD presented stimulus events. Relevant factors include the superior definition and crispness of the HUD stimuli that occupied a relatively large proportion of the HUD screen over the screen stimuli, which were comprised of fewer pixels and had less area. Also, the choice of shades of gray used to form the stimuli was limited to a discrete set, rather than a continuous range from which to choose.

This bias in favor of HUD detection partly contaminates the effects that the experiment was designed to examine. However, the results, though not fully conclusive, still support the experimental hypothesis. The finding that screen targets are detected more readily when attending to the screen than when attending to the HUD may be due to a combination of two effects: First, the relatively higher salience HUD primary task may inhibit detection of the relatively dull screen targets. This would account for the direction of the screen target effect—namely, screen targets are less perceptible when attending to the HUD. And, second, the direction of the effect also supports the proposed attentional allocation effect; detection in the alternative medium (the screen) is inhibited when attending to a primary task in a competing medium (the HUD).

The other significant result pertaining to the experimental hypothesis was that HUD targets were more perceptible when attending to a screen primary task. (This effect is in the opposite direction to the experimental hypothesis, and does not reach significance.) This may also be partially explained in terms of the relative salience of the HUD and screen primary tasks. If the relatively dull screen primary task is being monitored, a HUD target of relatively higher salience is easily perceived. Similarly, if a high salience HUD primary task is being monitored, then the high salience HUD targets are less perceptible due to interference from the primary task. In this case, this proposed HUD salience effect is in the opposite direction to the experimental hypothesis, that targets in the nonattended medium are less perceptible.

The relative strengths of these competing effects will determine the direction of the results. If the attentional allocation effect is stronger, then results confirming the experimental hypothesis would be expected. If, however, the HUD salience effect is stronger, results along the lines of those found would be expected: First of all, screen target detection is inhibited due to both effects when the primary task is in the HUD. This additive effect is reflected in the present findings of statistical significance in the appropriate direction for the number of targets detected. In addition, HUD target detection in the HUD primary task condition is inhibited due to the proposed salience effect, but aided due to the attentional allocation effect. The present findings suggest that the salience effect wins out, but only just. Statistical significance was not reached.

Implications for contact analogue HUDs in the automotive application arise from this finding that the higher salience image is more readily perceived. Design exploitation of this would favor a high salience HUD image that would replace rather than supplement missing information from the external road scene.

This solution is practical under impoverished viewing conditions where a relatively bright high contrast HUD image can replace missing information. However, to accommodate situations where the background illumination is relatively high (e.g., in daytime fog, or passing through intermittent street lighting), control over the luminance of the HUD image would be required. Mulvanny (1993) suggested automatic brightness level control as a preferable solution.

Finally, the issue of HUD focal plane distance needs to be addressed in a less ambitious design, with a greater number of observations for each focal plane setting.

CONCLUSIONS

The presentation of stimulus materials in the present study was artificially manipulated so that images appeared in either the HUD or as elements in the background. This replicated two possible scenarios that might occur in a real application: An object is imaged in the HUD but not visible to the naked eye, or an object is visible but either not picked up by the VES sensor, or not imaged in the HUD in a recognizable or useful format. A third, and potentially most common scenario, is that an object is both imaged and simultaneously visible or partly visible to the naked eye.

The present study found that HUD target detection was superior to screen detection even though only a relatively slight salience discrepancy existed between the HUD image and screen image (in favor of the HUD image). This would imply that in the contest between a clear HUD image and a partial naked-eye view, as implied in the third scenario, the HUD image would emerge superior. For automotive contact analogue HUDs it implies that perfectly matching the HUD image to the external scene is of secondary importance to providing a high quality HUD image with higher information content than the unenhanced forward view. However, care should be observed in ensuring that any scenario 2 type objects are adequately accounted for. The issue of integrating the central HUD image with the unenhanced peripheral external scene must also be addressed.

ACKNOWLEDGMENTS

This chapter forms part of the output of the vision enhancement strand of the BRIMMI Project, supported by UK Department of Trade and Industry, within the PROMETHEUS Program. The UK industrial partners are Pilkington Technology Management Limited and Jaguar Cars. Pilkington supplied the HUD unit used in the experiment. The assistance of co-workers Nic Ward and Andrew Parkes is gratefully acknowledged.

REFERENCES

Hills, B. L. (1980). Vision, visibility and perception in driving. *Perception, 9*, 183–216.

Khurana, B., & Kowler, E. (1987). Shared attentional control of smooth eye movement and perception. *Vision Research, 27*(9), 1603–1618.

Kowler, E. (1990). The role of visual and cognitive processes in the control of eye movement. In E. Kowler (Ed.), *Eye movements and their role in visual and cognitive processes* (pp. 1–70). Amsterdam: Elsevier Science.

Kowler, E., van der Steen, J., Tamminga, E. P., & Collewijn, H. (1984). Voluntary selection of the target for smooth eye movement in the presence of superimposed, full-field stationary and moving stimuli. *Vision Research, 24*(12), 1789–1798.

Mulvanny, P. (1993, July 8). *Vision enhancement.* Institute of Mechanical Engineers Vehicle Lighting Seminar, London.

Olson, P. L. (1988). *Visibility problems in nighttime driving* (Rep. No. 870600). Warrendale, PA: Society of Automotive Engineers.

Sperling, G. (1990). Comparison of perception in the moving and stationary eye. In E. Kowler (Ed.), *Eye movements and their role in visual and cognitive processes* (pp. 307–351). Amsterdam: Elsevier Science.

Weintraub, D. J., & Ensing, M. (1992). *Human factors issues in head-up display design: The book of HUD* (CSERIAC 92-2). Wright-Patterson AFB, OH: Crew Station Ergonomics Information Analysis Center.

Chapter 16

Drivers' Cognitive Process and Route Guidance

Tatsuru Daimon
Hironao Kawashima
Keio University, Yokohama, Japan

Motoyuki Akamatsu
National Institute of Bioscience and Human Technology, Tsukuba, Japan

In order to navigate (i.e., find their way through the road network to a particular destination), drivers must continually monitor their location relative to a given route and perhaps change route if and when circumstances warrant. A road map helps drivers in this task provided they orient themselves in relation to the map and maintain their orientation. Experience with passive navigation systems that do not provide route guidance (i.e., they display a digital map on a dash-mounted screen that may highlight a route and indicate the current position of the vehicle) reveals that drivers do not always reach their destinations without making navigation errors. Route guidance, in which the system provides en-route guidance in the form of instructions, is considered to be an important and desirable functional element of navigation systems.

In Daimon (1992), the characteristics and performance of drivers using an in-vehicle navigation system were compared with drivers using a road map. The results suggest that drivers' cognitive processes are influenced by the design of the navigation aid and its functional elements. Navigation errors and decisions leading to inefficient routes can arise if the navigation system (map or digital display) does not provide route guidance information about where and when to turn. The development of such route guidance functions, however, requires a thorough understanding of how and what kind of information best support the navigation task; that is, an understanding of the cognitive processes involved in wayfinding.

Wayfinding is composed of three principal elements: forming a cognitive map, planning the route, and making en-route navigation decisions. Navigation decisions require that drivers orient themselves by relating the information obtained from external sources, such as sign boards, to their cognitive map. This process leads to appreciation of their location relative to the destination. The cognitive map is, therefore, central to navigation decisions. The purpose of these studies is to gain further insight into the composition of the cognitive map and, in particular, to investigate the relation between guidance information and the cognitive map as the basis for decision making.

COGNITIVE MAP

Humans refer to internalized representations of environmental objects in order to know their position in space and to decide in which direction to move to arrive at a particular location. It is thought that these representations are structurally interlinked and stored in memory as a *cognitive map*. Cognitive maps enable us to relate our position to those of other objects even if we are in an open environment (i.e., in a situation lacking vision). Thus, a cognitive map permits "solving problems in space," such as self-orientation, estimating distances to objects, and so on. Drivers refer to their cognitive maps when they plan and follow routes; therefore, the study of the structure and role of cognitive maps in navigating in an unfamiliar area is important in the development of supportive navigation systems.

In the present study, drivers' cognitive maps were analyzed using the five components defined by Lynch (1960). The following components are depicted in Fig. 16.1:

Path: a course where people can move such as road, rail, airway, and so on.

Node: a principal point of entry to a path such as an intersection.

Landmark: a place or object that is recognizable and fixed to the path but not necessarily unique such as a building or signboard.

District: a distinct area such as a park or downtown, which bounds map elements contained within it.

Edge: a line that cannot normally be traversed such as a coastline, wall, etc.

The structural relation among these components in the cognitive map (representing real world knowledge) evolves through learning (i.e., expe-

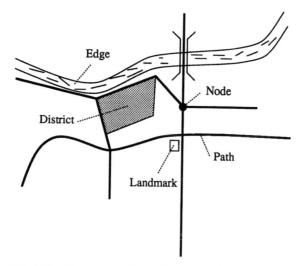

FIG. 16.1. The concept of cognitive map and its elements.

rience navigating in the area). In particular, Freundschur (1989) defined three levels of knowledge, corresponding to three steps in learning a geographical area, as shown in Fig. 16.2. The first level is *procedural knowledge*, comprising memories of discrete scenes that have been encountered. Although people who have procedural knowledge of an area have difficulty drawing a map of the area, they can often reach places in the area by recalling (or retracing) previous movements or decisions in relation to points along the route. The second level of knowledge, *network knowledge*, refers to topological knowledge of connected elements. Knowledge of dis-

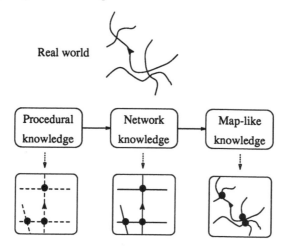

FIG. 16.2. The changing process of cognitive map and the categories.

tance and direction in the network knowledge is lacking or distorted with respect to the real world. The most advanced level is *maplike knowledge*, in which the cognitive map bears very accurate resemblance to the real world.

Characteristics of cognitive maps, such as the components identified by Lynch and the process of structural change described by Freundschur, were incorporated in the present study investigating the requirements for route guidance information.

STUDY 1: NAVIGATORS' INTERNAL REPRESENTATION

The aim of the first study was to identify the principal factors underlying wayfinding by observing the methods used by *navigators* to describe a specific route to drivers unfamiliar with the area. Navigators were drivers with good knowledge of the area and, therefore, had a well-developed cognitive map. It was postulated that the information they transmitted to other drivers about the route reflected the most salient aspects of their cognitive map. By conducting such analyses, it was possible to identify driver interface requirements.

Methodology

There is no method that can directly extract from humans their internalized representations of the real world. However, mental models or cognitive processes can be inferred from observations of related behaviors. In this study, the *map sketch method* and the *verbal description method* (Lynch, 1960) are utilized to extract the structural components of cognitive maps from navigators who have good local knowledge of an area. These methods have proved useful in the investigation of cognitive maps and extraction of basic structural information. The map sketch method uses maps drawn by subjects to extract internalized representations. The verbal description method uses written descriptions, in this case, about the route from origin to destination.

Procedure. The subjects, navigators, were required to describe the route they normally take between their homes and their university in a manner that unfamiliar drivers can understand and follow. The subjects used both the map sketch and the verbal description methods. The order of methods was randomized. The following descriptions were provided:

> **Instructions for the map sketch method:** Imagine that you want to invite your friend to your house. Your friend will go to your house from the university by car, but he/she does not have knowledge of the route or the

area. Explain to him/her the route between your house and university by drawing a map.

Instructions for the verbal description method: Imagine that you want to invite your friend to your house. Your friend will go to your house from the university by car, but he/she does not have knowledge of the route or the area. Explain to him/her the route between your house and university by written instructions. Do not use a picture or figure in your explanation.

After completion of these tasks (map sketch and written instructions), subjects were asked to draw a map of the area as accurately as possible.

Subjects

Subjects were 44 students (40 male and 4 female). All subjects have either driven the route from their university to their homes, or knew the route very well. Their driving experience was more than 10,000 km.

Results

Figure 16.3 shows an example of the results of the map sketch method and Fig. 16.4 shows an example of the results of the verbal description method. There were few differences among subjects in their reliance on the cognitive map components as defined by Lynch or the way they conveyed the information. Table 16.1 shows the total number of subjects who used the various Lynch components. Similarly, Table 16.2 shows the total number of cases in which the subjects utilized these components in their instructions. It should be noted that in these tables, "map" refers to the map sketch method and "verbal" refers to the verbal description method. These results indicate that subjects, in describing a route to people unfamiliar with an area, rely primarily on three of Lynch's components: namely, landmarks, paths, and nodes.

As an example, for traversing a river using a bridge, subjects tended to refer to the "bridge," which belongs to the landmark category, rather than to the "river" itself, which belongs to the edge category. It was noted that components that come directly into the driver's view—that is, landmarks, paths, and nodes—are more important than components, such as districts or edges, that may not be directly visible but are obviously essential for maps. In this respect, it might be inferred that districts and edges serve to orient the map reader, but that orientation is not essential for route guidance. This finding was not influenced by the routes or areas used by the subjects (each of the 44 subjects had different routes running through areas containing different features). In summary, effective route guidance messages should reference primarily landmarks, paths, and nodes.

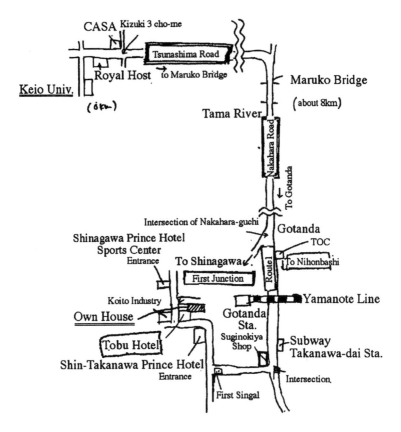

FIG. 16.3. An example for the results of the map sketch method.

First, after starting from Keio University, please turn to the right. Go straight ahead for a while. You will see a traffic signal and a main road (Tsunashima Road). At the intersection, please turn to the right. After a while, you will see Maruko Bridge. Cross the bridge and go straight ahead. After passing the first main intersection where Tsunashima Road and Loop 8 cross each other, please turn to the left at the second main intersection where Tsunashima Road and Loop 7 cross each other.

.
.
.

FIG. 16.4. An example for the results of the verbal description method.

TABLE 16.1
The Total Number of Subjects Who Used Lynch's Components

	Map	*Verbal*
Landmark	44	44
Path	44	44
Node	44	44
District	19	27
Edge	19	6

It was noted that the map sketch method can extract more information from subjects than the verbal description method. It was observed that, in most cases, the extracted information was structured according to the network knowledge representation of the cognitive map, as described by Freundschur (1989). Subjects found it easier to describe the route using maps rather than written descriptions because network knowledge can be more efficiently depicted in this way.

Figure 16.5 shows an example of the map that a subject drew when asked to draw it as accurately as possible. This map is almost equivalent to the real world. Although not all of the subjects were able to generate maps of this accuracy, these maps exhibited common trends with respect to distortions of the real world. In these maps, the routes and objects were connected to one another in a one-dimensional form. The directions and distances were distorted and curves were often flattened or replaced by a straight line. These distortions are consistent with the network knowledge representation of the area. The maps shown in Figs. 16.5 and 16.3 represent the same area and were drawn by the same individual. The difference between the two maps is a direct consequence of the different instructions given for drawing the map (i.e., drawing a particular route versus drawing a map as accurately as possible). Thus, even a navigator who is able to draw an accurate map of an area uses a distorted map to describe the route to unfamiliar drivers. Hence, the most useful information for route guidance is a representation of network knowledge of the area rather than an accurate map.

TABLE 16.2
The Number of Cases in Which Subjects Used Lynch's Components

	Map	*Verbal*
Landmark	660	511
Path	610	337
Node	494	303
District	65	61
Edge	20	8

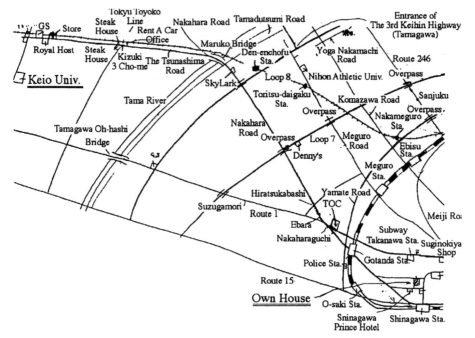

FIG. 16.5. An example for the results of the map drawn as exactly as possible.

In conclusion, a navigation system used for route guidance should be designed to provide information corresponding to the network knowledge representation of the route and rely primarily on landmarks, paths, and nodes as reference elements.

STUDY 2: INFORMATION THAT DRIVERS NEED

The aim of the second study was to investigate the information that drivers need to navigate in an unfamiliar area. The study used the *question-asking protocol method* to obtain information that could be analyzed within the construct of the cognitive map. In particular, the objectives of the study were to determine what kind of information the driver needs and when the driver needs the information.

Method of Analysis

The question-asking protocol method is similar to the thinking-aloud protocol method that is used in the field of cognitive science (Kato, 1986). The latter is often used in evaluations of human–computer interfaces or, more generally, to obtain human factors data relevant to the design of user interfaces (Newell & Simon, 1972). In this method, subjects are asked to "think aloud" as they perform given tasks (i.e., provide direct verbal reports). The verbal reports contain valuable information concerning the usability of the interface, thus permitting identification of user needs. One of the problems with the thinking-aloud protocol method, however, is that it is difficult to elaborate on the needs of users.

In the question-asking protocol method, a tutor is used to answer questions that subjects raise in the course of performing their tasks. However, under this limited assistance technique, the subject must ask the tutor for assistance when it is required and the tutor refrains from offering unsolicited assistance. In this way, the information that drivers need for wayfinding can be extracted from the recorded communications between subjects and tutors (e.g., questions and answers).

The verbal protocol analysis (VPA) method, as proposed by Kinoe (1989), was used to analyze the verbal protocol data. In the VPA method, a framework for categorizing verbal protocol data is developed before the data are collected. Data reduction is then simply a matter of categorizing the recorded messages in accordance with a preestablished schema. The analyses produce simple and contingent distributions of message categories.

Experimental Procedures

Subjects were given an origin and two destinations and asked to select the optimal route and drive to the destinations. During the drive, they were required to speak about their feelings and thoughts. The tutor sat in the passenger seat and answered questions about wayfinding only when asked. Drivers' performance and verbal protocol data were recorded using four small CCD cameras and a small tiepin-type microphone.

Subjects

Subjects were five male students. They were fully qualified drivers with a mean driving experience of 5 years and 25,000 km. They had no knowledge of the route used in the experiment. The tutor was a male graduate student who was raised in the area. He had very good knowledge of the area and detailed knowledge of the route. He had been driving for 9 years.

Results

The framework for the VPA is shown in Table 16.3 and the results in Table 16.4. The results indicate that subjects requested and relied primarily on information about landmarks, paths, and nodes to navigate to their destination. Specifically, the most frequently used landmark was the traffic

TABLE 16.3
Framework for Verbal Protocol Data in Experiment 2

Category 1 [Action for question]

  [Timing]   [Aim]

1. Just after turning   1. To inquire
2. Middle of two turn points   2. To confirm
3. Just before turning
4. Just before departure

  [Attention]

1. One intersection ahead
2. Two intersection ahead
3. More than two ones ahead

Category 2 [Situation]

  [Situation]

1. While running
2. While stopping

Category 3 [Content of information]

  [Pattern]   [Key]

1. Instruction (turn right or left)   1. Signal
2. Instruction + Landmark   2. Shop
3. Instruction + Path   3. Sign (include direction board)
4. Instruction + Node   4. Other landmark
5. Landmark alone   5. Distance
6. Path alone   6. Form
7. Node alone   7. Other path
8. Two-dimensional relationship   8. Node (intersection)
                                               9. District (the name of a place)
                                              10. Edge (river, etc)
                                              11. None

signal and the most frequently used path was the distance to the next
turning point. Landmarks that were generally not familiar to subjects, such
as the names of buildings or intersections, were found not to be useful
for wayfinding. Drivers therefore tended to ask information with reference
to highly familiar objects.

Table 16.5 shows the results of a content analysis of the questions asked
by the subjects. Each content category is intended to cover all the information sought in a single question. The term *instruction* refers to a request

TABLE 16.4
The Frequency and the Proportion of the Used Information

Component		Frequency	(%)
Landmarks	traffic signal	130	(20.12)
	shop	43	(6.66)
	traffic sign	41	(6.35)
	the others	61	(9.44)
Paths	distance	110	(17.03)
	form	33	(5.11)
	the others	8	(5.11)
Nodes		110	(17.03)
The others		110	(17.03)

for turning direction (i.e., right or left turn). An example of instruction is "turn left at the intersection named X after passing two intersections." It can be seen that, in addition to landmarks, paths and nodes, turning directions are frequently requested. Furthermore, subjects needed information concerning the distance from the current location to the next turn, and reference landmarks that can be used to confirm the location of a turning point.

Although subjects asked a variety of different questions, it was apparent that they were only interested in obtaining information about the location of the turn next. They did not want to receive too much information at any one time.

Clear trends emerged during the study concerning when the subjects asked questions in relation to the route and their purpose in asking these questions. The general scenario is depicted in Fig. 16.6. The figure shows a general route segment and indicates where along the route and why questions were asked. Most questions related primarily to the turning points. The data in parentheses indicates the proportion of questions at that particular point that pertained to the next turning point.

TABLE 16.5
The Contents of Questions for the Information

Content	Frequency	(%)
Landmark alone	179	(27.71)
Path alone	116	(17.96)
Instruction + Landmark	90	(13.93)
Instruction + Node	85	(13.16)
Instruction alone	68	(10.53)
Instruction + Path	39	(6.04)
Node alone	26	(4.01)
Two-dimensional relationship	12	(1.86)
The others	31	(4.08)

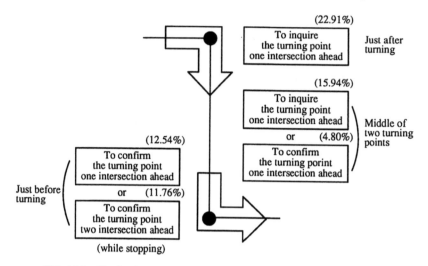

FIG. 16.6. The kind of the question and the location of the intersection.

The general wayfinding process can be described in terms of establishing subgoals. Just after turning at an intersection, the driver asks questions to determine where the next turn will be. Information is sought concerning the distance to the next turn and any identifying landmarks. When the target intersection appears in sight, they ask questions to confirm the name or location of the intersection. They direct their attention to the next intersection most often when waiting at a traffic light or when they have spare time. If there is sufficient distance between the current location and the next intersection, they might ask one of two types of questions: to inquire about the next intersection if they could not do so earlier due to workload, and to confirm that the intersection where they intend to turn is correct.

In this way, navigating to a destination can be regarded as stepping through a series of subgoals. That is, at any point in time, drivers set as the current subgoal the intersection at which they must turn next. Once the current subgoal is reached they determine the next subgoal and this process is repeated until they reach the final destination. Other intersections along the route are of minimal interest to the driver.

During this experiment, it was observed that subjects felt uneasy and their driving performance deteriorated when they came across unusual road configurations, such as intersections involving three or five roads or roads that have a low radius of curvature. Subjects normally expect to encounter standard four-way intersections and are sometimes confused when the real road configuration does not match their expectation. This suggests that it is important to determine what degree of distortion (of distance or form) can be tolerated in a network knowledge map without disturbing drivers' expectations.

DISCUSSION

The first experiment was aimed at identifying key route message elements that should be communicated to drivers by a navigation system. This was accomplished through analysis of information reflecting navigators' internal representations of the route. The second experiment was aimed at identifying route information needs of drivers by analyzing the information they request as they navigate a particular route. Both studies found that drivers primarily needed information about landmarks, paths, and nodes; and the components of the cognitive maps that navigators relied on, or unfamiliar drivers needed, were not necessarily accurate in terms of distance and direction. In other words, route guidance information is often more useful and easier to understand if it is distorted with respect to distance and direction. However, the distorted representation must be topologically consistent with objects in the real world. This implies that, for navigation, drivers use a cognitive map that corresponds to a network knowledge representation of the real world.

A model of the relation between the driver's cognitive map and the information required for route following is shown in Fig. 16.7. For route following, the driver constructs a cognitive map based on the network knowledge derived from maps, digital maps, and other external sources. Drivers select routes on the basis of the cognitive map. They proceed to navigate the route by establishing successive subgoals (next turning point)

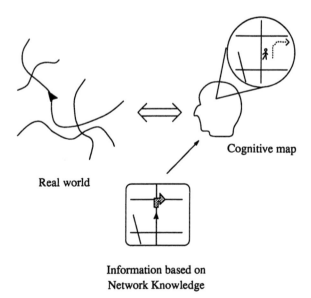

Cognitive map

Real world

**Information based on
Network Knowledge**

FIG. 16.7. The relation between cognitive map and information.

consistent with the overall route. During each subgoal, they seem interested primarily in information needed to reach the subgoal.

The information most useful to drivers in constructing a cognitive map is that which represents network knowledge. Network knowledge representations are very effective for navigation despite their distortion of the real world (in terms of forms and distance). Alternatively, road maps that are accurate representations of the real world may not facilitate internalization of the information in a manner useful for road navigation.

CONCLUSIONS

Navigation in an unfamiliar area, at least at the tactical level, can be regarded as a process of subdividing the task into a series of subtasks (or subgoals) demarcated by turning points. The driver needs information about unique buildings or objects (i.e., landmarks), specific route options (i.e., paths), and intersections along the route (i.e., nodes) in order to achieve each subgoal. Information such as landmarks, paths, and nodes are interconnected in one-dimensional form. That is, two-dimensional map information is not always required. It is easier for the driver to interpret the route if the path between a given intersection and the next is distorted to some extent. The optimal degree of distortion that facilitates route acquisition and does not result in disorientation is an important design issue that requires further research.

REFERENCES

Daimon, T. (1992). Driver's characteristics and performances when using in-vehicle navigation systems. *Third Vehicle Navigation and Information Systems Conference (VNIS'92)*, 251–260.

Freundschur, S. M. (1989). Does anybody really want (or need) vehicle navigation aids? *First Vehicle Navigation and Information Systems Conference (VNIS'89)*, 439–442.

Kato, T. (1986). What "question asking protocols" can say about the user interface. *International Journal of Man–Machine Studies, 25,* 659–673.

Kinoe, Y. (1989). The VPA method: A method for formal verbal protocol analysis. In G. Salvendy & M. J. Smith (Eds.), *Designing and using human–computer interfaces and knowledge based systems* (pp. 735–742). Amsterdam: Elsevier Science.

Lynch, K. (1960). *The image of the city.* Cambridge, MA: MIT Press.

Newell, A., & Simon, H. A. (1972). *Human problem solving.* Englewood Cliffs, NJ: Prentice-Hall.

Chapter 17

Automatic Versus Interactive Vehicle Navigation Aids

David M. Zaidel
Transportation Research Institute, Technion, Israel

Y. Ian Noy
Transport Canada, Ottawa

In-vehicle navigation and route guidance systems[1] are likely to be the most salient products emerging from technological developments within the domain of intelligent transport systems (ITS).

Like other in-vehicle transport information and control systems (TICS), navigation aids introduce auxiliary tasks that require some interaction with the driver. There is some concern among designers and researchers that auxiliary tasks may intrude on the "primary" driving tasks, such as vehicle control and obstacle avoidance (Noy, 1989) by distracting, confusing, or overloading drivers. Laboratory and simulation studies tend to show some decrement in performance of driving-related functions when drivers are given a concurrent task such as navigation (Noy, 1990). Although improvements in the ergonomic design of navigation and route guidance systems have been evident in recent demonstrations of new technologies, as yet there is no clear evidence that such systems are contributing to, or detracting from, safe driving (OECD, 1992).

The very nature of the driving task may change in fundamental ways with the use of different navigation systems. For example, a map display highlighting the recommended route may require drivers to orient themselves in relation to the map as well as to look for conventional street signs. Visual displays of simple route guidance instructions provide little orien-

---

[1]For purposes of this chapter, *navigation* provides geomatic information and *route guidance* suggests a course and/or provides instructions.

tation support but may be easier to follow. Visual displays, both spatial and verbal, may draw too much visual attention away from the road. On the other hand, a system providing oral instructions—for example, advising drivers when and where to turn—may eliminate the need for orientation and for reading street signs. However, voice displays may conflict with the acoustic environment, and provide information that must be kept in memory. It is important to understand the performance trade-offs of alternative driver interface designs in order to optimize safety and the performance of driving and navigation tasks.

Several laboratory and simulation studies have investigated the effects of display factors on driving. In a laboratory reaction time study, Verwey (1993) provided evidence for the advantage of auditory verbal instructions but recognizes the need to augment auditory instructions with simple, visual indications.[2] Such controlled studies can provide important principles for the design of appropriate driver interfaces.

However, there is a need for more ecologically valid field research to complement simulation research (Zaidel, 1991). Indeed, several large-scale field evaluations have been conducted and others are ongoing in an effort to advance TICS in Europe, the United States, and Japan. Results to date generally indicate driver satisfaction with the systems tested but fail to provide direct evidence for a safety benefit. In fact, a recent German study (Pohlmann & Traenkle, 1994) reported the possibility of negative safety impacts of the navigation system.

Driver interface designs of navigation systems are still at an early stage of development, though certain trends in design are emerging. Complex electronic map displays are giving way to simpler, visual–auditory display combinations, which appear to be more ergonomically sound. At the same time, new options, such as programmable features and custom settings, defer the selection of configuration to drivers, who may not understand system options or performance trade-offs sufficiently well to make appropriate choices. There is evidence, especially within the domain of human–computer interface design, that user preferences often lead to choices that actually hinder performance (Andre & Wickens, 1995; Bailey, 1993).

Many of the issues relating to message content, timing, display modality, and control of guidance information remain unresolved (Zaidel, 1991; Zaidel & Noy 1993). Their influence on the safety of driving could be subtle and difficult to observe or interpret, particularly in view of the compensatory mechanisms employed by drivers in normal driving.

The two experiments described in this chapter compared the quality of driving with different route guidance systems using a common experimen-

---

[2]The rationale for the redundant visual display is that simple auditory instructions may not be adequate in complex situations or may be partially missed by the driver.

tal technique that has been found to be reliable and sensitive to task/system manipulations.

OVERVIEW OF THE EXPERIMENTS

The two experiments used a paradigm involving open road driving in an instrumented vehicle equipped with video cameras and a data acquisition system. Drivers drove specially selected routes and searched for target addresses. An experienced observer rated the quality of driving along eight dimensions using a 9-point scale. Dependent variables included vehicle speed, looking behavior, quality of driving, subjective workload, and navigation performance.

The task in the first experiment compared two different types of navigation aids. The Guidance List provided basic visual guidance information in the form of a concise list of written directions. The more advanced Ideal Navigator provided, in addition to the visual list, timed voice information. The information contained contextual cues that were appropriate to the particular navigational decision point, including distance, road geometry, and landmarks. In both cases, drivers had to identify street names and addresses and make turning or stopping decisions. A third drive along a familiar route, with no address searching task, was included for experimental control.

The first experiment established the usefulness of the technique, the validity of the task environment, and the reliability and sensitivity of the dependent measures. The results also indicated that an intelligent navigation system can produce better driving performance than a paper-based list of instructions. However, the navigation system was different than the paper list in three fundamental ways: (a) the information was presented in the auditory modality as well as the visual modality, (b) the auditory information had additional contextual cues, and (c) information was automatically presented at "just the right time."

The second experiment was designed to delineate the factors confounded in the first study. Two display modalities, Screen and Voice, and two modes of control, Automatic and Interactive, were investigated in a mixed factorial experiment. The first group of drivers drove with the Screen system, which presented context rich verbal guidance information on a computer screen display. In one condition, the presentation was automatic (i.e., system driven) and, in a second condition, the presentation was interactive (i.e., under driver control). The second group drove with the Voice modality in which information was provided by digitized voice, without any visual navigation assistance. Here too, drivers drove under two conditions: an automatic and an interactive mode of control. The results obtained in the two experiments indicated the relative importance of each of the factors to successful navigation and quality of driving.

A brief description of the protocol used in both studies is described under Experiment 1. More detailed description of the two experiments can be found in Zaidel (1992, 1995).

EXPERIMENT 1: QUALITY OF DRIVING WITH ROUTE GUIDANCE ASSISTANCE

Purpose of Experiment 1

The purpose of the study was to develop and test a paradigm, based on measures of Quality Of Driving (QOD), which generally would be suitable for evaluating any TICS. The paradigm was tested in a comparison of two generic navigation aids. The two aids were designed to impose different demands on drivers' attention and were expected to produce noticeable differences in driving performance, looking behavior, navigation performance, and workload.

Methodology

Sixteen drivers drove an instrumented vehicle on predesigned city routes on three different occasions. In two of the drives, drivers searched for 12 address destinations using different navigation aids. In the third drive, drivers followed a familiar route with no requirement to search for addresses. Each of the three trips took 45 to 50 min to complete. During each drive, an observer in the front passenger seat entered QOD scores and codes representing navigation errors and related information, such as traffic conditions, into the onboard data acquisition system. Two video cameras recorded the front view of the road and the driver's eyes and head. At the conclusion of each trip, drivers rated overall workload for the trip and answered related questions.

The Test Routes. The use of a repeated-measures design required testing drivers on similar but not identical routes. Two similar test routes, 35 km long, were carefully designed utilizing a mixture of residential and commercial streets in downtown Ottawa. Each route included about 45 left and right turns.

A third test route was constructed from segments of the other two routes. This route was used as a control route in which no navigation task was required. This was always the last drive, so drivers were relatively familiar with the route.

The Navigation Task. The navigation task required drivers to follow a list of instructions to 12 address destinations. As they reached each destination, they identified it and proceeded to the next address. Drivers were asked to drive as they would normally. There were three navigation conditions and associated aids: Guidance List condition, an Ideal Navigator condition, and a Familiar Route condition.

In the *Guidance List condition* drivers received a concise list of written directions, grouped by destination, indicating street names where they had to make right or left turns. The list was typed in large typeface (24 pt.) and attached to a clipboard placed on the dashboard and in good view of the driver.

In the *Ideal Navigator* condition, in addition to the Guidance List, drivers were assisted by auditory messages from a "smart navigator." The Ideal Navigator, simulated by the experimenter in the back seat, provided advance information that included distance and landmark cues. Information was provided in a standard manner from a written text and repeated if requested. This aid, like GIDS (Michon, 1993), was expected to lead to improved quality of driving because the Navigator had "perfect knowledge" of the road network and, thus, able to time the delivery of the information in a contextually appropriate way. Also, it was hypothesized that auditory guidance would reduce the need to glance at the visual list, thus increasing the likelihood that drivers would keep their eyes on the road.

In the *Familiar Route condition,* a typed guidance list was provided for a route that was already familiar to drivers. It was composed of segments from the two previous routes. No destinations were included on the list.

Quality of Driving. Lack of attention to the driving task need not necessarily manifest itself immediately as gross deterioration of vehicle control and guidance, or in serious legal driving errors. The effects could be more subtle, affecting "higher order" traffic negotiation variables (Zaidel, 1991), such as situational awareness and anticipation.

QOD is a multidimensional construct, comprising performance variables as well as higher order strategic behaviors. Zaidel (1992) identified eight dimensions (not necessarily orthogonal) believed to underlie quality of driving as a composite indicator of safe driving. Many of them have also been used in other studies (e.g., Quimby, 1988; Risser, 1993). Definitions for the eight QOD dimensions follow:

1. *Speed maintenance.* The extent to which speed is kept within safe bounds, is adaptive to traffic conditions, and changes in a timely, consistent, and smooth fashion.

2. *Headway maintenance.* The extent that the longitudinal distance of the vehicle from other road users is within safe bounds, is adaptive, and allows good visibility.

3. *Lane position.* The extent that lateral placement of the vehicle within a lane and the choice of a lane are consistently appropriate, free of uncontrolled drifting, and do not infringe on the path of other road users.

4. *Turning and crossing.* The extent that drivers follow an appropriate sequence of visual checks, adjust spatial position, and speed prior to and during crossing or turning, maintain a margin of safety in gap acceptance and other conflict points resolutions.

5. *Traffic control devices.* The extent that drivers correctly interpret the meaning of signs, signals, markings, and other control devices; grasp the positions and intentions of other road users; and negotiate their course in accordance with the opportunities afforded by the controls.

6. *Vehicle handling.* The extent that drivers use vehicle's controls appropriately, with little overt attention, and achieve a consistently smooth ride.

7. *Dynamic space management.* The extent that drivers manage their time and space in a manner that optimizes the amount of space surrounding their vehicle without detracting from the safety of others; select the least risky course in complex situations and maintain maneuvering flexibility.

8. *Dynamic time management.* The degree to which drivers look ahead in time and space, use secondary cues to anticipate likely future positions of traffic participants, and avoid getting into situations requiring extra maneuvering.

The first three dimensions relate to the driver's control of the position of the vehicle in time and space. Dimensions 4–6 relate to interactions between the driver and the actual or potential presence of other road users, where conflicts and priorities have to be anticipated and resolved. Dimensions 6–8 relate to the drivers' management of the vehicle, the environment, and their own attention. In a previous study, a factor analysis demonstrated that QOD measures had good psychometric properties (Zaidel, 1992).

Ratings by Expert Observers. Each of these variables was rated by expert observers using a scale of 1 to 9, with 5 representing the typical behavior of the "average driver," and the higher and lower values representing better and poorer performance, respectively. A set of eight ratings was obtained for each of 12 segments of each test drive. Ratings were entered on average every 3 min at the conclusion of a predefined segment of the drive and before the new segment began. The observer was provided with a keypad for entering the ratings directly to a computer. The observer was not able to view previous ratings.

This procedure yielded a large number of scores for every driver, test drive, and segment. The composite QOD score for each drive was derived by averaging the ratings across the eight QOD scales and across the 12 segments of the route. Most previous studies that have employed expert ratings of driving performance used counts of illegal maneuvers or other driving errors, or rated the driver only once at the end of a drive (e.g., Fastenmeier, 1993; Oxley, Ayala, Alexander, & Barham, 1994; Pohlmann & Traenkle, 1994; Popp, Farber, & Schmitz, 1991; Quimby, 1988; Riedel, 1991; Risser, 1993).

The observer's task of rating quality of driving was not an easy one. The observer had to constantly look and assess the situation earlier than the driver in order to diagnose and understand the reasons for the observed behavior, and to determine its appropriateness given the traffic circumstances at the time.

Observers were selected from advanced driver trainers experienced in diagnostic evaluation and were given extensive training in the use of the QOD scales. There was one observer in Experiment 1, and three alternating observers in Experiment 2.

Workload Measure and Other Subjective Ratings. Drivers rated subjective workload using the NASA Task Load Index technique (TLX; Hart & Staveland, 1988), which was modified to suit the experimental protocol. Other rating scales were included to address issues of navigation and safety.

The ratings were provided four times by each driver. The first set of ratings, given after the initial briefing, was based on the driver's recollection of a "regular trip to work" (all test drivers drove to work). This provided familiarity with the rating process and produced "baseline" TLX data. The other ratings were obtained at the conclusion of each of the three experimental drives.

Other Performance Measures. *Navigation errors* were coded by the observer. A navigation error was defined as an unintended deviation from the course or a missed destination.

Speed was measured directly via a speed sensor. Speed modulation is an important mechanism for safe negotiation in traffic and for compensating for fluctuating demands of the driving task.

Glance frequency, glance duration, and direction were measured from video recordings. Only segments containing turns were analyzed because turns represent critical decision points requiring drivers to attend and react to visual cues outside the vehicle. Visual behavior prior to entering a turn can provide important insight into drivers' sharing of visual attention and task processing. Comparisons between navigation conditions provide insight into the "visual cost" (or attention demand) of driving with different devices. Video segments corresponding to 10 sec before and 5 sec after the turns were analyzed.

Drivers, Test Vehicle, Design, and Procedure of Experiment 1. *Drivers* were 16 volunteers, 10 men and 6 women. Their age ranged from 21 to 64. Their median experience of driving was 19 years. They drove on average 220 km per week, with a range of 80 km to 800 km. About 42% of their regular driving took place on city streets, 37% on highways, and 21% on residential areas. The drivers represented a variety of occupations including clerks, student trainees, skilled technicians, data analysts, and managers.

The *test vehicle*, a 1990 Dodge Spirit, was a regular production, midsize, four-door sedan with automatic transmission. It is a conventional car, requiring minimal practice for familiarization. The dynamic response of the vehicle was not affected by the instrumentation.

The *experimental design* was a repeated-measures design. The two navigation conditions (Guidance List condition and the Ideal Navigator condition) and the two routes were counterbalanced. The Familiar Route condition was always the last run.

The *procedure* included a briefing about the purpose and nature of the study, a pretest questionnaire that solicited task load ratings for a typical trip to work, a short drive to familiarize drivers with the vehicle, instructions specific to the condition of the first test drive, the test drive, a postdrive questionnaire, a brief rest period, instructions for the next condition, the second test drive, a rest period, instructions for last drive, the last drive, and debriefing.

Results

The Mental Cost of Navigation

Table 17.1 shows the mean ratings for subjective responses across all subjects for TLX and other variables. The TLX scores clearly differentiated among the three navigation conditions and the baseline condition. Other

TABLE 17.1
Ratings of Task Load, Navigation, and Safety

	Guidance List	Ideal Navigator	Familiar Route	Baseline (to work)
Task load (TLX)	3.98*	3.15*	2.18*	2.88
Similarity to normal drive	6.10*	6.65*	8.06*	na
Feeling well oriented	6.03	5.31	7.32*	7.44
How safe during drive	6.94	7.44	8.37*	na
Quality of navigation aid	6.36*	7.60	7.94*	na

*Denotes a $p \leq .05$ (two-tailed test) for pairwise comparisons within row.

differences in mean ratings in Table 17.1 are internally consistent and many are also statistically significant. (The differences were tested in both repeated-measures ANCOVA with base line scores as covariates, and in ANOVA's and pairwise t test.)

As expected, the three driving conditions produced three distinctly different levels of task demands. The Guidance List condition was the most demanding, the Ideal Navigator condition was less demanding, and the Familiar Route condition (with no address searching task) was the least demanding. Drivers' spontaneous comments about the task, during the drives and debriefings, and experimenters' observations support this finding.

The Ideal Navigator provided better navigation support and produced a greater feeling of safety than the Guidance List when navigating in an unfamiliar road network. All drivers mentioned that navigating with the Ideal Navigator aid was easier and for many it was also more relaxing. Four drivers, however, felt that this degree of support acted to reduce their sense of orientation and their sense of control.

The ratings for the Familiar Route condition indicate that subjects did indeed find the route to be familiar. Subjects' sense of orientation on the Familiar Route was much closer to what they reported for their (obviously familiar) trip to work, than for the first two drives. Their subjective task load in the Familiar Route condition was much lower compared with the other drives. In fact, subjective task load in the Familiar Route condition was even lower than in the Baseline condition, suggesting that just driving on a fairly familiar route may be more relaxing than driving to work during rush hours and under time pressure.

Quality of Driving With Navigation Assistance

Psychometric Properties of QOD Measure. Factor and reliability analyses examined the scale properties of the composite QOD measure. For the factor analysis, there were 576 cases (16 Drivers × 3 Conditions × 12 Segments), each having eight scores corresponding to different QOD dimensions. The results of the factor analysis lead to the conclusion that a composite QOD measure can be based on averaging the eight scales; the reliability coefficient, α, was .90.

In order to determine how the 12 road segments in each drive contributed to the overall QOD rating, the segments were treated as items of a questionnaire and subjected to a reliability analysis. The reliability coefficient ($\alpha = .96$) implies strong consistency between the segments. The correlation matrix provided no evidence for consistent sequential effects, which means that the observer rated each segment independently. Therefore, the composite QOD for a drive, the grand mean of all ratings during the drive, had good psychometric properties.

TABLE 17.2
Quality of Driving in Three Guidance Conditions

	Guidance List	Ideal Navigator	Familiar Route
QOD measure	5.00*	5.55	5.32
Navigation errors	2.56*	1.00	0.13
Speed (kph)	36.8	38.8	40.4
% Deceleration > .3 g	8.8	8.4	9.8*

*Denotes a $p \leq .05$ (two-tailed test) for pairwise comparisons within row.

Quality of Driving With and Without a Navigation Task. Table 17.2 summarizes the mean composite QOD scores, mean number of navigation errors, mean running speed, and proportion of decelerations that exceeded .3 g—for each of the three driving conditions, across 16 drivers.

QOD measures clearly distinguished between the three experimental conditions. QOD was best in the Ideal Navigator condition and worst in the Guidance List condition. Navigation errors were very rare in the familiar condition, infrequent in the Ideal Navigator condition, and relatively more frequent in the Guidance List condition. Travel speed (excluding long stops and navigation errors) was inversely related to navigation errors, reflecting the difficulty of the task. The Ideal Navigator produced the smoothest drive and the Familiar Route condition the roughest one. The data suggest that whereas the Familiar Route condition was the least demanding, it produced an intermediate level of safe driving; drivers tended to drive faster and decelerate at higher rates.

Correlations between individual QOD scores and navigation errors in the Guidance List condition ($r = .41$) suggest that under the high demands of this condition, there was a trade-off between driving well and navigating well. Under the relatively low demands of the other two conditions, drivers tended to drive and navigate well.

Looking Behavior While Driving and Navigating

Video segments containing 15-sec intervals during the approach to and execution of turns (excluding traffic light waiting) were coded with reference to the drivers' eye and head movements. Each glance was categorized as having one of four gaze directions: forward view, rear view (including all mirrors and shoulder check), in-vehicle display, and side view (through side windows). Glance frequency and duration were calculated from the time stamp data recorded on the tape.

Table 17.3 presents a summary of the glance data. The distributions of visual attention were generally similar for the two address search conditions: 76% to the front, 7% to the rear, 6% to the display, and 11% to the side.

TABLE 17.3
Glance Allocation by Direction and Guidance Condition

	Guidance List	Ideal Navigator	Familiar Route
Forward view	75.7%	75.6%	80.1%
Rear view	6.7%	6.6%	6.2%
In-vehicle displays	6.2%	5.8%	3.7%
Side view	11.4%	12.0%	10.0%

The Familiar Route condition produced only minor changes in the overall distribution of glances. The data in Table 17.3 are generally similar to distributions of visual attention obtained in other experiments (e.g., Oxley et al., 1994; Parkes, Ashby, & Fairclough, 1991).

Differences between navigation aids did not produce any noticeable effect on the allocation of visual attention while turning. In the control condition (Familiar Route), there was only a slight shift from in-vehicle displays to the forward view. It should be noted that in all three conditions drivers had a paper list on a dash-mounted clipboard. Thus, during complex maneuvers, such as turning, in which the demand for visual attention is high, drivers' looking behavior was relatively stable (e.g., unaffected by navigation condition). The data suggest that the attentional demands of driving had an overwhelming influence over the distribution of visual attention.

In summary, Experiment 1 demonstrated the usefulness of the QOD technique for evaluating navigation aids. It also demonstrated that the quality of driving can be adversely affected by the addition of a demanding navigation task. However, driving can also be improved by the provision of an appropriate aid, as evidenced by the fact that QOD improved and task load decreased under the Ideal Navigator condition as compared with the Familiar Route control.

EXPERIMENT 2: SCREEN VERSUS VOICE
AND AUTOMATIC VERSUS INTERACTIVE GUIDANCE

Purpose of Experiment 2

In Experiment 1, as in other field studies (e.g., Ashby, Fairclough, & Parkes, 1991; Fastenmeier, 1993), oral guidance instructions, especially when appropriately timed, produced the best QOD, the fewest navigation errors, and the lowest subjective workload. However, display modality (i.e., voice vs. visual information) and mode of control (i.e., automatic vs. interactive control of the guidance information) were confounded in Experiment 1, as they were in other studies. The Ideal Navigator was an automatic voice system that provided advance, contextually relevant information. The information included distance, geometry, and landmark cues chosen for their usefulness in the context of the specific routes and destinations selected. Any of these attributes of the Ideal Navigator could have contributed to its superiority over the alternative Guidance List—namely, the additional contextual information, the auditory modality, or the automatic generation of guidance messages.

The purpose of Experiment 2, therefore, was to isolate the principal factors that were confounded in the Ideal Navigator condition. Experiment 2 utilized both a Screen display and a Voice display. In the Screen modality, context-rich guidance information was presented on a dash-mounted computer display. In one condition, the presentation was *automatic* (system driven); in the second condition, the presentation was *interactive* (under driver control). In the Voice modality, guidance was provided by a digitized voice device, without a redundant visual display. Once again, guidance messages were presented under two conditions, *automatic* or *interactive*. The experimental procedure was similar to that used in the first study.

Methodology

Experiment 2 used a 2×2 mixed factorial design. The within-subjects factor was mode of control: Automatic versus Interactive. The between-subjects factor was display modality: Screen versus Voice. The experiment was run in two phases; the Screen study was run first, followed by the Voice study.

Drivers. Thirty-seven drivers were recruited by posting an advertisement in a local newspaper. They were paid $20 for their participation. Their ages ranged from 18 to 56 years. Half of the drivers wore glasses for driving and 40% needed glasses for reading as well. Their mean driving experience was 170 km per week. Most (69 %) preferred a standard map as an aid for navigation in an unfamiliar city, and their self-reported experience with navigating in a strange city ranged from 1 (very little) to 9 (a lot), with a median value of 4.5. Sixteen drivers participated in the Screen group, 12 males and 4 females. Twenty-one drivers participated in the Voice group, 12 males and 9 females.

Test Routes. The two test routes that were used in Experiment 1 were modified to reduce the number of destinations from 12 to 8. The resultant routes were 16 km and 21 km in length. (The longer route included a longer highway segment; driving time for each route was about 40 min.)

Procedure and Instructions. The procedure was similar to that used in the first experiment. Three expert observers were used in this experiment (including the observer who participated in Experiment 1).

For the Screen Navigator, the instructions for the Automatic condition included the following: "Directions along the route and address destinations will be shown to you on the screen. . . . As you execute the turn or point out the destination, the list on the computer screen will be updated for you." The instructions for the Interactive condition included the following variation: "You will be scrolling through the list of turns and destinations on the computer screen with the use of three buttons on a control box. These buttons are marked NEXT, ADVANCE, and HOME." (NEXT moved the current position to the next turn or destination; ADVANCE enabled a preview of directions down the list; HOME caused an immediate return to the current position). "You may press any of these three keys as often, and in any order that you like."

The Screen Navigator's instructions were digitally recorded by a male voice. In the Automatic Guidance condition, the voice delivered the information, via the radio's speakers, at predetermined positions along the route. Drivers were able to ask for a repeat of the last message. In the Interactive Voice condition, drivers had control over the delivery of the voice instructions by use of the keys on the control box. A REPEAT key was also provided.

Results

Ease of Navigating With Alternative Guidance Aids. Table 17.4 shows drivers' subjective ratings of task load and other dimensions, such as ease of navigation and feeling of safety. Drivers within each of the modality groups (e.g., Screen and Voice) received both Interactive and Automatic conditions, so more confidence can be attached to comparisons within display modality than to the comparisons between display modality. This limitation does not apply to QOD and other measures that do not rely on drivers' ratings.

TABLE 17.4
Ratings of Task Load, Safety, and Navigation

	Screen		*Voice*	
Variable	*Automatic*	*Interactive*	*Automatic*	*Interactive*
TLX (Mean)	3.39	3.29	3.95	4.62*
Similarity to normal	6.31	6.94	6.38	5.95
Feeling oriented	7.56	7.75	6.52	5.14*
Feeling in control	7.00	7.12	7.48	6.95*
Safety during drive	7.94	7.94	7.71	7.29*
Quality of navigation aid	7.12	6.37	7.05	7.38
Ease of navigation	na	na	7.62	7.00*

*Denotes a $p \leq .05$ (two-tailed test) for pairwise comparisons within row and Modality.

The differences in ratings between the Automatic and Interactive conditions in the Screen modality were small and statistically nonsignificant, indicating that the conditions were generally perceived as similar. The quality of the Automatic Screen aid was rated higher than that of the Interactive Screen, though the differences did not reach significance.

On the other hand, most of the differences between the Automatic and Interactive conditions in the Voice modality were significant. There was a consistent preference for the automatic mode of control. In comparison with the Interactive Voice condition, drivers in the Automatic Voice condition felt that their driving was more similar to normal driving, they felt better oriented and had a better sense of control, they felt safer during the drive, and they thought it was easier to navigate.

Quality of Driving With Alternative Navigation Aids. Table 17.5 presents mean performance measures for all conditions within the two experiments. The Ideal Navigator condition stands out as the one with the best driver performance. QOD scores in the Voice Navigator conditions came close to the Ideal Navigator, followed by the Screen Navigator conditions. The five alternative navigation aids produced better navigation performance compared to the Guidance List in Experiment 1.

Within the Screen and Voice display modalities, the Automatic condition tended to result in better quality of driving than the Interactive condition, though this did not reach statistical significance.

There was a marked difference in the number of REPEAT requests between the two Voice conditions. There were nearly twice as many requests in the Interactive condition than in the Automatic condition. This may be due to the uncertainty of the location of turns or other critical decision points along the route. It may also be due to the relative accessibility of the REPEAT key that was used in the Interactive condition. In the Automatic condition, drivers had to make a verbal request to have a message repeated.

Looking Behavior While Driving With Alternative Navigation Aids. Table 17.6 presents a summary of glancing behavior during the 15-sec intervals

TABLE 17.5
Quality of Driving, Navigation Errors, and Speed

	Screen		Voice		Paper	Mixed
	Automatic	*Interactive*	*Automatic*	*Interactive*	*Guidance List*	*Ideal Navigator*
QOD	5.22	5.07	5.43	5.37	5.00	5.55*
Navigation errors	2.13	1.81	1.19*	2.10	2.56	1.00*
Speed (kph)	29.94	29.70	30.62	30.34	36.80	38.30
No. repeat requests	na	na	7.67*	13.48	na	na

*Denotes a $p \leq .05$ (two-tailed test) for pairwise comparisons for every variable.

associated with negotiating the turns along the route. The criteria for selecting the video segments and the coding procedures were identical for both experiments.

Glance variables in Table 17.6 were first tested for normality using normal probability plots and the Shapiro–Wilks statistic. The data were normally distributed, thus justifying the use of the original values, without transformations, in further analyses. The effects of display modality (Screen, Voice) and mode of control (Automatic, Interactive) on glance frequency, glance allocation (% of total time), and average glance duration, were analyzed using multivariate analysis of variance (MANOVA) and paired t tests.

The frequency of glances inside the vehicle, and, indeed the mean glance duration, were lower under the Voice condition than under the Screen condition. Within the Voice condition, the Interactive mode of control led to slightly higher glance frequencies (these effects were significant at the .05 level).

The percentage time allocated to looking in the forward direction was higher under the Voice condition than under the Screen condition. The MANOVA analyses indicated significant main effects and no interaction. Mean glance durations exhibited similar trends—mean glance durations in the forward direction were longest in the Voice condition and shortest in the Screen condition.

Not surprisingly, the opposite trends can be seen for glances at the display. The mean glance duration at the display in the Screen conditions was higher than in the Voice conditions. Glances to the rear were very short and consistent across all conditions, including the two conditions of Experiment 1. Glances to the side windows were also relatively consistent in frequency and duration across conditions.

TABLE 17.6
Glancing Behavior With Four Navigation Aids

	Screen		Voice	
	Automatic	*Interactive*	*Automatic*	*Interactive*
Glances to display/15 sec	10.31	10.56	7.45	8.50
Glance Allocation Time				
Forward view	64.2%	61.9%	75.6%	73.3%
Rear view	2.8%	3.0%	2.3%	1.6%
In-vehicle display	13.0%	16.6%	~0.0%	3.8%
Side	20.0%	18.5%	22.1%	21.3%
Mean Glance Duration(s)				
Forward view	1.99	1.88	3.00	2.95
Rear view	0.58	0.62	0.56	0.50
In-vehicle display	1.02	1.07	0.13	0.62
Side	1.34	1.32	1.23	1.26

The Automatic conditions for both Voice and Screen displays produced longer forward glances (and larger overall allocation of visual attention) than the Interactive conditions.

SUMMARY: PERFORMANCE WITH ALTERNATIVE, GENERIC TYPE, NAVIGATION AIDS

Figures 17.1–17.5 illustrate the main results of the two experiments. Means and standard error bars are shown for quality of driving, time load index, percent attention allocation to the forward road scene, percent attention allocation to the in-vehicle display, and navigation errors as functions of navigation task conditions. The conditions S/A, S/I, V/A, V/I, PL, IN, and Control refer to Screen-Automatic, Screen-Interactive, Voice-Automatic, Voice-Interactive, Paper List, Ideal Navigator, and Familiar Route navigation conditions, respectively.

The figures give a consistent picture of the influence of different attributes of a navigation system on task demands and driver behavior. Following a familiar route without searching for addresses (the control condition) was the least demanding task for drivers, as expected. Their visual attention

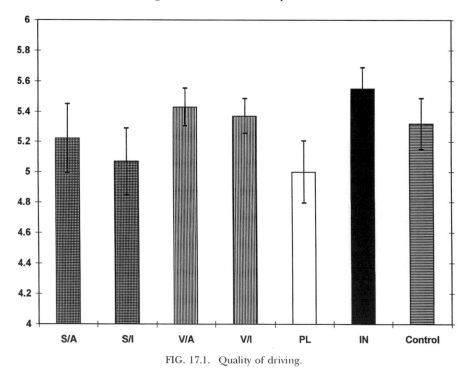

FIG. 17.1. Quality of driving.

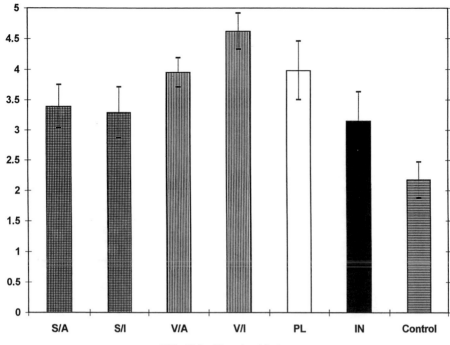

FIG. 17.2. Time load index.

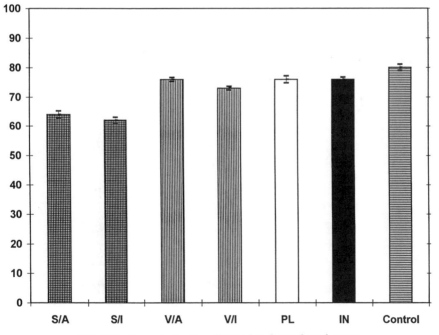

FIG. 17.3. Percent attention allocated to forward road scene.

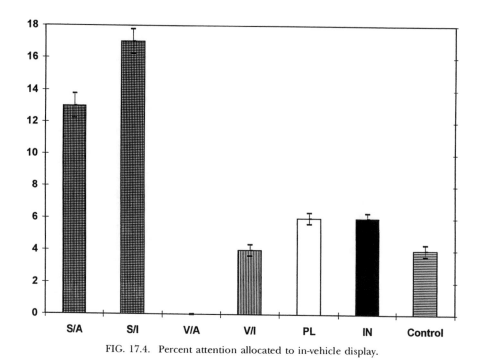

FIG. 17.4. Percent attention allocated to in-vehicle display.

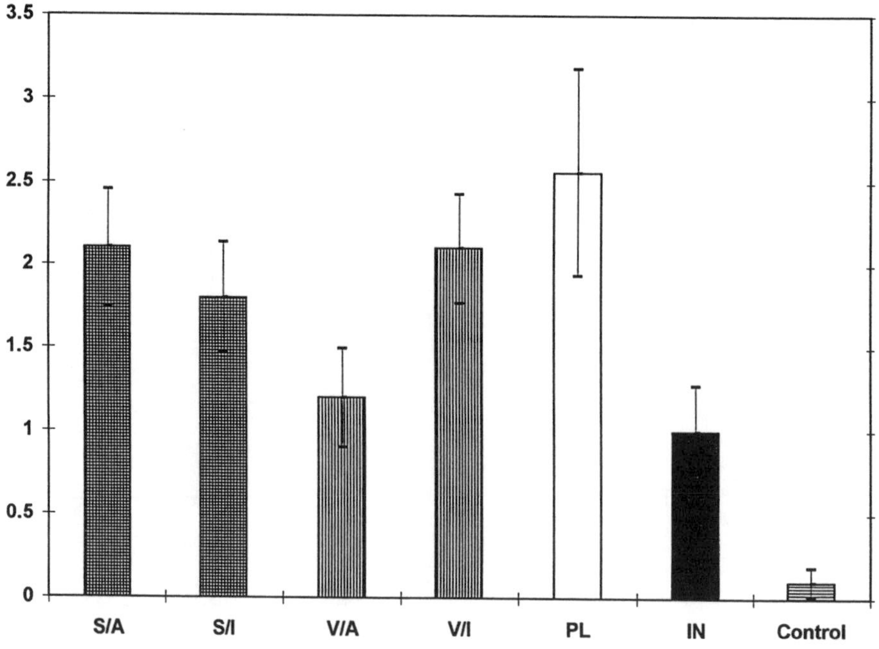

FIG. 17.5. Navigation errors.

was directed mostly to the road (80%), and minimally (4%) inside the vehicle. Interestingly, the proportion of attention allocated to the in-vehicle display was very similar across the three conditions of the first experiment, probably due to the presence of the paper display of instructions.

An address searching task in an unfamiliar city environment, regardless of the type of navigation aid used, was indeed experienced as a more difficult task than just following a route. Under conditions of minimal information provided by the Guidance List, the QOD was degraded relative to the Familiar Route condition. However, under the more supportive Ideal Navigator condition, the task was made easier, resulting in fewer navigation errors and better quality of driving compared with both the Guidance List and the search-free Familiar Route conditions. Most drivers expressed a preference for the Ideal Navigator as a navigation assistance device.

The second experiment demonstrated that a Voice guidance aid can lead to better driving performance, less visual attention directed to the in-vehicle displays, and more attention directed to the forward scene than a Screen guidance system displaying the same verbal information. In fact, the Voice-Automatic condition produced quality of driving that was equivalent to the Ideal Navigator. However, drivers using the Voice modality reported higher task load than drivers using the Screen modality. This may be due to between-group differences in TLX ratings and/or the need to keep instructions in short-term memory under the Voice modality.

Figure 17.1 shows that an Automatic display resulted in consistently better performance than an Interactive system. The Interactive Screen condition produced the poorest overall quality of driving, not much different than navigating with the basic paper list. It should be noted that although all mean QOD values fell within the range of 5.0 to 5.6, the differences between conditions were significant at the .05 level of confidence.

For the Voice condition, driving with an Automatic system was also subjectively less demanding than driving with an Interactive system perhaps because the latter imposed additional demands associated with deciding when to request information and the need to depress the repeat key.

It was noted that drivers' personal preferences were consistent with the system's relative advantage as indicated in these figures. For both display modalities, the Automatic mode of control was preferred over the Interactive mode of control, though the preference was more pronounced in the Voice modality.

CONCLUSIONS

The results of these experiments reaffirm the sensitivity and reliability of the QOD technique for evaluating the safety of TICS. Further efforts should aim to refine and validate the QOD methodology.

The data were used to compare features of navigation system interfaces in terms of their support of drivers' needs for guidance and contribution to safe driving. Certain design principles (which apply while driving, not necessarily for trip planning) emerge from these analyses.

Contextual guidance information on a screen display (including landmarks and other orientation cues) is potentially helpful, provided it is automatic, or system driven. However, the potentially greater allocation of visual attention to the display increases the risk of insufficient attention being directed to the road.

Voice guidance, with context-rich information, and without additional or concurrent visual information, is more effective than displaying guidance information on a visual screen.

Automatic Voice guidance appears to be the most effective and safest. Possible explanations for the superiority of the automatic voice system are that the system can better anticipate upcoming decision points because it "knows" the road network, it frees drivers from having to decide and initiate interactions, and it causes drivers to drive at a pace established by the system. There is growing evidence that voice instructions with a redundant screen display is the optimal configuration for a navigation interface (Parkes & Burnett, 1993; Verwey, 1993). Indeed, the Ideal Navigator condition (i.e., automatic voice messages and a redundant visual display) produced the best QOD scores.

REFERENCES

Andre, A. D., & Wickens, C. D. (1995). When users want what's not best for them. *Ergonomics in Design*, October 10–14, pp. 10–14.

Ashby, M. C., Fairclough, S. H., & Parkes, A. M. (1991). *A comparison of two route information systems in an urban environment* (DRIVE Rep. No. 49). England: HUSAT Research Institute.

Bailey, J. H. (1993). Performance vs. preference. *Proceedings of the Human Factors and Ergonomics Society 37th Annual Meeting*, 282–286.

Fastenmeir, W. (1993). Traffic safety evaluation using behavioural criteria. *Proceedings of the 26th International Symposium on Automotive Technology and Automation*, 373–380.

Hart, S. G., & Staveland, L. E. (1988). Development of NASA-TLX (Task load index): Results of empirical and theoretical research. In P. A. Hancock & N. Meshkati (Eds.), *Human mental workload* (pp. 139–183). New York: North-Holland.

Michon, J. A. (1993). *Generic intelligent driver support.* London: Taylor & Francis.

Noy, Y. I. (1989). Intelligent route guidance: Will the new horse be as good as the old? In D. H. M. Reekie, E. R. Case, & J. Tsai (Eds.), *Proceedings of Vehicle Navigation & Information Systems Conference* (pp. 49–55). New York: IEEE Publishing Services.

Noy, Y. I. (1990). *Attention and performance while driving with auxiliary in-vehicle displays.* Transport Canada Report TP 10727 (E).

OECD (1992). *Intelligent vehicle highway systems: Review of field trials.* Road Transport Research Programme, Organization for Economic Cooperation and Development, Paris.

Oxley, P., Ayala, B., Alexander, J., & Barham, P. (1994). *Evaluation of route guidance systems. Parts 1&2* (DRIVE II Rep. No. 20, 20A). Cranfield University.

Parkes, A. M., & Burnett, G. E. (1993). An evaluation of medium range "advance information" in route-guidance displays for use in vehicles. *IEEE Vehicle Navigation and Information Systems Conference*, 238–241. Warrendale, PA: Society of Automotive Engineers.

Parkes, A. M., Ashby, M. C., & Fairclough, S. H. (1991). The effects of different in-vehicle route information displays on driver behavior. *IEEE Vehicle Navigation and Information Systems Conference* (Paper No. 912734), 61–70. Warrendale, PA: Society of Automotive Engineers.

Pohlmann, S., & Traenkle, U. (1994). Orientation in road traffic. Age related differences using an in-vehicle navigation system and conventional map. *Accident Analysis and Prevention, 26*(6), 689–702.

Popp, M. M., Farber, B., & Schmitz, A. (1991, November). *Guiding drivers through a metropolis: Traffic safety aspects of the guidance and information system Berlin (LISB).* Paper presented at the 13th International Technical Conference on Experimental Safety Vehicles, Paris, France.

Quimby, A. R. (1988). *In-car observation of unsafe driving actions* (Research Rep. ARR 153). Australian Road Research Board.

Reidel, W. J. (1991). Eye-movements, expert ratings, weaving and time-to-line-crossing as measures of driving performance and driving performance impairment. In A. G. Gale et al. (Eds.), *Vision in vehicles* (Vol. 3, pp. 299–306). Amsterdam: Elsevier Science.

Risser, R. (1993). A study of the behavior of drivers of electric cars. *Proceedings of the 26th International Symposium on Automotive Technology and Automation*, 529–538.

Verwey, W. B. (1993). Further evidence for benefits of verbal route guidance instructions over symbolic spatial guidance instructions. *IEEE Vehicle Navigation and Information Systems Conference*, 227–231.

Zaidel, D. M. (1991). *Specification of a methodology for investigating the human factors and safety of advanced driver information systems.* Transport Canada Report TP 11199 (E).

Zaidel, D. M. (1992). *Quality of driving with route guidance assistance: An evaluation methodology.* Transport Canada Report TME 9201.

Zaidel, D. M. (in press). *Performance vs. preference for alternative route guidance systems.* Transport Canada Report.

Zaidel, D. M., & Noy, Y. I. (1993, January). *Ergonomic issues in the evolution of advanced driver interfaces.* Paper presented at the 72nd Annual Meeting of the Transportation Research Board, Washington, DC.

Chapter 18

Ergonomic Issues on Entering the Automated Highway System

James R. Buck
Anil Yenamendra
University of Iowa

Many people regard Automated Highway Systems (AHS) with the same credulity as James Hilton's (1933) *Shangri-La.* Others believe it is a system that will appear soon. The realization of AHS lies somewhere between these extremes and depends on solving technical, economic, and political problems. This chapter is only about some of the technical ergonomic problems and some possible solutions.

Other related dimensions where the beliefs of people differ widely is about the configuration and operating procedure of a future AHS. Some view AHS as an expressway system purely dedicated to automated vehicles (Varaiya, 1993). Others, at the opposite extreme, see AHS occurring on existing highway systems with minimum alterations, at least one lane for automated cars and manually controlled cars in the other lanes. Other opinions differ with regard to having a strong central AHS control in contrast to a modest central authority and strong local control by and between automated vehicles. Some of this diversity was indicated by Alicandri and Moyer (1992); Hedrick, Tomizuka, and Varaiya (1994); Fenton (1994); and Shladover, Desoer, et al. (1991). Such was the case when the authors first attempted to identify ergonomic issues about the illusive and visionary AHS in the United States. Clearly, those issues depend on both the system configuration and modes of operation, but those dependencies are fuzzy at best. Accordingly, the development of this study was exploratory from both an ergonomics and a systems point of view and the first actual AHS seen may differ substantially from the one described later. Nevertheless, the outcome of the study provided a surprise, even to the authors.

AHS are not just the embellishments of technical visionaries. Increasing traffic in the United States is expected to overstress the capacities of the current expressways by the year 2020, especially expressway linkages around and through urban areas. However, it is precisely those areas where there is no space to add greater capacity using additional conventional roadways. AHS offer the prospect of greatly increasing traffic throughput with even greater safety than conventional manually controlled vehicles without needing significantly greater space (Saxton, 1980).

There are a variety of ergonomic issues that need to be resolved before a safe and successful AHS can be designed. However, those issues change with the configuration and operating mode of AHS. There are few ergonomic issues in an AHS dedicated solely to automated vehicles and operating in a totally automated mode, but such systems are enormously expensive to build. Accordingly, most experts believe such systems are unlikely for years to come because of political and economic problems. Hence, a configuration toward a lower end was investigated. A first consideration was to see if a less costly configuration was adequate. If not, then more sophisticated and expensive configurations could be examined. In the assumed configuration and operating mode, the driver manually controlled the automated vehicle until the vehicle was in the AHS lane of the expressway and then control was transferred over to the AHS. The following are three important ergonomic issues for the assumed configuration:

1. How well can drivers best enter the AHS in the prescribed manner?
2. How can control be transferred from the driver to the AHS?
3. What design speeds and vehicular gaps should be maintained on the AHS?

An experiment was designed and executed to address several features associated with these issues.

There has been controversy in the ergonomics community about the need for vehicle simulators. This study exhibits a case in favor of simulators because the study could not have been undertaken without simulator capability (Stadden, 1991).

METHOD

Subjects

The 24 paid volunteers used in this experiment were recruited through newspaper advertisements. They ranged from 25 to 34 years age. Each had a valid state driving license without restrictions other than to wear eyeglasses during driving. Half of these drivers were female and half were male in order to maintain a representative sample of drivers in this age group. Driving experience of these subjects appeared to be well within the limits typically reported for this age group. As there was no way to accommodate for driver disabilities in our testing equipment, all applicants who needed special driving devices were rejected. Every subject was administered a series of Titmus vision tests and a peripheral visual test for acuity and other abilities. However, the results of these visual tests neither qualified nor disqualified subjects. None of the subjects exhibited simulator sickness and none were excused from the experiment due to this or visual reasons.

Experimental Equipment

Driving behavior and performance data were recorded while the subjects were driving in the Iowa driving simulator (IDS). This simulator has a moving base with a sedan car body mounted to it, a screen about 154 cm tall around the edge of this base, and a dome over the top. A hydraulic hexapod drove the base with 6 degrees of freedom to impart realistic sensations of acceleration, braking, and turning. After each base maneuver, there was a very low frequency neutral centering movement of the base. A Harris Nighthawk 440 computer controlled these base movements, as well as performing other activities of event scheduling and data collection. An additional computer, the Alliant FX-2800, performed all the Newtonian mechanical computations for the suspension and drive systems for a vehicle performing in the same manner. Due to parallel processing, those computations were performed in real time. An Evans and Sutherland CT-6 visual display system was used to generate the visual scenes. Four channels of chromatic visual projection were used, three projecting forward on the screen to cover over 190 degrees in front of the driver and one projected backward providing a 65-degree field to be seen in the rearview and the driver's sideview mirrors. The resolution of each projector was about 785 K pixels. Buck, Stoner, Bloomfield, and Plocher (1993) and Kuhl, Evans, Papelis, Romano, and Watson (1995) provided further details on this simulator.

AHS Configuration in the Experiment

In the development of this research project, it was assumed that the initial version of an AHS would occur in urban areas because those areas have the highest cost land and the least land available. A low-cost minimum-change configuration was envisioned as one of the alternatives during the initial development as part of the highway system. Part of the basis for this assumption was the fact that the initial demand for automated highways would follow the manufactured volume of automated vehicles. It was also realized that vehicles in at least one automated lane of traffic would be in strings of several cars with small headway's between adjacent vehicles in the string but larger gaps between successive strings. This configuration could carry heavy traffic densities safely to and from central urban business areas.

A minimum AHS configuration consisted of a three-lane expressway for each direction of travel. The left-most lane was dedicated to automated vehicles, whereas the other two lanes contained manually controlled vehicles. Some of the vehicles in the manually controlled lanes would be vehicles capable of operating in the automated lane and were leaving that lane, about to enter it, or remained in the manually controlled lanes because of driver choice. The center lane of this configuration served two purposes: It allowed faster vehicles to pass slower ones in the right-most lane, and it provided an AHS entry/exit lane. However, there was no segregation of the entry and exit functions. Also, all three lanes of the expressway was 3.66 m wide (i.e., 12 ft), which meets current design standards. No barriers were used. The speed limit in the non-AHS lanes was the current speed limit of 88.6 kph (55 mph). Vehicles in the AHS were assumed to travel at constant velocities in strings of three or four cars with a gap of $\frac{1}{16}$ sec between successive AHS vehicles within a string (i.e., about 1.5 m from the back bumper of the car ahead to the front bumper of the folowing car). Three different AHS speeds were examined in this experiment: 104.6 kph (65 mph), 128.7 kph (80 mph), and 152.9 kph (95 mph).

A variety of other vehicles played parts in the scenario of the driving task. Both cars and trucks occupied the rightward two lanes at an average density of 6.2 vehicles/km in each lane (i.e., about 10 vehicles/mile/lane). These vehicles traveled at an average speed of 88.6 kph (55 mph), but the speed of individual vehicles varied according to a normal distribution with a standard deviation of about 6.3 kph (3.9 mph). The initial distances between successive vehicles that were introduced to the visual scene were randomly distributed following a Pearson Type III distribution independently of the speed variation in order to create a realistic scene of traffic. Only automobiles were generated for the AHS lane. These vehicles traveled at one of the three AHS design speeds in groups of two to four cars to a string. Interstring distances within a string were varied experimentally, as is explained later.

The Test Vehicle and Experimental Personnel

The test vehicle was a Ford Taurus body with four doors and the standard left-hand steering wheel. The inside remained as built, but there were a few modifications outside. The hood was standard, although without an engine, but it was necessary to cover it with a dull black cloth to minimize reflections from the projected imagery. Also, the trunk at the rear was removed and covered over in order to fit the moving base. The removed part of the rear end could not be seen from the driver's seat.

During the experimental procedures, one of the experimental personnel always introduced the drivers to this test vehicle so that they knew the locations of all instruments and controls. A safety button was installed so that subjects could activate it any time during the testing and the test would be terminated as quickly as safely possible. Also, one of the experimental personnel sat in the back seat during the testing to record unusual events, prevent drivers from leaving the vehicle while the base was still in motion, watch for signs of simulator sickness, and serve as a backup for shutting down testing (if need be).

AHS Entry Maneuvers and Procedures

It was assumed that drivers who had appropriately equipped vehicles and wished to enter the automated lane would request such entry while they drove in the center lane at a prescribed velocity. AHS would verify that the car was adequately equipped and, if so, then instruct the driver when to make the needed lane change. Both automatically controlled entry and manual control were considered during the initial planning of this experiment. Only manually executed lane changes were considered in this experiment. However, it was uncertain if the automatic-entry technology was low enough in cost or high enough in reliability. Consequently, because of these and other reasons, only manually controlled steering was examined. If safety or efficiency appeared to be impaired during the tests, the alternative form of entry could be recommended.

At the start of an experimental trial, the driver was instructed to move safely from the initial entry-ramp location to the center lane and then maintain a speed of 88.6 kph (55 mph). So long as the driver held a speed between 80.6 kph and 96.5 kph (i.e., 50.1 and 60.0 mph) for 15 sec, the AHS issued a verbal "Enter" command. On hearing the "Enter" command, the driver was expected to steer quickly into the left lane behind the last of a passing string of cars. Two methods of transferring control were used after entering the left-most lane. With the total-manual control method, the driver pressed the cruise control button on the vehicle's steering wheel. The other method automatically transferred vehicle control to AHS when the last wheel of the vehicle crossed into the automated lane.

Entering cars started the entry procedure at speeds about 88.6 kph (55 mph) within an envelope of about 2 kph. Entry was made between successive strings of automated cars travelling at speeds at least 16 kph greater than the entering vehicle. Accordingly, the string of vehicles that just passed the entering driver continued to get farther away while the string behind approached the driver's vehicle. Once under AHS control, the entering car was commanded to accelerate as rapidly as possible to the AHS design velocity. Accelerations in this study were equivalent to that of a standard U.S. sedan, such as a Ford Taurus. Consequently, the time it took to bring the entering vehicle to the AHS design speed increased with the designated AHS design velocity exponentially. In a theoretically perfect entry, the gap between successive strings of cars should be large enough to allow a safe entry to be made behind the last car in the passing string and so that the entering car could accelerate to the AHS design speed precisely as the first vehicle in the trailing string joins up .0625 sec behind it.

The minimum interstring gaps used in the experimental AHS traffic for each design velocity were based on the time required by the simulator vehicle, which has the acceleration characteristics of a modern U.S. sedan on a straight, level, and dry road to the AHS design velocity. Those times were about 2.0 sec for both 104.6 and 128.7 kph (65 and 80 mph) and 5.6 sec for 152.9 kph (95 mph). Minimum gaps were set at approximately those values. Those minimums ignored reaction times and other sources of delay in entering the AHS lane, so it was expected that those cars following the entering car would in fact be delayed by the entering vehicle. The larger gaps were arbitrarily set at 2.4, 4.0, and 7.5 sec to provide slack for the behavioral causes of a nonoptimal entry, respectively, at the three design AHS speeds. Losses in that slack would indicate the magnitude of those behavioral causes.

When the driver was in the center lane for the prescribed duration at an acceptable speed, the AHS instructed the driver, "After the count-down, enter the automated lane—four, three, two, one, enter." Previous to the testing, each subject was informed that if they drove too fast or too slow to enter they would receive a "Don't Enter" command and in that case they would be given a later entry opportunity. During a successful entry, a driver would need to turn the wheels toward the automated lane and make the lane-change maneuver. If a manual method of control transfer was appropriate, then the driver had to additionally press the button on the steering wheel after entering the AHS lane.

Experimental Variables Examined in the Study

There were three experimental variables in this experiment: AHS design speeds, intrastring gap sizes, and the method of transferring control from the driver to the AHS. It was determined that the random switching between methods of control transfer would be confusing to a subject. Hence, the "method of control transfer" was a between-subject variable. Each subject performed under all six conditions obtained by three AHS design speeds and two intergap sizes. The values of the variables at those six experimental conditions are described in Table 18.1. These six conditions were assigned to each subject such that the order in which each condition appeared was equal in frequency over the subjects who always used each

TABLE 18.1
AHS Design Velocities and Interstring Gaps
Used in the Experiment Given in Multiple Units

			Interstring Gaps			
AHS Design Velocities			Shorter Gap		Longer Gap	
kph	[m/sec]	(mph)	sec	[m (ft)]	sec	[m (ft)]
104.6	[29.1]	(65)	2.0	[58.1 (190.7)]	2.4	[69.7 (228.8)]
128.7	[35.8]	(80)	2.0	[71.5 (234.7)]	4.0	[143.1 (469.3)]
152.9	[43.5]	(95)	5.5	[233.6 (766.33)]	7.5	[318.5 (1045.0)]

method of control transfer. Also, half of the conditions under each mode were assigned to the males and females. This gender requirement not only assured better subject representation. It also provided a basis for observing if any behavioral differences occurred due to gender.

Behavioral and Performance Measurements

Figure 18.1 describes driver actions starting with the entry to the expressway from the ramp, traversing the right lane, and entering the center lane to start the "AHS entry" maneuver. Prior to the start of that maneuver, a request to enter the AHS lane would be made in practice, but it was not part of the experimental protocol. If the vehicle was in the center lane and traveling within the proper speed envelope, then AHS would give the "Enter" command when the gap in the AHS lane was positioned appropriately to the entering car. That command started the AHS entry maneuver. Several events before and during the maneuver are shown.

Events occurring during the entry maneuver are also reported in Fig. 18.2, starting with the "enter" command and ending when control was transferred to the AHS. During this time, the steering and speed of the entering car was under the control of the driver. The events in this figure partition the total maneuver time interval into component intervals. Data were collected on these component times as behavioral measurements along with the velocities at the events. Drivers who attempted to steer sharply over each component of the maneuver were clearly distinguished from those taking a more gradual approach. Those component time intervals were measured with the precision of 60 Hz. Speeds at the events also consisted behavioral measurements to distinguish drivers who accelerated strongly over the maneuver from those who did not. In the planning of this experiment, it was suspected that higher AHS velocities, shorter entry gaps may induce some behavioral changes or that some unsafe situations during the entry may be preceded by sharp or fast entry behaviors.

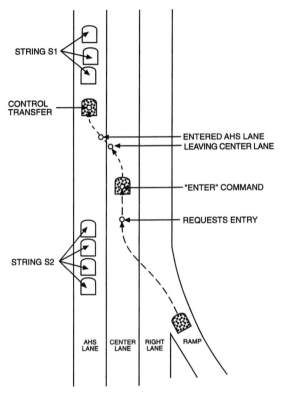

FIG. 18.1. A schematic of the entering maneuver in the experiment with events that provided behavioral data.

The final period of importance in this AHS joining maneuver, which is not part of Fig. 18.2, is the "string joining time." This was the time period from the event where the AHS takes control until the entering car becomes the lead vehicle in the string that joins it. Because vehicle is under AHS control during string joining, this time period is not ordinarily of importance to ergonomics. However, delays during the earlier part of the entry maneuver impacts the effectiveness of the AHS system, delaying vehicles that follow.

A direct measurement of system performance was "delay distance," which was defined for the vehicle immediately behind that of the entering car after the entering maneuver was completed. The delay distance is the distance the vehicle should have traveled had the car not entered, less the distance it actually traveled. An associated measure of "delay time" was the delay distance divided by the AHS design speed. The greater delay distance or time, the greater the impact of an operating mode on system effectiveness, and the lower the vehicle throughput rate on the AHS lane.

Figure 18.3 illustrates the the relation between the entering car and the strings of cars immediately ahead and behind during entry maneuver and

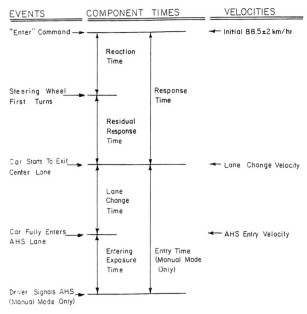

FIG. 18.2. Events, component times, and velocities during the entry maneuver.

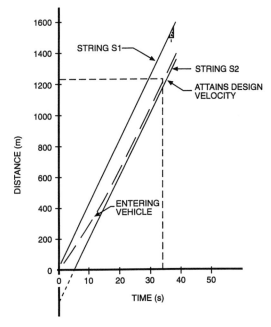

FIG. 18.3. The AHS lane at the time of entry and changes over time assuming a 152.9 km/hr AHS design velocity and no time losses during entry other than acceleration effects. The strings S1 and S2, respectively, precede and follow the entering vehicle.

the string joining time that follows. The specific example associated with this figure is the AHS design velocity of 152.9 kph (i.e., 95 mph or 43.5 m/sec). This figure shows the entering car is behind the last car in the leading string (denoted as S1) during the beginning of the entry. At the end of the "string joining time," it is the first car in the following string (denoted as S2). It is assumed that the entering car starts to accelerate at entry and continues to do so until the following string of cars is within $\frac{1}{16}$ sec (or 2.7 m at 95 mph, 1.82 and 2.23 m at 65 and 80 mph). Figure 18.3 shows the trailing string of cars is far behind when the car enters AHS. The two parallel lines of S1 and S2 shows constant distances over time by the two successive strings of AHS cars and the curvature in the distance achieved by the subject's car over time is the acceleration curve from approximately 88.6 kph (or 55 mph) to 152.9 kph (95 mph or 42.5 m/sec). The acceleration curve of the entering car in Fig. 18.3 reaches 152.9 kph in about 34 sec (5.3 sec at 104.6 kph and 15 sec at 126.7 kph). Hence, the distance between strings S1 and S2 can be reduced with a greater acceleration curve. Those S1 to S2 distances denote the minimum size of gaps needed in the AHS lane between successive strings of vehicles.

Typical measurements associated with accidents or near accidents or other unsafe conditions (e.g., collisions or incursions into other lanes) were included in order to describe the apparent safety of the AHS under these conditions and operating procedures. The experimental personnel were instructed to note any occurrences of unsafe or potentially unsafe behavior (e.g., sudden braking, close tailgating, or wide swaying). Video footage was also recorded on the four visual channels as a secondary record for identifying unsafe situations.

An exit questionnaire was developed to examine subjective perceptions of the drivers' viewpoints on system design, simulator realism, perceptions of AHS safety, preferences among conditions, and related concerns. These questions and their significance are discussed later.

RESULTS

The experimental variables already described were subjected to analyses of variance (ANOVAs) to identify those variables displaying statistical significance. Although the selected level of significance was 5% level or smaller, the actual levels reported here are much smaller, which was in keeping with the large number of criteria tested. These ANOVAs accounted for differences between the within- and between-subjects variables. For example, Table 18.2 shows the summary of the ANOVA conducted on the response times, which is described in Fig. 18.2 as the elapsed time from the command until the AHS lane is just penetrated. When in the partially

TABLE 18.2
A Summary ANOVA Table for the Response Time

Source of Variation	df	Sum of Squares	Mean Square	F Ratio	Probability P
Mode of Control Transfer (T)	1	2.011	2.011	2.64	0.1184
Subjects within Mode (S within T)	22	16.754	0.762		
Size of Interstring Gap (G)	1	0.020	0.020	0.10	0.7500
$T \times G$	1	0.221	0.221	1.15	0.2962
$G \times S$ (within T)	22	4.240	0.193		
AHS Design Velocity (V)	2	0.099	0.050	0.34	0.7110
$T \times V$	2	1.224	0.612	4.24	0.0207*
$V \times S$ (within T)	44	6.353	0.144		
$G \times V$	2	0.133	0.067	0.21	0.8142
$T \times G \times V$	2	0.292	0.146	0.45	0.6387
$G \times V \times S$ (within T)	41	13.202	0.322		

automatic method of transferring control, the response time increased from 2.2 sec to 2.3 sec and to 2.4 sec over the three AHS design velocities of 104.6, 128.7, and 152.9 kph. But in the manual method of control transfer, response time dropped from 2.6 sec while at the 104.6 kph speed down to 2.5 sec at the other two AHS design speeds.

Other ANOVAs were analyzed in a similar way. A summary of those ANOVAs reaching statistical significance are reported in Table 18.3.

An ANOVA was also performed on the "delay time." Table 18.4 reports the summary ANOVA results. There was a minor loss of data reflected by the difference in degrees of freedom in this table compared to the previous summary ANOVA. The delay time was found to differ significantly when the AHS design velocity and the method of control transfer varied. Delay times increased with the AHS design velocities, but those delay times were almost uniformly greater with the manual control transfer method than the other method, as Fig. 18.4 shows. That figure also verifies the results in Table 18.4 that the control transfer methods and the AHS design velocity exhibited virtually no interaction.

In regard to safety–accident-related statistics, it should be stated that there were no collisions, no lane incursions, or other unsafe conditions observed. Data from the questionnaires regarding perceptions of safety and other issues are reported in Table 18.5. The actual data were marks made on a line where one end represented total disagreement and the opposite end complete agreement. The data reported in Table 18.5 are percentage points from the negative end toward the positive end, which were proportional to the average of the subjects' mark.

TABLE 18.3
Summary Results of Time and Velocity Criteria

Criterion	Significant Variance & Probability	Comments A	Comments B
Response Time (RT)	Interaction of the transfer method (T) and AHS design Velocity (V) Interaction (T × V) probab. .0207	with partial automatic method, RT increased from 2.17 to 2.33, and to 2.42 sec, respectively, over velocities 104.6, 128.7, and 152.9 kph	with manual method, RT changed from 2.7 sec at 104.6 kph to 2.5 sec for 2 higher speeds
Lane Change Time (LCT)	AHS design velocity, V, probab. .022	LCT decreased from 1.22 sec at 104.6 kph down to 1.14 sec at 128.7 kph	LCT rose to 1.28 sec at the velocity of 152.9 kph
Exposure Time (ET)	Transfer mode T, probab .0001	In the auto mode, ET was constrained at zero	In the manual mode, ET averaged 1.17 sec
String-Joining Time (SJT)	AHS design velocity V, probab .0001	SJT was 5.7 sec at 104.6 kph, 14.6 sec at 128.7 kph, & 36.1 sec at 152.9 kph	
Minimum Velocity in the Center Lane (MVCL)	Interstring gaps G, probab .0006, & the Interaction of T × V, probab. .0485	MVCL was 97.62 kph at the smaller gap & 95.66 kph @ the larger gap	In partial auto mode, MVCLs were 51.32, 50.94 & 50.32 kph over 3 AHS design speeds. In manual mode, MVCLs were 49.55, 51.47, & 50.32 kph

TABLE 18.4
A Summary ANOVA Table for the Delay Time

Source of Variation	df	Sum of Squares	Mean Square	F Ratio	Probab. P
Mode of Control Transfer (T)	1	2.00	2.00	6.73	0.017
Subjects within Mode (S within T)	22	6.55	0.30		
Size of Interstring Gap (G)	1	0.05	0.05	0.57	0.457
T × G	1	0.00	0.001	0.00	0.981
G × S (within T)	22	1.93	0.09		
AHS Design Velocity (V)	2	716.58	358.28	2604.9	0.0001
T × V	2	0.49	0.25	1.79	0.1785
V × S (within T)	43	5.91	0.14		
G × V	2	0.30	0.15	1.28	
T × G × V	2	0.23	0.12	1.00	
G × V × S (within T)	40	4.72	0.12		

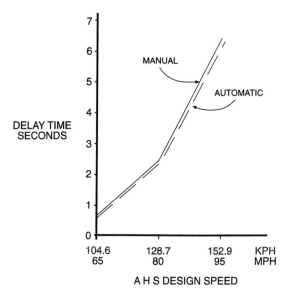

FIG. 18.4. Time delay as a function of design velocity for the partial auto-matic and manual modes of transferring control from the driver to the AHS.

DISCUSSION OF RESULTS

Before discussing the implications of the behavioral and performance data collected, it is important to recognize that the drivers in this experiment gave every appearance of respecting the realism of these physical simulations, as indicated by the questionnaire data shown in Table 18.5. The first six questions dealt with issues of realism, simulator driving enjoyment, and a comparison of the simulator and the driver's own car. Responses from these questions were graded from a totally negative extreme at a zero value to a totally positive extreme near 100 points. Average subject responses from each of the first six questions were over the midpoint. Although responses to Question 2, comparing their personal car to the simulator car, and Question 4 on sound (i.e., audio) realism, were only 59% and 60% averages. The remainder were 73% and above. These results indicate that most subjects perceived the simulator as exemplifying realism even though it drove a bit differently from their own or did not sound quite correct.

Prior to entering the AHS lane, each driver needed to get into the center lane and hold a set speed about 88.6 kph (55 mph) for around 15 sec. That behavior was a condition of entry and every driver was able to perform it. After the "Enter" command, it took time for the driver to react to the command and steer the vehicle toward the white line separating the center and AHS lanes. This response time was shorter for some drivers and it varied

TABLE 18.5
Summary of Questionnaire Results

Question	Scores
Simulator Realism	
1. How much did you enjoy driving the simulator?	90.0
2. How did driving in the simulator compare to driving in your car?	59.0
3. How realistic was the view out of the windshield in the simulator?	73.0
4. How realistic were the sounds in the simulator?	60.0
5. How realistic was the vehicle motion in the simulator?	74.0
6. While driving in the simulator, did you feel queasy or unwell?	83.0
Message Understanding and Timing	
7. Was the message giving you the command to enter the automated lane easy to understand?	97.0
8. Did you have enough time to react to the message telling you to enter the automated lane?	91.0
Feelings of Safety and Control	
9. How safe did you feel when you drove into the automated lane?	84.0
10. Did you control your car poorly or well as you changed lanes from manual to the automated lane? (left very poorly controlled and right very well controlled)	85.0
11. To what extent did you feel in control of the situation when you drove into the automated lane and transferred control of your vehicle to the Automated Highway System?	75.0
Gap Size and AHS Design Velocity	
12. When your car was under automatic control, the distance between you and the cars in front and behind was varied from trial to trial—which separation did you prefer? (left, preferred longer distance; right, preferred longer distance)	36.0
13. When your car was under automatic control, were you comfortable with the speed, or would you have preferred to have traveled faster or slower? (left, much slower; right, much faster)	69.0
Attitudes Toward AHS	
14. You spent some time in the manual lanes and some in the automated lane—which did you prefer? (left, strongly preferred manual; right, strongly preferred automated lane)	74.0
15. Was it more challenging to be in the automated lane or the manual lanes? (left, manual lanes more challenging; right, automated lanes more challenging)	14.0
16. How would you feel if an AHS was installed on I-380 between Iowa City and Waterloo? (left, very unenthusiastic; right, very enthusiastic)	82.0
17. If an AHS was installed, would you feel safer driving on I-380 than you do now without the system? (left, much safer on current highway; right, much safer with AHS)	65.0
18. How will the installation of an AHS affect the stress of driving? (left, will greatly decrease stress; right, will greatly increase stress)	30.0

between conditions of the experiment The ANOVA of the response time data indicated that the first-order interaction between the transfer method and the AHS design velocity was the only statistically significant factor ($\alpha <$ .05). Response times of those subjects making partial automatic transfers increased with greater AHS design speeds but those making manual transfers showed a decreasing response time from the lowest AHS design velocity to a near-constant response time with the two higher AHS design speeds. The main difference in response times occurred between the two methods of control transfer at the 104.6 kph velocity, but not at the other AHS design velocities. This result appears to the authors as an anomaly.

The lane-change times were found to differ with statistically significance ($\alpha < .05$) over the three different AHS design speeds. Average lane-change performance times over the 104.6, 128.7, and 152.9 kph (i.e., 65, 80, and 95 mph) were respectively 1.22, 1.14, and 1.28 sec. These deviations are very small and, consequently, are unimportant.

It was not surprising to find a long average exposure time with the manual transfer of control. The partially automatic method performs the transfer at the first possible time when the rear-right tire crosses the lane marker separating the center and AHS lanes. During the lane-changing maneuver, the driver's attention is focused on making the steering actions, and it is not obvious to the driver exactly when that wheel crossing occurs. Also, signaling the relinquishment of control undoubtedly seems secondary. So it is to be expected that drivers who make manual control transfers delay 1.17 sec in actuating the signal. It also follows that some form of automatic signaler should be recommended.

String-joining time was found to increase significantly with increasing AHS design speeds. Because the mean string-joining time for the car was 5.7 sec when traveling at 104.6 kph, that time interval is equal to about 165 m (540.5 ft) in distance at that design speed. At the next AHS design speed of 128.7 kph (80 mph) the mean string-joining time interval was 14.6 sec, which is equivalent to 422.8 m. With the speed increase of 23%, string-joining times and distances increase 2.6 times. When at the highest AHS design speed, the string-joining time was 36.1 sec, which is 6.4 times that at the lowest speed, although the speed of 152.9 kph (95 mph) is only 46% faster. These times occur because of the particular acceleration curves of simulator cars and the actual gaps set in the experiment.

The delay time and distance is of extreme importance because of its implications for the effectiveness of the AHS. If the interstring gap distance used for AHS is smaller than the delay distance plus the length of the entering car and the intrastring gap distance, then a delay equal to the difference will be experienced by the string of vehicles following the entering car. For the conditions in this experiment, minimum gaps without delays should be the delay distances plus 6.24 m to 7.08 m for design speeds from

104.6 to 152.9 kph plus the minimum delay required for the vehicle to enter. Estimates of the delay required, based on the empirical data from this experiment, varied from 11.6 m (38.1 ft) at 104.6 kph to 51.9 m (170.3 ft) at 152.9 kph. These minimum gap estimates are 33, 121, and 311 m (108.3, 397.0, and 1020.4 ft), respectively, for the three AHS design speeds tested. Accordingly, the gaps used in the this experiment that exceeded those approximate minimum gap estimates would not have produced delays to the cars following the one entering. Only the shorter gaps at the two higher AHS design speeds would require enlargement to avoid delays.

Clearly, solutions to our overcrowded expressways require either some new technology, such as the AHS concept, or a radical increase in the expressway infrastructure. Whereas an increase in the infrastructure may be feasible—although costly—in rural settings, there are difficult constraints in many urban setting. This study shows that delay times can be quite substantial, particularly with higher speeds. The results for the string-joining times show that acceleration curves are the major factor. Of course, newer engine designs and transmission designs could substantially change the acceleration characteristics. An alternative to improved acceleration curves would be the capability of accelerating in a special lane before entering the AHS lane. This alternative creates an added burden on the infrastructure design. Whichever way the ultimate decisions are made, the prospects for higher AHS design speeds look dim without changing the acceleration curve impact on the tested AHS configuration.

Traffic Throughput Implications

One of the more compelling aspects of this study is revealed by comparing the maximum traffic capacity that can be achieved with the three design velocities. Using these velocities and the estimated minimum interstring gaps 33, 121, and 311 m (108.3, 397.0, and 1020.4 ft), and assuming that all strings are of four cars with intrastring gaps of .0625 sec, the maximum capacities that can be achieved are shown in Table 18.6. This table also gives the calculations of minimum gaps, the length of a string of four cars plus the interstring gap, the number of four-car strings per kilometer, and the number of cars per hour for the three AHS design speeds. Because of cumulative error, the values in Table 18.6 were computed in higher precision but rounded back in the table. The average minimum gap between successive strings of cars was computed using the AHS design velocity in meters per second less 24.6 m/sec (88.5 kph or 55 mph) times the sum of the mean response time plus the mean lane-change time as two components. Two other components of this gap are the mean delay distance (i.e., the product of the mean delay time and the design velocity), and the mean length of vehicles plus the within-string gap size. The length of a car was assumed here to be 4.4 m (14.5 ft) and the within-string gap size was 1.8 m, 2.2 m, and 2.7

TABLE 18.6

Statistics of String Distances, Lane-Change Distances, and
Acceleration Distances in Determining the Interstring Distance
and the Number of Cars in the AHS Lane at Maximum Densities
(based on 29.0576 m/sec at 65 mph as rounded back to $\frac{1}{10}$ precision)

	AHS Design Velocities		
	104.6 kph 65.0 mph 29.1 m/sec	128.7 kph 80.0 mph 35.8 m/sec	152.9 kph 95.0 mph 42.5 m/sec
[1] Average Response Time Automatic Method	1.4 sec	1.5 sec	1.6 sec
[2] Average Lane-Change Time	1.2 sec	1.1 sec	1.3 sec
[3] Distance S1 Traveled During Car Entry {[1] + [2]}{V − 24.6 m/sec}	11.6 m	29.1 m	51.9 m
[4] Mean Delay Time (Distance)	0.5 sec (14.5 m)	2.4 sec (85.8 m)	5.9 sec (250.6 m)
[5] Car Length + Intrastring Gap = 4.4 m + 0.0625 V	6.2 m	6.6 m	7.1 m
[6] Four Car String Length 3 × [5] + 4.4 m	23.0 m	24.3 m	25.7 m
[7] Minimum Interstring Gap = [3] + [4] + [5]	32.3 m	121.5 m	309.5 m
[8] Minimum String-to-string Distance with 4 cars each = [6] + [7]	55.4 m	144.3 m	335.0 m
[9] Maximum Number of Strings per km 1000 ÷ [8] =	18.1 strings	6.9 strings	3.0 strings
[10] Maximum Hourly Capacity {[9] × 4 cars/string × V*} =	7,551 cars/hr	3,531 cars/hr	1,825 cars/hr

*V in kph.

m, respectively, for the three AHS design velocities. The assumptions behind this calculation of the minimum required gap between strings are that the entering car comes in immediately behind a string of vehicles referred to here as S1, that it accelerates immediately after the lane change, and that the string of cars following the entering car, referred to as S2, joins the entering car just as that car achieves the AHS design velocity. Figure 18.1 describes the movement of the entering car over time in order to portray those events. Accordingly, the estimated maximum hourly capacities of the AHS lane are theoretical upper limits rather than reasonable estimates of actual maximum capacities. Note that those are 7,551 cars per hour at 104.6 kph, 3,531 cars at 128.7 kph, and 1,825 cars at 152.9 kph. This result clearly shows that higher AHS design speeds produce lower traffic capacities. It should be understood that it is not the use of the higher design velocities itself that causes this effect. It is the higher differential speeds between it and the speed of the entering car that necessitates the long acceleration times that causes the larger gaps between strings with a greater design velocity and reduces capacity. Also, these estimates are overestimates because they fail to account for population variations in either drivers or vehicles. Because the

AHS must serve all qualified drivers and vehicles, actual highway capacities will be much less. However, the maximum capacity of a conventional lane would be 1,712 vehicles/hr based on typical gaps between manually controlled vehicles and otherwise similar assumptions to those for the AHS estimates. Note that the theoretical maximum with AHS at 104.6 kph has a capacity that is about 4.4 times that of conventional highways with manually controlled vehicles. Moreover, higher highway speeds for conventional highways offer little advantage. At 128.7 and 152.9 kph (80 and 95 mph), maximum theoretical capacities on conventional expressways are respectively only 1,736 and 1,752 vehicles/hr. Accordingly, a version of AHS with low velocity differentials between the automated and unautomated lanes appears to provide considerably improved capacities over conventional highways.

Control Transfer Mode, Safety, and Public Acceptability Indications

One should note that the portions of the vehicle-entry maneuver that depends on the driver are the successive time periods of response and lane-change maneuver as described by Fig. 18.2. The exposure time, immediately following, is also under the driver's control but only in the case of the manual method of transferring control. An average driver using the manual method of transfer requires about 4.3 sec between the "Enter" command and the control transfer. Drivers using the automatic method required 2.6 sec. Either way, the time periods of human control in taking a vehicle from manual control to AHS control is very short.

Although the fact that no collision or lane incursion occurred during the current experiment is a positive indication, accident rates relative to driving times or distances are such that it would be unexpected to observe an accident in less than 24 highway driving hours. However, there was additional support for the safety of these entry maneuvers in the responses to three questions (i.e., Questions 9, 10, and 11 in Table 18.5) were asked on the questionnaire relative to the degree of safety perceived by the subjects during the lane change into the AHS lane. These responses indicated that the drivers felt safe in the AHS lane, had control of the vehicles as they changed lanes, and felt they were in control of the situation.

Numerous people have expressed skepticism to the authors about public acceptability of an automated highway system, but the drivers in this experiment expressed several positive statements in the questionnaire. These responses show that the average driver preferred the automated lane over the manual lane from a moderately strong to a very strong degree and they felt that the degree of challenge and stress of driving in the manual lane was also stronger than they experienced when traveling in the automated lane. Questions 14 and following, in Table 18.5, address some concerns about the preferences of the drivers from this experiment. These results show that, at least the 25- to 34-year-old drivers tested liked what they saw about the AHS. However, while traveling in the AHS lane, drivers generally prefer longer distances between successive cars within a string of cars than the $\frac{1}{16}$ sec intrastring distances, and they prefer the faster AHS design speeds over the slower one, although both preferences were at best mild (Questions 12 and 13 on Table 18.5, respectively, show average scores of 36% and 69%, which are closer to the middle than the extremes).

CONCLUSIONS

A major conclusion from the ergonomic viewpoint is that the maneuver for entering the automated lane with this AHS configuration used here appears to be easy, effective, and safe. This conclusion needs to be qualified to say that it is difficult to adequately assess automotive safety from these limited simulation results. The conclusions in Buck, Yenamandra, and Bloomfield (1994) and Yenamandra (1994) are essentially the same.

Considering system effectiveness, the use of a low velocity differential for vehicle entry to the AHS and some automatic form of control transfer are recommended. The reason for recommending the low velocity differential is that higher differentials necessitate much larger gaps between successive strings of cars and those larger gaps greatly reduce traffic capacity. Added delays due to the manual method of transferring vehicular control reduce traffic capacity even more.

ACKNOWLEDGMENTS

The authors are greatful to the U.S. Department of Transportation for its sponsorship of this research effort. However, the published materials represent the position of the authors and not necessarily that of the Department of Transportation. Also, the authors recognize the extensive contributions of John R. Bloomfield and J. Marty Christensen, both of the Center for Computer Aided Design of the University of Iowa, Engineering Research Facility, Iowa City, Iowa.

REFERENCES

Alicandri, E., & Moyer, M. J. (1992). Human factors and the automated highway system. *Proceedings of the 56th Annual Meeting of the Human Factors Society*, pp. 1064–1067.

Buck, J., Stoner, J., Bloomfield, J., & Plocher, T. (1993). Driving research and the Iowa driving simulator, Contemporary Ergonomics 1993. In E. J. Lovesey (Ed.), *Proceedings of the Ergonomics Society's 1993 Annual Conference* (pp. 392–396). London: Taylor & Francis.

Buck, J. R., Yenamanda, A., & Bloomfield, J. R. (1994). Ergonomics of vehicle entry for automated highway systems. *Proceedings of the 12th Triennial Congress of the International Ergonomics Association* (Vol. 4, pp. 209–211). Human Factors Association of Canada.

Fenton, R. E. (1994). IVHS/AHS: Driving into the future. *IEEE Control Systems, 14*(6), 13–20.

Hedrick, J. K., Tomizuka, M., & Varaiya, P. (1994). Control issues in automated highway systems. *IEEE Control Systems, 14*(6), 21–32.

Hilton, J. (1933). *Lost horizons.* New York: W. Morrow.

Kuhl, J. G., Evans, D. F., Papelis, V. E., Romano, R. A., & Watson, G. S. (1995). The Iowa driving simulator: An immersive environment for driving-related research and development. *Computer* (IEEE), *28*(7), 35–41.

Saxton, L. (1980). Automated highway system—Considerations for success (Paper D4.5). *Vehicular Technology Society* (IEEE).

Stadden, K. (1991). Simulators: Tools or Toys? *Heavy Duty Trucking,* June, 78–81.

Shladover, S., Desoer, C., Hedrick, J., Tomizuka, M., Wlrand, J., Zhang, W., McMahon, D., Peng, H., Sheikholeslam, S., & McKeown, N. (1991). Automated vehicle control developments in the PATH program. *IEEE Transactions of Vehicular Technology, 40*(Feb.), 114–130.

Varaiya, P. (1993). Smart cars on smart roads: Problems of control. *IEEE Transactions on Automatic Control, 38*(2), 195–207.

Yenamandra, A. (1994). *Human factors issues of entry of vehicles into the automated highway system.* Unpublished master's thesis, Department of Industrial Engineering, College of Engineering, University of Iowa, Iowa City, IA.

Chapter 19

Driver Fatigue:
An Experimental Investigation

John Richardson
Stephen H. Fairclough
Simon Fletcher
HUSAT, Loughborough University, Leics., UK

John Scholfield
Ford Research and Engineering Centre, Essex, UK

The work described here was part of the European PROMETHEUS program launched in 1986 to encourage precompetitive collaborative research within the automotive industry. The objectives of the program were to develop and apply new technologies in order to increase traffic safety and road system capacity and to reduce vehicle emissions. Its aims were therefore consistent with other IVHS programs.

In the program's "safe driving" stream, research effort has been invested in areas such as vision enhancement, collision avoidance, and proper vehicle operation. The Ford Motor Company (UK) and the HUSAT Research Institute have undertaken research on driver alertness monitoring within this last area.

Although driver fatigue is now being accepted as a major factor in causing road accidents, there is still considerable difficulty in quantifying the problem (O'Hanlon, 1978). Harris (1977), for example, found a clear relation between time spent driving and the likelihood of accident involvement. But other investigators, such as Hamelin (1987), have found stronger effects for time of day and sleep pattern disruption. This latter effect is consistent with typical findings regarding feelings of drowsiness and unintentional sleep episodes during the day. These show pronounced peaks between 2 a.m. and 7 a.m. and, to a lesser extent, 2 p.m. and 5 p.m. (Carsakadon, Littel, & Dement, 1985). Experimental investigations of driver fatigue have proven surprisingly contradictory, despite a large body of literature. There is certainly no simple causal relation between time spent driving, fatigue, and driving impairment.

The current project was conceived to provide a diagnostic tool that could be used to investigate driver fatigue in field research and, if successful, form the basis for a reliable in-vehicle warning system. The potential benefits provided by such a system are increased if current developments in IVHS technologies gain widespread implementation.

Many of the intelligent in-vehicle systems currently under development could result in decreased arousal and increased vigilance requirements. The decreased arousal may result from the provision of automated functions (e.g., intelligent cruise control and lane support), which were previously under the driver's active control. The greater vigilance demands may result from increased information presentation (e.g., navigational support and medium range pre-information) and the need to monitor automated functions. The driver's ability to respond appropriately to any given eventuality will depend on their maintaining a sufficient level of arousal. However, the problem of maintaining high levels of arousal while an individual is undertaking a prolonged and undemanding activity is well established.

In order to ensure that a driver is capable of responding when required—to either a normally occurring, but infrequent, driving demand, a critical incident or a new automated support system—it is essential that the driver's own arousal status is monitored and they are warned whenever this falls below the necessary level. It is also conceivable that output from a driver status monitor could be used directly to adapt the parameters employed in certain intelligent subsystems. For example, the minimum headway for an intelligent cruise control system might be increased, or the auditory feedback from a lane support system raised in volume if the driver was judged to be under aroused.

The project's initial goal, however, was simply to investigate whether a reliable alertness monitor was technically achievable.

BASIC STRATEGY

In the phased development program, the initial stage required the direct measurement of driver fatigue, along with measures of driver performance, during extended driving sessions. In the subsequent *training* stage, a neural net would be trained offline to identify patterns in the driving performance data sets that were consistent with the behavior of tired drivers, as indicated by the direct measures of fatigue. In the final evaluation stage, the net would be given driver performance data in real time during an extended drive and be expected to predict driver state. The current work is largely concerned with the first stage.

The original conception of the system involved continuous psychophysiological driver monitoring based on the measurement of eye blink *events*.

A rearward facing video camera mounted on the test vehicle's dashboard would capture an image comprising the driver's head and immediate surrounds. Image-processing software operating in real time would then identify those portions of the captured image corresponding to the driver's eyes and track them. Changes in the image's contrast level in the eye region could then be used to determine whether the eyes were open or closed on a frame-by-frame basis and thus derive an estimate of eye blink frequency. This approach enjoyed the advantage of being continuous and completely unobtrusive to the driver.

In conjunction with the psychophysiological driver measurement, an instrumented vehicle would be developed with autonomous data recording capabilities. Quantitative data relating to the driver's performance is relatively easy to gather from contemporary motor vehicles. Sensor systems around the vehicle already monitor inputs such as throttle position, brake activation, and road speed to control antilock brakes, adaptive damping, engine management, and other systems. A Ford Scorpio was therefore equipped with additional sensors and recording equipment to monitor the use of all vehicle controls and the vehicle's dynamic performance. The variable of primary interest was judged to be steering correction based on a well-established correlation between driver steering performance and driver status (see, e.g., Elling & Sherman, 1994; De Waard & Brookhuis, 1991).

The final components of the system were to be computational algorithms, which could receive data from the vehicle sensors and detect patterns consistent with those produced by a driver suffering from levels of fatigue sufficient to cause or lead to impairment.

The development strategy required the collection of data from drivers completing extended journeys to validate the image processing and "train" the neural net-based algorithms. However, a major review of current and likely progress resulted in a number of significant changes to these plans at both the practical (image-processing capability) and strategic level (choice of fatigue indicator).

The initial video recording system was capable of capturing usable images under relatively limited operating conditions. The wide variance in ambient lighting exceeded the camera system's ability to automatically accommodate them. Strong sunlight also encouraged drivers to wear dark glasses, and given the extended nature of the planned sessions, it was felt unreasonable to prevent their use. Whereas the system could cope with moderately tinted glasses, the frequently overexposed video rendered the images unsuitable for subsequent analysis—although efforts were made to salvage the situation by way of manual analysis. The camera position also led to the capture of images with rapidly changing backgrounds (e.g., buildings, trees, passing vehicles, etc.), as well as changes in illumination, and this exacerbated the problem of locating the face and then subsequently tracking the eyes. It proved necessary for the software to interrupt the image analysis in order to confirm that the system was indeed tracking the eyes and not some other component of the image. In addition, the number of processing cycles required placed upper limits on the video sampling rate that could be achieved.

The review of the fatigue detection criteria established that simple measures of eye blink frequency alone would be insufficient to determine fatigue satisfactorily because many factors, other than fatigue, are known to influence blink rates (Stern, Boyer, & Schroeder, 1994). A more robust estimation would require the incorporation of eye blink duration data but the available video frame rate (25 Hz) would be inadequate. As a consequence, a decision was made to adopt alternative psychophysiological measures of fatigue until the image-processing system's performance could be sufficiently improved. (Significant progress was subsequently made. This was largely a result of increasing the computer's processing speed and power and with a switch to color image processing, which allowed skin tone data to be utilized. In what has been described as a "virtuous circle" effect, a faster system produces greater accuracy and makes the task easier because fewer frames are skipped between processed frames and there is less chance of the tracked targets being lost. The system is currently capable of giving a continuous estimate of the driver's eyelid separation—a substantially more reliable indicator of fatigue than eye blink frequency.)

The revised strategy involved using more direct laboratory-oriented methods to measure fatigue: cortical activity, heart rate, muscle tone, and eye movements. It was accepted that these direct contact measures would be appropriate in the system's development even if they could not be part of a future system. The justification for such an approach is well established; see, for example, Torsvall and Akerstedt's (1987) work on train drivers and existing research program on drowsy driver detection (Artaud et al., 1994; Knipling & Wierwille, 1994).

An initial program of field trials was devised to provide the vehicle control and driver data necessary to develop the neural net. For future systems to be capable of recognizing instances of fatigued driving from a continuum of driver states it would be necessary to collect data from drivers who themselves were attempting to drive with markedly different degrees of arousal (i.e., from fully alert to very tired). This requirement for training data was a major determinant of the project's initial field work program.

In designing the trials, two apparently opposing requirements had to be reconciled. First, there is the need to collect data from highly fatigued drivers under realistic driving conditions. Second, this must be accomplished without jeopardizing the safety of the test drivers and research staff involved, which is a familiar dilemma in driver fatigue research. As a consequence, a series of three trials were designed entailing the manipulation of fatigue under controlled conditions but not on the public road. This compromise allowed the use of highly fatigued drivers but did not place individuals at risk.

The first trial used a basic laboratory driving simulator to collect baseline psychophysiological data from drivers experiencing levels of fatigue, which could not be safely induced on public roads. The second trial used a real driving task, which was completed at low speeds on a private airfield and again employed sleep deprived drivers. The final trial used a 480 km motorway journey completed at normal speed but without any sleep deprivation.

THE FIELD TRIAL PROGRAM

Laboratory Trials

In addition to investigating the onset of drowsiness and sleep during a simulated driving task, the first trial established a common trial procedure that was adopted with relatively minor variations in the subsequent trials.

A total of 20 subjects were recruited from the local university community, with the majority being staff and postgraduate students. The limited capability of the simulator required only limited driving skills (i.e., basic steering control), so only minimal previous driving experience was required. However, health history and current medication, if any, were checked for all subjects because this was likely to have a bearing on the interpretation of the psychophysiological measures.

The simulator comprised a video display cabin in which the stationary test vehicle could be positioned and a forward view of a road scene projected on to 5 m × 3 m screen in front of the driver. The video footage was recorded during a 4-hr daylight drive with a "lead" vehicle constantly positioned some 70 m in front of the "test" vehicle. The driver's task in the simulator was to "track" the lead vehicle by way of a small dot of red light projected onto the video image from a laser light source mounted on one of the simulator vehicle's front wheels. This arrangement allowed the driver to adjust the horizontal position of the projected dot using the steering wheel. The task was therefore designed to mimic the relatively low steering demands imposed on a driver completing a motorway journey under light traffic conditions.

The subjects were invited to attend the test site in pairs in the early evening and were then given a full briefing on the procedure, task, and measurements to be made. The subjects were then prepared and fitted with sensor electrodes by an electrophysiology technician recruited from a local hospital. The montage supported four electroencephalogram (EEG) channels for the recording of electrocortical activity (brain waves), two periorbitally placed electrooculogram (EOG) channels to record eye movements, and single channels for electrocardiogram (ECG) and Electromyogram (EMG) to record heart rate and muscle activity, respectively. The data was recorded on to an 8-channel Medilog 9000 ambulatory recorder with a data capture life of some 24 hr.

The subjects were then asked to complete a 30-min familiarization session in the test vehicle. This preliminary session also provided baseline electrophysiological data from the subjects when in an unfatigued state. The subjects were then required to complete a sleep latency test. The procedure was a simplified form of the multiple sleep latency test (see Carsakadon et al., 1986) because the subjects were asked to attempt to fall asleep only once during the 30-min period allowed. This test was repeated during the course of the trial and allowed a further longitudinal assessment of the subjects' underlying state of fatigue (subjects typically fall asleep more quickly as they get more tired). The subjects were then allowed to return home after being instructed to stay awake all night and refrain from consuming alcohol or caffeine.

The subjects returned to the test facility the following day at 9.00 a.m. or 1.00 p.m. for their full test session. The subjects were instructed to attempt to track the lead vehicle for as long as they could (up to a maximum of 4 hr) but to stop if they felt too distressed to continue (e.g., excessive nausea, etc.). They were not allowed to listen to the radio or chew gum, but they were allowed to adjust the ventilation.

The subjects were videotaped and their task performance and behavior was monitored. If the subjects appeared to have gone to sleep (eye closure for 30 sec and absence of task performance) they were roused. This was repeated a second time. After this, the subjects were left undisturbed until they appeared to have achieved 10 min of uninterrupted sleep. The session was terminated if this occurred. All the subjects then completed a further sleep latency test before the electrophysiological recording sensors were removed and they were escorted home.

Only one of the subject's data proved to be impossible to analyze, and this was due to critical sensor electrodes becoming detached during the trial session. The data from the remaining subjects' Medilog recorders were viewed and analyzed polygraphically by an experienced analyst.

The analyst completed an initial, and relatively high level, review to establish a profile for each subject. This confirmed that the data set was complete, the subject's EEG response was "conventional" (some 5% of the population show an atypical response), and the subject had lost the previous night's sleep. With one or two exceptions, they all had.

The second and more detailed analysis of the test session data involved the examination of successive 16-sec epochs from the polygraph with a single global score for level of arousal being given to each epoch (see Alford et al., 1992). Although the six-level scoring system takes account of all the electrophysiological data types, the primary defining characteristics for the more aroused levels (Levels 1–3) are the eye movements and muscular activity, and EEG data are a more significant determinant of levels where drowsiness is apparent (Levels 4–6). Table 19.1 provides a summary of the drowsiness scale and defining characteristics.

The drowsiness scale can be said to compliment Rechtschaffen and Kales' (1968) standard scale for sleep stages. The Rechtschaffen and Kales' scale identifies 5 stages of sleep and the waking state, whereas the drowsiness scale defines 6 waking stages prior to the onset of sleep (thus, drowsiness Level 7 = sleep Stage 1). Both systems are based on 3 electrophysiological measures: EEG, EOG, and EMG.

Comprehensive electrophysiological data was successfully recorded from 19 of the 20 participating subjects. The results indicated that subjects exhibited significant levels of fatigue behavior within a short time in the simulator. One subject asked to withdraw after 117 min because they found the temperature inside the vehicle uncomfortably hot but the remaining 18 subjects completed a full session in the simulator (8 managed to stay awake for the duration of their participation). Table 19.2 shows the elapsed time from the beginning of the session to when subjects first exhibited significant levels of drowsiness and the total time spent in the simulator. Three key stages are included: the onset of drowsiness (Level 4), the appearance of significant impairment (Level 6), and the onset of sleep (Level 7).

Although there was considerable intersubject variation, all 19 subjects showed a clear increase in recorded drowsiness during their test sessions. The effect was so strong that 17 of the subjects reached Level 7 (Stage 1 sleep) after an average of just 53.6 min. The onset of the effect was equally

TABLE 19.1
The Drowsiness Scale

Index	Category	Characteristics
1	Active	EEG active/alert, > 2 eye movements per epoch, definite body movements
2	Quiet waking (plus)	EEG active/alert, > 2 eye movements per epoch, minimal body movements
3	Quiet waking	EEG alert, < 2 eye movements per epoch, absence of body movements
4	Waking with intermittent alpha	Definite bursts of alpha but less than half the epoch
5	Waking with continuous alpha	Continuous alpha; i.e., > half the epoch
6	Intermittent theta	Definite bursts of theta (+ alpha if present)
(7)	Sleep stage 1	Continuous theta; i.e., > half the epoch

TABLE 19.2
Simulator Trial: Elapsed Time (min),
Level of Drowsiness, and Total Session Length

Subject No.	4	6	7	End
1	13.60	30.13	55.20	61.33
2	0.53	5.33	11.20	64.27
3	2.40	8.53	33.07	77.33
4	3.20	28.53	113.33	124.27
5	17.87	38.40	40.53	51.73
6	20.00	21.60	22.13	182.13
7	2.40	12.80	0.00	180.53
8	7.47	10.13	28.00	181.33
9	20.00	20.00	39.47	95.20
10	12.00	13.87	24.00	111.73
11	1.33	13.07	65.60	85.07
12	38.13	43.20	74.93	116.80
13	11.20	33.87	58.13	181.07
14	16.80	17.07	0.00	179.20
15	10.13	13.87	47.20	58.93
16	2.13	6.40	73.07	181.60
17	12.00	18.40	83.20	181.07
18				
19	12.53	80.27	90.40	181.33
20	7.47	7.47	54.67	111.20
M	11.12	22.26	48.11	126.64
SD	9.15	17.88	30.85	51.50

rapid, with the mean time taken to exhibit Level 4 effects being just 11.12 min and Level 6 being 22.3 min.

The second most significant characteristic revealed in the analyzed data was the episodic nature of the drowsiness response. None of the subjects showed a simple linear response with an increase in fatigue directly linked to length of time on task. The graphed output of each subject session clearly show that, although an underlying trend toward increased drowsiness may have been present, subjects typically experienced a succession of brief periods of reduced arousal (often only one or two epochs long) followed by a return to a more alert status. The video recordings of the subjects' faces showed this process very clearly. After an initial period of alertness, they experienced an extended period when they alternated between states of pronounced and more moderate drowsiness. The extreme episodes were characterized by head nodding, periods of eye closure lasting several seconds, and intermittent attempts to pursue the tracking task.

Figure 19.1 shows a typical plot from a subject in the simulator trials. Characteristic features are an initial period when the subject settles to Level 2 (quiet waking plus) but with occasional and temporary transitions

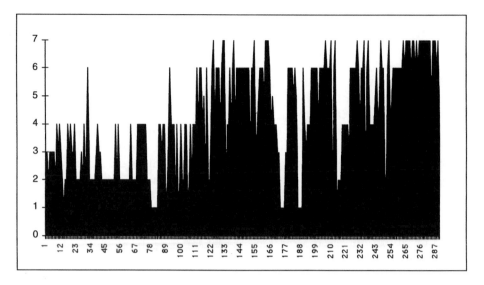

FIG. 19.1. Simulator trial, Subject 3, 8:48 a.m.–10:05 a.m. Changes in drowsiness with time (16-sec epochs).

to a more drowsy Level 4 (waking with intermittent alpha). This pattern is maintained for about the first 100 epochs and then repeated sleep episodes occur (Levels 6 and 7) and relatively little time is spent in the active states. Over the course of the subject's participation, there is a clear increase in the duration and magnitude of the episodes of drowsiness. The subject's session was terminated when they were judged to have been asleep at the wheel for 10 min. From the video record and the subject's subsequent comments, it appeared that continuous sleep onset would have been even sooner were it not for the uncomfortable upright sitting position.

A major limitation of the first trial was the lack of an incentive for the subjects to maintain their arousal, despite being apparently well motivated. The conditions were very conducive to sleep (sleep deprivation, warmth, dark, and little task demand) and with no element of risk the subjects did not appear to resist sleep onset. This reduced the usefulness of the steering control data because, for extended periods, the drivers appeared to maintain a very reduced state of arousal with little effort to maintain task performance but without actually progressing to stage one sleep. Many subsequently claimed that they frequently felt they were falling asleep but reawakened when their heads "dropped" suddenly. The use of a more interactive simulator would have enabled performance-related reward schedules to have been used (see, e.g., Cook, Allen, & Stein, 1981). The second trial was able to overcome this problem through the introduction of a greater element of task realism.

Closed Circuit Trials

The second trial was conducted on a partially used airfield and involved both day and night sessions. Two similar L-shaped courses of 3.5 km and 2.4 km were marked out with traffic cones. Although they both incorporated U turns, the runway width (some 50 m) was sufficient to allow these turns to be made with minimal loss of speed. The drivers were instructed to maintain a speed of 50 kph and their safety was further secured through the presence of a co-driver, who was instructed to rouse the driver or take control of the steering wheel, if necessary. If a vehicle did run off the circuit there were no obstacles that could have caused a hazard. The co-driver also made notes regarding the drivers' performance, self-reports of fatigue, and any incidents that might assist the interpretation of the electrophysiological data.

A total of 22 subjects took part in the trial and the procedure was similar to that used in the first trial. The subjects were invited to attend the test facility in the early evening and were fitted with the recording apparatus after a full briefing. The subjects completed a 10-min familiarization session on the track and then completed a sleep latency test. The subjects remained on site overnight in a private hotel and stayed awake accompanied by a supervisor. A nighttime session approximately 2 hr long was completed by each subject starting at either 1:00 a.m. or 3:00 a.m. Upon completion, the drivers returned to the hotel, took a second sleep latency test, and then stayed up for the rest of the night. The following day, they completed the major test session—a 4-hr drive starting at either 9:00 a.m. or 1:00 p.m. After a final sleep latency test, the drivers' sensors were removed and they were driven home.

The procedure proved highly effective in generating instances of severely fatigued driving, which was evident from both subjective and objective sources of data. Examination of the daytime video recordings showed instances of subjects driving with their eyes closed for several seconds at a time; on a number of occasions, the co-driver was obliged to intervene to prevent a possible loss of control incident. In some cases, the intervention involved the co-driver simply speaking the subject's name loudly or placing a hand on the steering wheel to make a corrective action. However, occasionally a more pronounced intervention was required (i.e., an abrupt steering correction or knocking the transmission into neutral).

Further evidence was drawn from comments noted on the co-drivers' record sheets. These included observations made by the co-drivers regarding the drivers' state of apparent arousal and driving performance, as well as verbal comments volunteered from the subjects and their responses to regular questioning from the co-driver regarding their fitness to drive. Table 19.3 contains excerpts from a co-driver's comment sheet.

TABLE 19.3
Co-Driver Comment Sheet From Closed Circuit Trial (Excerpt)

Time	Co-Driver's Comments
08.35	Start
08.40	Q—? "*Pretty Good*"
08.55	Steering showing signs of drifting already, also told him to change into 4th gear after doing 2 laps in 3rd gear.
08.59	Subjects eyes closing, speed increasing.
	Q—? "*A bit tired but getting better.*" He then turned heating off.
09.05	Subject asleep 10 seconds, almost missed corner.
09.08	Asleep again, speed increases.
09.11	Driver asleep, I had to wake him up, said he "*felt really tired now.*"
09.20	Eyes closing again, sound of (*event recorder*) button woke him up, subject seems to be really struggling to keep awake.
<<	<< << << << << << << << <<
11.03	Noticed speed starting to get erratic, braking very harshly at ends of runway, no gear changing anymore.
11.07	Speed very erratic, driver asleep.
	Q—? "*My reactions are getting very slow but I wish to continue.*" I am not too sure though (*co-driver comment*).
11.16	Car wandering again.
11.22	Driver falling asleep again—I think we have just about reached our limit (*co-driver comment*).
11.24	Spoke to driver. He didn't hear me initially.
11.26	Q—? "*Not too bad, but very tired.*"
11.30	Driver drifting again, eyes becoming very heavy.
11.32	Fighting to keep eyes open, seems very jumpy, speed fluctuating.
11.34	Driver asleep and speeding.
11.35	Driver asleep almost full length of runway.
11.37	Driver asleep until reaching corners, hands sliding on wheel.
11.40	Driver almost asleep, passed end of runway. I feel unhappy about continuing much longer. I've decided to give it another 10 minutes.
11.44	Driver asleep full length of runway.
11.47	Driver blinking a lot.
11.48	Driver asleep passed end of runway, shouted at him. . . .
11.49	End.

Although driving performance varied significantly, a striking finding was the ability of some subjects to continue driving while in an extremely fatigued state. This often included driving the length of the runway with eyes closed and only regaining apparent control at the co-driver's intervention prior to the turn. In these cases, although there was a tendency for the vehicle to slowly drift off course, vehicle speed was often maintained and, in some cases (to the alarm of the co-drivers), steadily increased. As with the simulator trials, there was a marked tendency for subjects to experience episodes of almost overwhelming drowsiness interspersed with periods of greater arousal (see Fig. 19.2). The plot differs from that shown

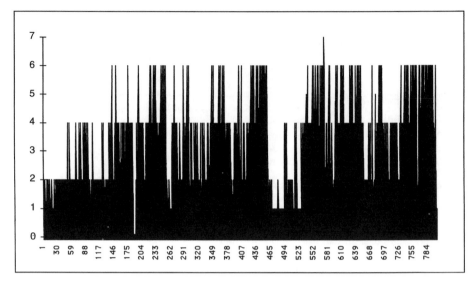

FIG. 19.2. Closed-circuit trial, Subject 21, 8:56 a.m.–12:29 p.m. Changes in drowsiness with time (16-sec epochs).

in Fig. 19.1 in a number of ways. The session duration is considerably longer and the drowsiness episodes, while just as severe, are briefer. After a brief period spent at Level 2, the subject spends the majority of their session at Level 4 with brief descents to Level 6, which increase in frequency and duration as the session proceeds. The single epoch rated at Level 7 coincided with a loss-of-control incident. The differences in driver response can be explained in terms of the greater arousal provided by the task.

Road Trials

The experience of the second trial clearly ruled out any possible manipulation of fatigue in the road trials, but apart from allowing the subjects to sleep prior to the daytime sessions, the procedure was again very similar to that employed in the second trial.

The subjects were drawn from a database of volunteer drivers maintained by HUSAT. Only those with 10 years driving experience, an annual mileage of at least 7,000 miles (12,000 km), and a high proportion of this distance completed on motorways were recruited.

A 120-km section of the M1 motorway between Loughborough and Luton was used in the study. Drivers had to complete one return trip (240 km) in the late evening and two successive trips (480 km) the following day. The drivers arrived at the test site in the early evening and were fitted with the electrophysiological recording equipment before a brief familiarization drive in the car and a first sleep latency test. The drivers then completed their evening session starting at either 9:00 p.m. or 11:00 p.m. followed by a second sleep latency test. They then went home for a night's sleep before returning to the test site to complete a 4-hr drive the following day.

The daytime drive was interrupted by a vehicle refueling break after approximately 2 hr and the drivers were allowed a light refreshment at this point. Although no co-driver was employed, the drivers were accompanied by a member of the research team; this person was primarily present to operate the data logging equipment and assist if a technical problem arose with the vehicle. However, the researcher kept a discrete watch on the drivers' apparent alertness and performance.

Most of the drivers reported an increase in the level of subjective fatigue over the course of the trial, although very few admitted to feeling drowsy. There was only one incident, which appeared to be the result of a loss of control resulting from underarousal (a near lane departure during a nighttime session). One subject's fatigue response is shown in Fig. 19.3. Again, there are significant differences between the road trial plots and those generated in the previous trials. The subject spends the majority of the session at Level 2 but with frequent, brief episodes of Level 4 behavior. The midpoint break is clearly evident. Although the underlying level of drowsiness is much less than that shown in the two previous trials, there are two obvious characteristics with a potentially serious safety implication.

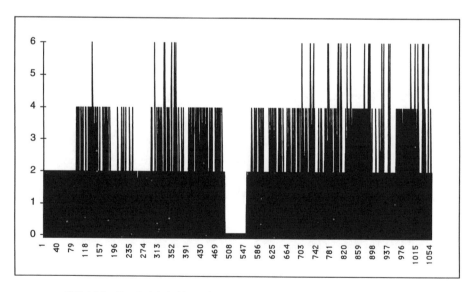

FIG. 19.3. Road trial, Subject 12, 1:14 p.m.–5:55 p.m. Changes in drowsiness with time (16-sec epochs).

For a considerable portion of the journey, the subjects remained in a relaxed state (Level 2) but frequently switched in and out of a more drowsy state (Level 4) with occasional episodes of pronounced drowsiness (Level 6). Most of the subjects drove both sessions at, or above, the speed limit (110 kph) and their ability to respond to a critical incident when experiencing the highest levels of drowsiness must presumably have been reduced.

A second common finding was the limited restorative value of the midpoint break. Whereas drivers are typically encouraged to take regular breaks when undertaking lengthy journeys, the data from this trial suggests that the relief from fatigue may be very short lived. For some drivers, evidence of drowsiness was only found after the break.

CONCLUSIONS

At the time of this writing, the offline processing of the fatigue and vehicle data to enable net generation had yet to be completed. However, once this has taken place, a further session of field trials is required to assess the accuracy of the net's predictive output. This trial may also be able to assess the effectiveness of the revised eye blink event monitoring system. Recent results published by Wierwille and his colleagues (1994) give strong encouragement. Wierwille reported successful detection of driver fatigue in a simulator using driver eye closure, independent observers' ratings of driver facial images, and additive models incorporating eye closure and a variety of EEG-based measures.

The initial validation of the current detection system is likely to be compared once again with a direct psychophysiological measure of fatigue. However, once the net is capable of achieving a reliable output, a second, equally important type of validation will be required. This is a correlative investigation of the net's output with the drivers' self-assessments of their own arousal. If drivers are unaware of subjective feelings of fatigue, then this is likely to have a significant impact on their readiness to respond to a warning device. It is assumed that the optimal timing for a warning output will largely depend on a comparative assessment of the drivers' objectively measured, and subjectively experienced, state of fatigue.

ACKNOWLEDGMENTS

The authors gratefully acknowledge the contributions of Chris Alford, University of the West of England, who provided expert assistance with electrophysiological recording.

REFERENCES

Alford, C., Rombaut, N., Jones, J., Foley, S., Idzikowski, I., & Hindmarch, I. (1992). Acute effects of hydroxyzine on nocturnal sleep and sleep tendency the following day: A C-EEG study. *Human Psychopharmacology, 7*, 25–35.

Artaud, P., Planque, S., Lavergne, C., Cara, H., Lepine, P., Tarriere, C., & Gueguen, B. (1994, May). *An on-board system for detecting lapses of alertness in car driving.* Paper presented at the 14th International Technical Conference on the Enhanced Safety of Vehicles (ESV), Munich, Germany.

Carsakadon, M. A., Dement, B., Mitler, M. M., Roth, T., Westbrook, P., & Keenan, S. (1986). Guidelines for the multiple sleep latency test (MSLT): A standard measure of sleepiness. *Sleep, 9*(4), 519–524.

Carsakadon, M. A., Littel, W. P., & Dement, W. C. (1985). Constant routine: Alertness, oral body temperature and performance. *Sleep Research, 14*, 293.

Cook, M. L., Allen, R. W., & Stein, A. C. (1981). Using rewards and penalties to obtain desired subject performance. *Proceedings of the Seventh Annual NASA–University Conference on Manual Control.* Pasadena, CA: Jet Propulsion Laboratory.

De Waard, D., & Brookhuis, K. (1991). Assessing driver status: A demonstration experiment on the road. *Accident Analysis and Prevention, 23*, 297–394.

Elling, M., & Sherman, P. (1994). Evaluation of steering wheel measures for drowsy drivers. *Proceedings of the 27th International Symposium on Automotive Technology and Automation (ISATA)* (pp. 207–214).

Hamelin, P. (1987). Lorry drivers' time habits in work and their involvement in traffic accidents. *Ergonomics, 30*, 1323–1333.

Harris, W. (1977). Fatigue, circadian rhythm and truck accidents. In R. R. Mackie (Ed.), *Vigilance* (pp. 133–146). New York: Plenum.

O'Hanlon, J. F. (1978). What is the extent of the driving fatigue problem? *Driving fatigue in road traffic accidents* (EUR 6065 EN). Brussels: Commission of the European Communities.

Knipling, R. A., & Wierwille, W. W. (1994, April). *Vehicle-based drowsy driver detection: Current status and future prospects.* Paper presented at IVHS America, 1994 Annual Meeting, Atlanta, GA.

Rechtschaffen, A., & Kales, A. (1968). *A manual of standardized terminology, techniques and scoring system for sleep stages of human subjects* (Publication No. 204). Washington, DC: U.S. Government Printing Office.

Stern, J. A., Boyer, D., & Schroeder, D. (1994). Blink rate: A possible measure of fatigue. *Human Factors, 36*(2), 285–297.

Torsvall, L., & Akerstedt, T. (1987). Sleepiness on the job: Continuously measured EEG changes in train drivers. *Electroencephalography and Clinical Neurophysiology, 66*, 502–511.

Wierwille, W. W. (1994, May). *Overview of research on driver drowsiness definition and driver drowsiness detection.* Paper presented at the 14th International Technical Conference on the Enhanced Safety of Vehicles (ESV), Munich, Germany.

Chapter 20

Navigational Preference and Driver Acceptance of Advanced Traveler Information Systems

Kathryn Wochinger
Science Applications International Corporation, McLean, VA

Deborah Boehm-Davis
George Mason University, Fairfax, VA

The successful implementation of Advanced Traveler Information Systems (ATIS), a component of Intelligent Transportation Systems (ITS), depends on user acceptance of ATIS products and services. The focus of much ITS and ATIS research to date, however, has been on technical issues rather than on user acceptance. To fill the void in knowledge regarding user acceptance, the U.S. Department of Transportation (USDOT) is currently conducting a research program to provide broad-based, publicly available information on potential consumer reaction to ITS (Zimmerman & Elliot, 1995). Such information could then be applied to the design of ITS products and services, as well as to the development of an ITS implementation strategy.

Information on consumer reaction to previously introduced technological innovations may shed light on potential user acceptance of ITS. For example, several years after automated teller machines (ATMs) were introduced in Seattle, less than half of all bank account holders generated ATM transactions, and customers over 50 years of age rarely used ATMs (Kantowitz et al., 1993). The implication of this finding is that many drivers, particularly older drivers, may resist using technically innovative products (such as ATIS).

User acceptance is particularly important to the successful implementation of ATIS because the accuracy of traffic information it conveys is dependent on the number of ATIS-equipped vehicles. As predicted by Muir's model of user trust in automated systems (Muir, 1987, 1994), the initial trust users have in a system's performance is strong due to users' high expectations of innovative products. However, when a system fails in some way, user trust diminishes and can be difficult to rebuild. During the initial implementation of ATIS, there will be a relatively low number of users and system inaccuracies or failures will likely be higher, as compared to when ATIS is more widely adopted. Receiving inaccurate information from an ATIS device may break the trust a driver has in the system and lead to user rejection. Consumer rejection of ATIS, in turn, may lead to decreased system reliability and accuracy. ATIS is unique in that the degree of consumer use affects system effectiveness. Thus, to optimize ATIS accuracy, initial acceptance of ATIS should be maximized.

Clearly, the success of ATIS would be enhanced if older drivers were to embrace it. Older drivers constitute the most rapidly growing segment of the driving population in terms of the number of drivers licensed, miles driven, and proportion of the driving population (Barr & Eberhard, 1991; Waller, 1991). Further, each succeeding cohort is likely to adopt technology used by the preceding generation. One challenge facing ITS implementation is the development and marketing of products and services that are especially useful and appealing to the older driver.

A recent study of user acceptance of ATIS showed that drivers rated its features and functions favorably after learning about them from a training videotape (Campbell et al., 1995). Also, the results of an ATIS field evaluation found that drivers who actually used ATIS route guidance provided higher willingness-to-pay ratings than drivers who only had access to predrive functions that did not include route guidance (Perez, Fleishman, Golembiewski, & Dennard, 1993). This finding suggests that drivers who have "hands-on" experience with ATIS functions are more likely to adopt ATIS than are drivers who learn about ATIS secondhand.

One factor that may influence user acceptance of ATIS is navigational preference. Individuals show strong variations in their preferences for maps or text directions and they vary in competence in reading maps (Streeter & Vitello, 1986). Drivers who prefer maps to text directions and who have good map-reading skills may feel little need for route guidance; they also may be unwilling to use simplified turn-by-turn displays. Similarly, drivers who are uncomfortable using maps may resist using a route guidance system that primarily displays spatial information.

This study obtained usability data on drivers' initial preferences for maps or text directions from a questionnaire; performance with a simulated ATIS display, maps, and text directions in a part-task driving simulator; and subjective evaluations of those aids (as shown in Fig. 20.1). The data were examined to determine whether initial navigational preference influenced performance and/or subjective evaluations of the specific navigational aids used.

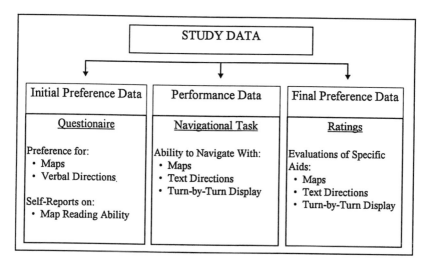

FIG. 20.1. Study data.

METHOD

Design

The study used a 2 (age group) × 4 (navigational aid) mixed experimental design. The type of navigational aid was the within-subjects variable. Participants performed a navigational task using each of the following aids in a random order: text directions; an enlarged, mounted paper map; a standard paper map; and a simulated ATIS turn-by-turn route guidance display. The measures of navigational performance were accuracy and decision time. Participants also performed a secondary tracking task during the navigational task.

Participants

Participants were 28 younger (30–45 years) and 28 older (65–75 years) licensed drivers. Each age group was balanced by gender. Each participant had a minimum of 20/40 corrected vision, and was paid $20.00 for participating in the 2-hr session.

Materials

Navigation Questionnaire. The questionnaire, shown in Fig. 20.2, had 16 questions that measured self-appraisal of navigational abilities and habits, as well as preferences for verbal directions and maps. The questions used a

Circle the number from 1 to 5 that best fits your answer for each question.

How much do you rely on street names when driving someplace new?	Not Much	Somewhat	Very Much
	1　　　2	3　　　4	5
On a trip with others, do you like doing the map-reading and navigating?	Not Much	Somewhat	Very Much
	1　　　2	3　　　4	5
To what degree do you like using maps as a source for directions?	Not Much	Somewhat	Very Much
	1　　　2	3　　　4	5
How easy is it for you to find different ways of getting to the same place when driving without a map?	Difficult	Average	Easy
	1　　　2	3　　　4	5
When driving someplace new, would you rather someone gave you verbal directions OR drew you a map?	Words	Either One	Map
	1　　　2	3　　　4	5
How important to you are landmarks, such as shopping centers and gas stations, when you are driving someplace new?	Not at All	Somewhat	Very Important
	1　　　2	3　　　4	5
How would you rate your ability to follow verbal directions when driving or navigating to unfamiliar places?	Poor	Average	Excellent
	1　　　2	3　　　4	5
When giving directions to someone in person, do you usually give directions in words OR draw a map?	Words	Either One	Map
	1　　　2	3　　　4	5
Do you like to find different ways of getting to a place you drive to frequently?	Not at All	Sometimes	Like A Lot
	1　　　2	3　　　4	5
How would you rate your ability to select a route from a map in a new area?	Poor	Average	Excellent
	1　　　2	3　　　4	5
How easy is it for you to locate your position on a map?	Difficult	Neither	Easy
	1　　　2	3　　　4	5
When stuck in a serious traffic jam, how likely are you to attempt an alternate route to your destination?	Not at All	Somewhat	Very Likely
	1　　　2	3　　　4	5
Do you find it easy or difficult to learn the roads in a new neighborhood?	Difficult	Neither	Easy
	1　　　2	3　　　4	5
If you get lost while driving in an unfamiliar area, how difficult is it for you to find your way back?	Difficult	Neither	Easy
	1　　　2	3　　　4	5
How easy is it for you to give verbal directions off the cuff to someone who is driving to your home from out of town?	Difficult	Neither	Easy
	1　　　2	3　　　4	5
How would you rate your sense of direction?	Poor	Average	Excellent
	1　　　2	3　　　4	5

FIG. 20.2. Navigation questionnaire.

Likert-type 5-point rating scale (Likert, 1932). Some questions were based on previous research (e.g., Streeter & Vitello, 1986).

Experimental Apparatus. The apparatus was a part-task driving simulator located at the Turner–Fairbank Highway Research Center (TFHRC). The simulator included a driving buck (with a steering wheel, dashboard, and adjustable seat) facing a 36-in. Mitsubishi color television monitor. A 9-in. JVC monitor was placed to the right of the steering wheel and was parallel to the dashboard. Two computer-operated NEC PC-VCRs were used; one played a videotape of a real-world route on the Mitsubishi monitor and the other played a simulated ATIS display on the JVC monitor.

A modified keypad with three arrow keys was centered on the steering wheel within easy reach of the participant. To indicate a left turn, the participant hit the key in the 9:00 position; for a right turn, a key in the 3:00 position; and for straight, a key in the 12:00 position.

Routes. Four 5-min routes were videotaped in a suburban area of Northern Virginia from the driver's perspective. The roadways were two-lane streets in residential neighborhoods. The routes were matched by the time of day (about noon), traffic density (typically no more than 1–2 cars were on the roadway), and the number ($N = 11$) and type of turns (right, left, straight-throughs).

Navigational Aids. For each route, the following four types of navigational aids were created:

A *simulated ATIS turn-by-turn display* based on the TravTek system (see Rillings & Lewis, 1991). The display was computer-generated using Animator Pro, an animation drawing program, and transferred to videotape. The lettering was ⅛″ and the intersection display was 3″ × 2½″ (see Fig. 20.3).

A *standard street map.* This consisted of one page (9½″ × 8″) of a common map of Northern Virginia (ADC, 1994). The route was highlighted. The entire route covered a space of about 4 square inches. The size of the lettering for the street names was ¹⁄₁₆″. The map was either hand-held, or placed on the dashboard or lap.

A *"big" paper map.* This was created by enlarging the standard map by 225%. The route was highlighted. The entire route covered an area of about 5″ × 5″. The size of the lettering was ⅛″.

Text directions. These consisted of a linear list of turns, the number of blocks before a turn, and street names. The directions were printed in a 25-point font, with a letter size of approximately ¼″ (see Table 20.1).

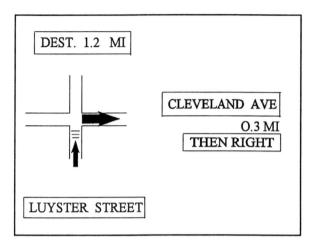

FIG. 20.3. Simulated ATIS turn-by-turn display.

TABLE 20.1
An Example of the Text Directions

You are heading southeast on Holly Avenue.
Take the second left onto Aspen Avenue N.
Take the second right onto Ash Road W.
Take the second left onto Alder Avenue.
Take the fourth left onto Derby Avenue W.
Take the second right onto York Road.
Take the first right onto Laurel Avenue W.
Take the second right onto Vernon Street N.
Take the first left onto Vernon Court W.
Stop at the end of the street.

Both the big paper map and the text directions were fastened to the front of the JVC monitor to ensure that they were viewed from the same visual angle as the ATIS display.

Navigational Task. The navigational task required each participant to follow a route depicted by one of the four navigational aids while the videotape of the route was shown on the front monitor. The videotape was paused at selected intersections during which the participant indicated the proper direction that the car should take by pressing the appropriate arrow key (straight, left, or right). After the participant pressed the arrow key, street names were superimposed on the front monitor because the actual street signs were illegible on the videotape. The participant then made a second response, which could be the same as or different from the first response. The tape resumed play immediately after the participant's second response and followed the proper route, regardless of the participant's response. Performance was measured by error rate and decision time. The error rate was the average number of incorrect responses during a route made before and after the street names were displayed, and decision time was the average time taken for all the responses made during a route.

Tracking Task. While performing the navigational task, the participant performed a tracking task via a computer controlled servo-laser (Mini-Tracer Laser Pointer, model B38-585). The goal of the task was to keep a semi-randomly moving laser within fixed boundaries on the bottom of the front monitor. The laser projects a continuous red laser dot .75 by .5 cm in size, approximately 14 cm from the bottom edge of the monitor. The laser moved randomly from left to right (range was 9.32 degrees left and right of center) at a rate of 3.22 degrees per second. A warning tone was triggered when the laser crossed either boundary, and sounded continuously until it was returned to the lane. The computer paused the laser when the videotape paused at intersections, and it resumed the laser when the videotape resumed play (after the participant completed the second navigational response). The rate the laser moved was based on previous work (Walker, Sedney, & Mast, 1991), in which five active drivers (ages 65–70) were able to maintain the laser within bounds at least 80% of the time. The results from the tracking task have been presented in a previous report (see Wochinger & Boehm-Davis, 1995).

Usability Ratings. Participants rated each aid immediately after using it by answering four questions on a 5-point scale, as shown in the example in Fig. 20.4.

Please circle the number from 1 to 5 that best fits your answer for each question.

	Strongly Liked		Neutral		Strongly Disliked
How did you feel about using the paper map?					
	1	*2*	*3*	*4*	*5*
How easy or difficult was it for you to use the paper map?	Very Difficult		Neither		Very Easy
	1	*2*	*3*	*4*	*5*
Did you ever feel lost?	Never		Sometimes		Always
	1	*2*	*3*	*4*	*5*
If you had a choice, would you choose to use a paper map like this one?	No		Maybe		Yes
	1	*2*	*3*	*4*	*5*

FIG. 20.4. Navigation aid evaluation form.

Procedure

Each participant signed an informed consent form, took a visual acuity test, and completed the navigational questionnaire. The participant was given 2 min to practice the tracking task only, and then four 3-min sessions to practice the navigational task using each aid. While practicing the navigational task, participants also performed the tracking task.

For the test session, the order of the presentation of the navigational aids and the routes was randomized. The participant had 3 min to study the navigational aid before beginning the task. When using ATIS, participants reviewed an ATIS instruction sheet for 3 min. Immediately after using each aid, the participant rated it using the four questions listed in Fig. 20.4.

RESULTS

First, the questionnaire data were analyzed to identify drivers' initial preferences for text or maps. Second, the data were analyzed to determine whether this preference influenced navigational performance in the simulated task. Third, the ratings data were examined to determine how drivers evaluated the aids, and whether initial preferences affected final preferences for the specific aids.

Factor Analysis

A factor analysis was conducted to determine if the questionnaire tapped underlying dimensions of navigational preferences. A principal-axis extraction analysis was used with a varimax rotation on the 16 questions. The results showed that six questions had weak contributions (loading factors less than .40) to a factor, or loaded only on factors with eigenvalues less than 1.00; these questions were dropped from subsequent analyses. A second factor analysis was conducted with the remaining 10 questions, which resulted in a two-factor solution. Table 20.2 shows the two-factor solution and the factor loadings for each age group.

One factor, labeled the "map factor," pertained to a preference for maps and consisted of four questions about maps. The other factor, labeled "wayfinding," pertained to general navigational habits and self-reported ability, but not specifically to preference. The map factor summarized how much a driver liked "using maps as a source for directions" and liked "map-reading and navigating." The factor also included a self-report of the ability to "locate your position on a map." Drivers with above-average self-reports of map-reading ability liked using maps more than did the drivers

TABLE 20.2
Factor Loadings for a Two-Factor Solution Using
Principle-Axis Extraction and Varimax Rotation

Questions	Wayfinding Factor	Map Factor
Do you find it easy or difficult to learn the roads in a new neighborhood?	.82	.16
If you get lost while driving in an unfamiliar area, how difficult is it for you to find your way back?	.82	.01
How would you rate your sense of direction?	.79	.28
How would you rate your ability to select a route from a map in a new area?	.59	.54
Do you like to find different ways of getting to a place you drive to frequently?	.58	.13
How would you rate your ability to follow verbal directions when driving or navigating to unfamiliar places?	.53	−.09
To what degree do you like using maps as a source for directions?	.06	.83
How easy is it for you to locate your position on a map?	.56	.71
When driving someplace new, would you rather someone gave you verbal directions OR drew you a map?	−.38	.64
On a trip with others, do you like doing the map-reading and navigating?	.18	.59

with below-average self-reports. The questions "To what degree do you like using maps as a source for directions?" and "How easy is it for you to locate your position on a map?" were strongly correlated ($r = 0.64$, $p < .001$), as shown in Fig. 20.5.

A 2 (age group) × 2 (gender) analysis of variance (ANOVA) found an effect for age, $F(1, 52) = 8.85$, $p = .004$, but not for gender, $F(1, 52) = 2.86$, $p = .097$, on the map factor scores. The older group had a significantly higher score than did the younger group, showing that the older drivers on average rated maps more favorably than did the younger drivers. Figure 20.6 shows the percent of drivers who preferred maps to those who preferred verbal directions, by age and gender.

Preference, Type of Navigational Aid, and Performance

The objective of this analysis was to evaluate the influence of navigational preference on performance with the different navigational aids. Toward that end, the drivers were grouped by their map factor scores derived from the factor analysis. Drivers with above average map factor scores were placed into a higher "map preference" group ($N = 33$), and drivers with below average scores were placed into a lower "map preference" group ($N = 20$).

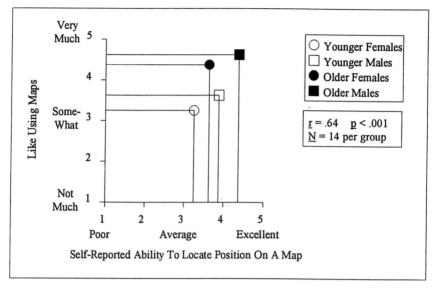

FIG. 20.5. Correlation between self-reported ability to locate a position on a map and map preference ratings by age group and gender.

To determine the influence of age, map preference, and navigational aid on performance, a 2 (age) × 2 (gender) × 2 (map preference—higher and lower) × 4 (navigational aid) repeated-measures multiple analysis of variance (MANOVA) was performed with error and decision time as the dependent variables. Data from three subjects (two younger males and one younger female, all with lower map preference scores) were not in-

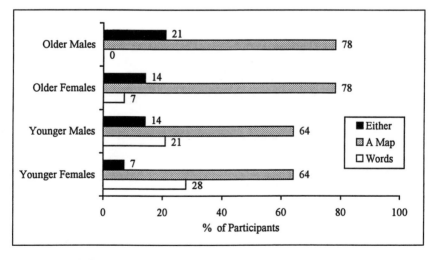

FIG. 20.6. Participants' preferences for navigational information.

cluded in the MANOVA because their error rates and decision times were more than three standard deviations above the overall mean. The MANOVA showed significant main effects for navigational aid, $F(7, 49)$ = 43.98, $p < .001$; age group, $F(1, 49)$ = 41.51, $p < .001$; and map preference group, $F(1, 49)$ = 11.00, $p < .01$, but not for gender. Further analyses were conducted to clarify the specific effects of the variables omitting gender as an independent factor due to its lack of effect in the MANOVA.

Navigational Error. A 2 (age group) × 2 (map preference group) × 4 (navigational aid) repeated-measures ANOVA was performed with error rate as the dependent measure. Significant main effects were found for the between-subjects factors of age group, $F(1, 49)$ = 42.22, $p < .001$, and map preference group, $F(1, 49)$ = 6.86, $p < .05$, and for the within-subjects factor of navigational aid, $F(3, 147)$ = 18.53, $p < .001$. There was an interaction effect between age group and navigational aid, $F = (3, 147)$ = 3.38, $p < .05$.

Results showed that the higher map preference group made significantly fewer errors ($M = 1.88$) than did the lower map preference group ($M = 2.69$), and the older group made significantly more errors ($M = 3.04$) than did the younger group ($M = 1.22$). Drivers, on average, made fewer errors when they used the ATIS display ($M = 1.04$) and the text directions ($M = 1.57$), than when they used the big map ($M = 2.79$) and the standard street map ($M = 3.34$). Newman–Keuls post-hoc pairwise comparisons showed that there were significant differences in the error rate between each type of map and the ATIS display, and between each type of map and the text directions, but there were no differences in error rate between the big map and the standard map, or between the text directions and the ATIS display. The younger group made significantly less errors using the ATIS display ($M = .32$) than when using the text directions ($M = 1.00$). In contrast, the older group made statistically the same number of errors with the ATIS display ($M = 1.68$) as with the text directions ($M = 2.07$). Figure 20.7 illustrates error rate by age group, map preference group, and the type of navigational aid.

Decision Time. A 2 (age group) × 2 (map preference group) × 4 (navigational aid) repeated-measures ANOVA was performed with decision time as the dependent measure. Significant main effects were found for the between-subjects factors of age group, $F(1, 49)$ = 17.40, $p < .001$, and for the within-subjects factor of navigational aid, $F(3, 147)$ = 60.66, $p < .001$. There was an interaction effect between age group and navigational aid, $F = (3, 147)$ = 3.93, $p < .05$. The F value for the map preference group approached significance, $F(1, 49)$ = 3.12, $p = .08$.

The pattern of results for decision time was similar to that for error rate in that the higher map preference groups made quicker decisions (M

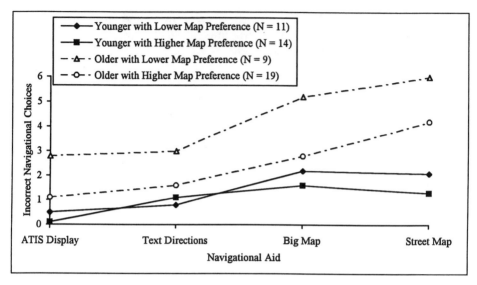

FIG. 20.7. Error rate by navigational aid, age group, and map preference group.

= 3.96 sec) than the lower map preference group ($M = 4.67$ sec), and the older group took longer to make a response ($M = 5.18$ sec) than the younger group ($M = 3.16$ sec). Overall, drivers made quicker decisions when using the ATIS display ($M = 2.48$ sec) and the text directions ($M = 2.74$ sec), than when they used either the big map ($M = 5.37$ sec) or the standard street map ($M = 6.32$ sec). The interaction effect was due to the fact that the average decision time for the younger group using the big map ($M = 4.15$ sec) was not significantly different than when using the street map ($M = 4.69$ sec). The older group, however, had a significantly slower average decision time with the street map ($M = 7.77$ sec) than with the big map ($M = 6.46$ sec). Figure 20.8 illustrates average decision time in seconds by age group, map preference group, and navigational aid.

Ratings of Navigational Aids

The ratings data were analyzed to determine how the drivers rated the aids in terms of the four questions listed in Table 20.3, and whether ratings were influenced by age group or by map preference group. Thus, a 2 (age group) × 2 (map preference group) × 4 (navigational aid) repeated-measures ANOVA was conducted on the ratings data for each question. Results showed that the ATIS display was rated significantly higher than each of the other aids for each question, as shown by Newman–Keuls post-hoc tests. Neither the age group nor the map preference group had main

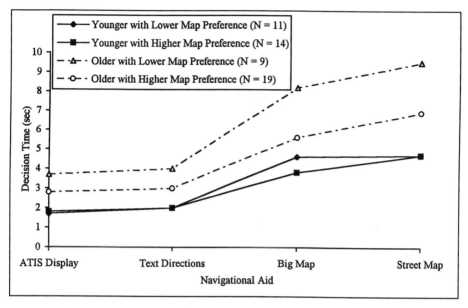

FIG. 20.8. Decision time by navigational aid, age group, and map preference group.

effects. The ratings data show that the drivers liked the ATIS display the most, found it easiest to use, were least likely to feel lost when using ATIS, and would choose ATIS over the other aids. The average ratings and the F values for the interaction effects for each navigational aid are shown in Table 20.3.

There were interaction effects between the map preference group and the type of navigational aid for ratings data from the question "How did you feel about using the aid?", $F(3, 156) = 2.92$, $p < .05$. Although the two map preference groups "felt" the same about the ATIS display, the big map, and the text directions, the higher map preference group liked the street map significantly more ($M = 3.5$) than the lower map preference group ($M = 2.7$), as shown in Fig. 20.9.

The ratings data from the question "How difficult or easy was it to use the aid?" showed an interaction effect between the age group and the type of navigational aid, $F(3, 156) = 3.87$, $p < .05$, in that the older group did not rate the ATIS display as easy to use ($M = 4.3$) as did the younger group ($M = 4.9$), as shown in Fig. 20.10.

Two interaction effects were found in the ratings data from the question "Did you feel lost when using the aid?" (Fig. 20.11). There was an interaction between map preference group and the type of navigational aid, $F(3, 156) = 4.61$, $p < .01$, in that the lower map preference group felt significantly more lost when using the big map ($M = 2.8$) or the street

TABLE 20.3
Average Ratings and F Values for the Navigation Aids

	How did you feel about using the aid?	How difficult or easy was it to use the aid?	Did you feel lost when using the aid?	Would you choose to use this aid?
ATIS Display	1.29 (0.65)	4.62 (0.78)	1.27 (0.75)	4.45 (1.01)
Text Directions	2.11 (0.95)	3.96 (1.06)	1.68 (0.96)	3.82 (1.28)
Big Map	2.50 (1.24)	3.32 (1.28)	2.36 (1.14)	3.16 (1.44)
Street Map	2.82 (1.40)	2.98 (1.34)	2.18 (1.05)	3.09 (1.50)
Navigational Aid, $df = 3, 156$	$F = 10.12, p < .001$	$F = 26.00, p < .001$	$F = 15.84, p < .001$	$F = 11.59, p < .001$
Map Group × Aid, $df = 3, 156$	$F = 2.92, p < .05$	n.s.	$F = 4.61, p < .01$	n.s.
Age Group × Aid, $df = 3, 156$	n.s.	$F = 3.87, p < .05$	$F = 2.62, p < .05$	n.s.

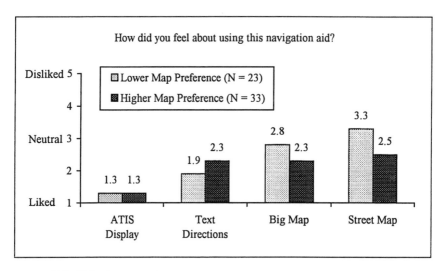

FIG. 20.9. Average ratings for "How did you feel about using this navigation aid?" by navigational aid and map preference group.

map ($M = 2.6$) than the higher map preference group when using the big map ($M = 2.1$) or the street map ($M = 1.9$). Both the higher and the lower map preference groups rarely felt lost when using the ATIS display or the text directions. An interaction effect was also found between the age group and the type of navigational aid, $F(3, 156) = 2.62$, $p = .05$. The older group felt just marginally more lost with ATIS ($M = 1.5$) than did the

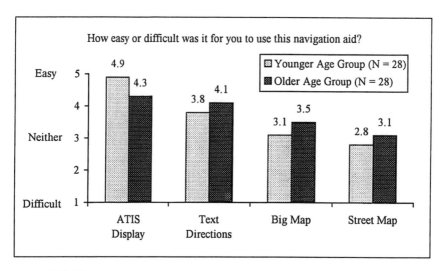

FIG. 20.10. Average ratings for "How easy or difficult was it for you to use this aid?" by navigational aid and age group.

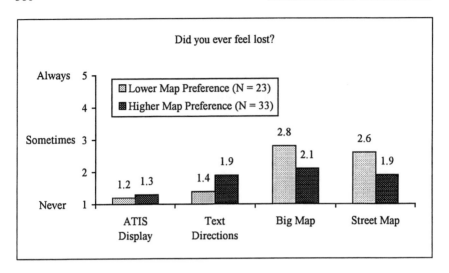

FIG. 20.11. Average ratings for "Did you ever feel lost?" by navigational aid and map preference group.

younger group with ATIS ($M = 1.0$). The older group did not feel any more lost when using the big map ($M = 2.4$) or the street map ($M = 2.0$) than did the younger group using the big map ($M = 2.3$) or the street map ($M = 2.4$). When using either map, however, the older group made more navigational errors than did the younger group, which suggests that, on average, the older drivers underestimated their "lost" time.

DISCUSSION

The results from this study suggest that drivers are aware of the type of navigational aid most suitable to them. Also, drivers generally perform better with the type of aids they prefer. For example, drivers with a strong preference for maps tended to work better with maps than those who did not prefer maps. Regardless of their initial preference, however, navigational performance was typically better with the ATIS display than with either type of map.

The drivers showed strong differences in their initial preferences for maps and text directions. However, most of the participants rated ATIS higher than the other aids after a "hands-on" experience with it. This finding supports previous research on consumer reaction to ATIS and implies a great potential for user acceptance of ATIS route guidance.

Of course, consumers may nonetheless be reluctant to purchase ATIS products and services for various reasons. For example, drivers with confidence in their navigational ability may feel no need to use ATIS. Similarly, drivers with strong preferences for maps may resist switching from paper maps to ATIS and relying on a nontraditional system that is not in their immediate control. Older drivers in particular may be unlikely to embrace a technologically innovative system, as suggested by the data on ATM use described previously. Further, the older driver group in this study rated maps more highly than the younger group, suggesting that older driver cohorts may be particularly reluctant to forego using maps. Finally, the simulated ATIS display used in this study was completely accurate and reliable. In contrast, real-world systems, particularly during early implementation, will undoubtedly have flaws and potentially damage the development of user trust in the system. Despite these caveats, however, the data presented here strongly suggest that drivers may be willing to use ATIS.

An ITS implementation strategy can facilitate user acceptance by presenting information to positively influence consumer reaction to ATIS. For example, descriptions of ATIS that are targeted at older drivers could emphasize the capacity of ATIS to enhance the safety and mobility of travelers. Likewise, educating users about potential system weaknesses might lead to realistic expectations of ATIS technology and the maintenance of trust in the system even in the face of less-than-perfect performance. Further research on potential user reaction can be applied to the design and implementation of ATIS products and services.

REFERENCES

ADC of Alexandria. (1994). Northern Virginia Area Map. Alexandria, VA.

Barr, R. A., & Eberhard, J. W. (1991). Special issue preface. *Human Factors, 33,* 497–498.

Campbell, J. L., Kinghorn, R. A., & Kantowitz, B. H. (1995). Driver acceptance of system features in an advanced traveler information system (ATIS). *Proceedings of the 1995 Annual Meeting of ITS America,* 967–973.

Kantowitz, B. H., Lee, J. D., Becker, C. A., Bittner, A. C., Kantowitz, S. C., Hanowski, R. J., Kinghorn, R. J., McCauley, M. E., Sharkey, T. J., McCallum, M. C., & Barlow, S. T. (1993). *ATIS and CVO components of IVHS: Task H.* Washington, DC: National Technical Information Service.

Likert, R. (1932). A technique for the measurement of attitudes. *Archives of Psychology,* No. 140.

Muir, B. M. (1994). Trust in automation: Part I. Theoretical issues in the study of trust and human intervention in automated systems. *Ergonomics, 37*(11), 1905–1922.

Muir, B. M. (1987). Trust between humans and machines, and the design of decision aids. *International Journal of Man–Machine Systems, 27,* 527–539.

Perez, W., Fleishman, R., Golembiewski, G., & Dennard, D. (1993). TravTek field study results to date. *Proceedings of the IVHS America 1993 Annual Meeting,* 667–673.

Rillings, J. H., & Lewis, J. W. (1991). TravTek. In *Conference Record of Papers at the 2nd Vehicle Navigation and Information Systems (VNIS '91) Conference,* Dearborn, MI.

Streeter, L. A., & Vitello, D. (1986). A profile of drivers' map-reading abilities. *Human Factors, 28*, 223–239.

Waller, P. F. (1991). The older driver. *Human Factors, 33*, 499–505.

Walker, J., Sedney, C. A., & Mast, T. (1991). Older drivers and useful field of view in a part-task simulator (Paper No. 920702). Washington, DC: Transportation Research Board.

Wochinger, K., & Boehm-Davis, D. (1995). The effects of age, spatial ability, and type of navigational information on navigational performance (Tech. Rep. FHWA-95-166). Washington, DC: Federal Highway Administration.

Zimmerman, C. A., & Elliot, C. A. (1995). ITS user acceptance research at the U.S. Department of Transportation. *Proceedings of the 1995 Annual Meeting of ITS America*, 941–945.

Chapter 21

Monitoring Driver Fatigue Via Driving Performance

Stephen H. Fairclough
HUSAT, Loughborough University, Leics., UK

A cursory examination of the relevant literature provides ample evidence that fatigue is a problem for the driving public. Estimates of the contribution of driver fatigue to total traffic accidents vary from 4% (Treat, 1980) to 25% (Maycock, 1995). A questionnaire survey conducted by Tilley, Erwin, and Gianturco (1973) illustrated drivers' awareness of the fatigue problem. Survey results revealed that 64% of respondents had experienced fatigue on the road. In addition, 7% of the subject sample claimed to have had a traffic accident due to fatigue.

However, the issue of driver fatigue has not found equivalent expression in public awareness as drunk driving, overspeeding, and other violations of road safety. According to Brown (1994), this is due to three factors:

1. Evidence of the contributory role of fatigue to traffic accidents is usually of the circumstantial variety (i.e., time of accident, pattern of injury). Therefore, the contribution of fatigue to accidents tends to be underestimated.

2. A lack of political will to research into the limitations of work hours for professional drivers. Brown (1994) claimed this is due to perceived negative effects on the competitiveness of alternative forms of transport.

3. A public perception that accidents due to the influence of fatigue are a matter of individual liability. It is reasonable to expect drivers

to behave responsibly and to respond appropriately to the symptoms of impending fatigue, however, the innate capability of the driver to perceive and self-diagnose may be impaired by an increase in fatigue (Brown, 1994).

This latter point is crucial to this chapter and central to the development of systems designed to monitor, diagnose, and inform drivers of their level of fatigue. McDonald (1989) presented the same argument as Brown (1994) and particularly emphasized self-monitoring as a means of managing fatigue: "Anything that impairs self-monitoring, particularly in terms of over-estimating one's psychological resources or under-estimating the demands that have to be met, is likely to give rise to unpredictable or uncontrollable effects of fatigue" (p. 88). The implication of McDonald's (1989) assumption is that self-monitoring provides subjective basis for the prediction of one's ability to continue driving. Similarly, Brown (1994) defined fatigue as "a subjectively experienced disinclination to continue performing the task in hand because of perceived reductions in efficiency" (p. 299). Therefore, self-monitoring drivers must perceive "reductions in efficiency" before they can experience "a disinclination to continue" (p. 299) the driving task. The blunted sensitivity of drivers' perceptions and decision-making abilities under conditions of fatigue is at the root of the driver fatigue problem.

DRIVER IMPAIRMENT MONITORING SYSTEMS

The aim of a driver impairment monitoring (DIM) system is to supplement the process of self-monitoring (as described previously). Increasing levels of fatigue reduce the precision of drivers' ability to self-monitor performance, to predict the ability to continue, and to estimate the appropriate point to discontinue. The purpose of DIM feedback based on actual performance is to increase the fidelity of these processes under conditions of fatigue.

In order to produce feedback, DIM technology is dependent on a group of suitable fatigue indicators, that is, real-time measures of fatigue used as input to system diagnostics. The caliber of the system diagnosis is dependent on the quality of these fatigue indicators. According to Mackie and Wylie (1991), fatigue indicators must fulfil the following three criteria:

1. *Reliability.* Do fatigue indicators reliably detect the presence of fatigue rather than another aspect of behavior, the driving environment, or some interaction between the two?
2. *Sensitivity.* Are the indicators sufficiently sensitive to respond rapidly to increases in fatigue and to permit rapid system diagnosis?

3. *Acceptability.* Can fatigue indicators be collected unobtrusively without either affecting the driver's ability to perform or introducing any hindrance or inconvenience for the driver?

Candidate fatigue indicators may be grouped into three measurement categories. The first of these measures is psychophysiology (PP), which has been used extensively for the quantification of sleep, stress, and drug effects in the scientific literature. PP variables have been demonstrated as both sensitive indicators of driver impairment on the road (i.e., Brookhuis & De Waard, 1993). The problem of PP measures is that they involve the placement of electrodes on various sites of the body, which is naturally unacceptable to the majority of the driving public.

A solution to the acceptability problem is to remotely monitor physiological properties without the physical attachment of electrodes to the driver. This approach has been pursued by Skipper and Wierwille (1986; see also Wierwille, 1994), who measured the percentage of the eye closure during driving via video observation. One of the symptoms of increasing fatigue is that the eyelids feel heavy and begin to droop. These authors demonstrated how eyelid closure was correlated with both psychophysiological and vehicle-based measures of driver drowsiness. Several systems are currently under development to automatically monitor this variable. Ueno, Kaneda, and Tsukino (1994); Aoki, Katahara, and Nakai (1994); and Tock and Craw (1992) have described image-processing technology specifically designed to measure eye activity as a proxy measure for driver drowsiness.

Similarly, Wierwille and Ellsworth (1994) noted that tired drivers demonstrated a number of observable symptoms during experimental studies. For example, sleepy drivers would yawn, stretch, and rub their eyes. Wierwille and Ellsworth (1994) formalized these behaviors into an observation scale. They reported a validation of the scale where they found a significant association between psychophysiological variables and the observable symptoms of fatigue.

The second category of measures describes those secondary or subsidiary tasks that may be introduced as both alarms and alertness maintainers. The "deadman switch" used by train drivers is example of a secondary task, where a cue (i.e., a tone) is presented that demands a response (i.e., depress a switch). These measures have been demonstrated as sensitive to driver fatigue (i.e., Drory, 1985; Riemersma, Sanders, Wildervanck, & Gaillard, 1977), but unfortunately comprise a discrete and rather continuous source of data (i.e., the update of system information is limited by the maximum number of cues presented); also, the act of responding to the cue creates an additional burden on the driver. Wierwille, Wreggit, and Knipling (1994) suggested that a secondary task may function as the second stage of a fatigue detection process. Their conception of a DIM system included the detection of driver fatigue based on vehicle measures as a first stage followed by the introduction of a secondary task as a second stage. The purpose of the secondary task is twofold: to provide supplemental information on the detected level of driver fatigue and to stimulate the (presumably) drowsy driver.

The current study is concerned with measures of primary vehicle input as indicators of driver fatigue. Candidate measures of vehicle input may be categorized as either concerned with lateral control of the vehicle (i.e., steering control, lane deviation) or speed control (i.e., headway-keeping, speed variability, control of accelerator pedal).

A large amount of research has focused on steering control as a means of detecting driver drowsiness (i.e., Allen, Parseghian, Kelly, & Rosenthal, 1994; Artaud et al., 1994; Brookhuis & De Waard, 1993; Elling & Sherman, 1994; Fairclough, 1994; Gabrielsen & Sherman, 1994; Riemersma et al., 1977; Skipper & Wierwille, 1986; Wierwille, 1994). The basic hypothesis of these studies was that psychological impairment increases the variability of steering control, that is, steering adjustment becomes increasingly erratic. Most authors are in agreement with this central hypothesis, but there is little consensus on the exact details of how steering activity is captured and quantified. Steering behavior may be measured in terms of velocity or amplitude or subjected to a Fast Fourier Transform analysis, which collapses both into a single dimension (i.e., Artaud et al., 1994). Steering activity is usually analyzed in terms of time "windows" (i.e., averaged across units of time). Wierwille et al. (1994) reported research demonstrating advantages (in terms of reliability) for 6-min time windows when compared to smaller time windows. By contrast, Allen et al. (1994) opted for a smaller time window of 15 sec. Obviously, an optimal measure of steering would combine both characteristics of sensitivity (i.e., small window size) and reliability (i.e., not influenced by confounding variables within the environment). A critical review of these factors is provided by Brekke and Sherman (1994).

Impaired driving may also be assessed by measuring the sensitivity of speed control. This behavior may be quantified either in terms of global variability (i.e., Riemersma et al., 1977) or as a response to speed changes of a lead vehicle (Brookhuis & De Waard, 1993).

Measures of primary vehicle control are potentially more acceptable to the driving public as they do not involve any overt monitoring (i.e., PP electrodes) or responding (i.e., secondary tasks). Although the experimental studies already listed provide evidence of the sensitivity to these measures to fatigue, it is difficult to assess if this category of measurement is sufficiently sensitive to provide input to a working DIM system.

The cumulative result of sensitivity and reliability characteristics is the accuracy of the DIM system. System accuracy will be a key factor in shaping

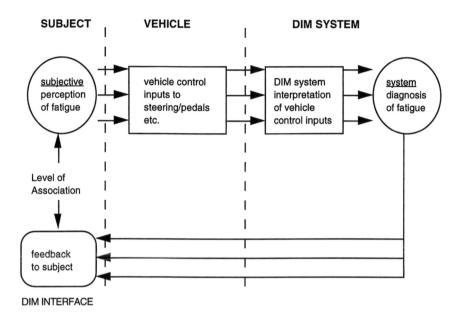

FIG. 21.1. Feedback loop between driver and DIM system.

user acceptance of DIM technology. The system collates data from primary vehicle input and produces a diagnosis of fatigue, which may subsequently be presented to the driver. The principle is identical with a biofeedback paradigm where vehicle-based indicators are substituted for physiological data. This flow of information through system components is presented in Fig. 21.1.

When drivers receive feedback from the system, they will naturally match their subjective estimation of fatigue against the system diagnosis (i.e., actively self-monitor). Drivers may agree or disagree with system output based on strength of association or agreement between system and subjective estimates of fatigue (see Fig. 21.1). Repeated exposure to a system will allow drivers to construct their estimate of system accuracy, which, in turn, determines the degree of trust drivers invest in the system. In order for these systems to provide an effective countermeasure against fatigue, it is essential that the trust of drivers is earned by the system.

The drivers' subjective estimates of fatigue are central to the matching process described in Fig. 21.1. However, the rationale for the development of a DIM device has rested primarily on the relation between psychophysiology (PP) and behavior. Subjective measures of fatigue are relegated to a position of secondary importance. There are several reasons why this should be so: First, subjective measures are difficult to capture concurrently with other data sources without confounding (i.e., alerting the driver) and, second, PP measures are the "scientific standard" for the study of sleepiness.

However, as DIM technology develops and moves closer to the marketplace, it is important to consider the role of subjective impairment within the driver–system interaction.

The purpose of the current study was to investigate aspects of sensitivity and reliability of primary vehicle indicators of fatigue. Several measures of steering control and speed control are presented as candidate indicators of fatigue. One aim of the study was to evaluate the relative sensitivity of these measures. In addition, the study includes both psychophysiological and subjective measures of fatigue. Both domains of measurement are included to assess the reliability of primary vehicle control by association with both psychophysiological and subjective correlates of impairment.

METHOD

Subjects

The subjects were British police drivers who regularly worked on motorway patrol duty. Nine male subjects participated in the study. The subject group were all qualified at police advanced level and had from 18 months to 11 years of professional experience ($M = 5.1$, $SD = 3.5$).

Experimental Design

The subjects participated in two experimental sessions. The "short-shift" condition functioned as a sleep deprivation manipulation. Subjects performed seven consecutive nightshift duties on the seven nights prior to the experimental session. Each nightshift lasted from 11:00 p.m. until 6:00 a.m. On the day of the trial, the nightshift schedule switched to an early afternoon schedule where officers returned to duty at 2:00 p.m. Officers reported to the institute at noon on the day of the short shift. The subjects were asked to report early in order to avoid the late afternoon rush-hour traffic on the motorway. The control condition took place when officers had worked either a standard dayshift, between the 9:00 a.m. and 8:00 p.m., or had received at least two normal nights sleep, between 9:00 p.m. and 9:00 a.m., prior to the control session.

Apparatus and Experimental Route

The subjects completed each experimental journey in a specially equipped Vauxhall Cavalier. This vehicle contained dual controls and a range of sensors to monitor vehicle variables that were captured and stored on a lap-top PC. Psychophysiological data was captured via an 8-channel MacLab™ and a PowerBook™ portable computer, supported by bioamplifiers and Chart™ software.

The experimental route took the form of a 281 km round trip between Loughborough and Leeds. The initial stage of the journey took place on the northbound section of the M1 motorway between junctions 21 and 42, at the latter junction, the drivers joined the M62. The drivers left the M62 at junction 33 and joined a two-lane A road (A6) for a short spell, before switching to the M18 and rejoining the southbound section of M1 at junction 32. The journey was concluded at junction 23 where the subjects returned to Loughborough. On average, the journey took between 2 hr and 30 min and 3 hr to complete.

Experimental Measures

The experimental variables used in the study were separated into three classes: (a) psychophysiological, (b) vehicle, and (c) subjective self-report.

A single bipolar channel of electroencephalogram (EEG) data was recorded between sites C_3 and O_2 (Jaspers, 1958). The EEG data was subsequently analysed via Igor™ waveform analysis program. The EEG was subjected to a Fast Fourier Transform (FFT) algorithm, which produced a spectral resolution of .78 Hz. Three principal bandwidths of EEG were analyzed: theta (3.9–7.8 Hz), alpha (8.6–12.5 Hz), and beta (13.3–20.3 Hz). This data was averaged over 100 sec windows to produce a single ratio score (i.e., alpha + theta/beta).

The vehicle variables constituting group (b) were steering wheel variability and standard deviation (SD), accelerator pedal SD, mean speed, speed SD, acceleration/deceleration SD. Also included in this group were four vehicle variables developed specifically for the detection of driver fatigue. These were the three Y variables and the S function developed by Khardi (1992) and described in the DETER project report by Fairclough (1994).

The S function described in Fairclough (1994) combines speed data and steering data. The S function concentrates on the relation between the magnitude of steering change and speed. It is given by the following formula:

$$S = \left[\frac{\dfrac{1}{N} \sum_{i=1}^{N} \left| \dfrac{dsi}{dt} \right|}{\dfrac{1}{N} \sum_{i=1}^{N} V_i} * 10^4 \right]^2$$

where
N = the number of points in the sample
si = the steering wheel angle in degrees
Vi = the instantaneous speed of the vehicle

The Y functions also describe both steering wheel movement and speed. The frequency and magnitude of steering wheel adjustments (i.e., $N1$ = number of positive and negative crossings at 1 degree) are transformed by constants and added to the area beneath the steering curve (STOT), and divided by the speed of the vehicle. The Y functions measure different aspects of steering behavior; $Y1$ quantifies the underlying trend of steering change, whereas $Y2$ and $Y3$ focus on the number and magnitude of steering adjustments. The $Y3$ variable provides weights for higher magnitude steering adjustments. The equations for the Y functions are described here:

$$Y1 = \text{STOT}$$

$$Y2 = \frac{\text{STOT} + 5/6(N1 + N1.5 + N2 + N2.5)100}{\text{vehicle speed}}$$

$$Y3 = \frac{\text{STOT} + 5/6(N1 + 2N1.5 + 3N2 + 4N2.5)100}{\text{vehicle speed}}$$

Subjective self-assessment (SSA) data was quantified over six 9-point self-rating scale administered at the beginning and end of each experimental journey and at 30-min intervals during the journey. The subjects were asked to verbally rate themselves over six bipolar adjective scales: interested–bored, relaxed–stressed, energetic–lethargic, awake–fighting sleep, attentive–inattentive, and no effort–much effort. Each bipolar scale contained 9 points, where 9 was always contained the most negative connotation (i.e., bored, stressed, lethargic, fighting sleep, much effort).

Experimental Protocol

The subjects were briefed and asked to complete consent forms. While the EEG electrodes were attached, the subjects familiarized themselves with the verbal rating scales and a paper-based description of the experimental route. When electrode attachment was complete, the subjects were familiarized with the vehicle controls and dashboard layout in the stationary vehicle. Seat position and mirror position were adjusted by each individual driver. The subjects were introduced to the co-driver, who was a trained driving instructor. The subjects were instructed not to speak to either the experimenter or the co-driver unless responding to the verbal self-reports. Co-drivers had been instructed not to use the dual-control pedals or to speak to the subject unless they judged safety to be compromised.

The subjects completed a 10-min drive to the motorway intersection. They were told that they were allowed a break of approximately 10 min duration at any point during the journey. The subjects performed a short (i.e., 25 min) familiarization journey on the southbound section of the motorway. A paper description of the route was attached to the dashboard. This description contained a list of the four navigation instructions (i.e., M1 Junction 42—exit). This list also contained the names and approximate locations of available service stations. The drivers were instructed to drive normally and to pull over if they felt sufficiently impaired to jeopardize safety.

The entire journey took approximately 3 hr. Subjective ratings were administered at the beginning and end of the journey, and at 30-min intervals in-between. On return to the institute, the EEG electrodes were removed and the subjects were debriefed.

RESULTS

One set of vehicle data was lost during experimentation, hence only eight subjects data was used in the analysis. The following sections of each experimental journey were edited out of the analyzed data record:

1. The 5-min period preceding and following each of two roundabout junction and associated entry and exit roads.
2. The 5-min period including elicitation of the subjective self-assessment data.
3. The 5-min period prior to and following the subjects' rest breaks.
4. The initial 20-min familiarization period.

The goal of this procedure was to ensure that the analyzed data was generic to straight stretches of road where the driver was habituated to the driving task. The data editing process yielded six sections of "clean data," each of which retrogressively corresponds to an on-route elicitation of subjective data; that is, SSA data is paired with mean values for the preceding data chunk.

Vehicle Parameters

Nine vehicle parameters were captured the study, this data was subjected to multivariate (MANOVA) statistical analysis. Two main effects were tested, one for experimental condition (CONDITION) and one for route section (SECTION). The MANOVA revealed an effect for both CONDITION and SECTION (Wilks' Lambda = .48 and .28, respectively, $p < .01$) across all nine vehicle parameters, several of which demonstrated significance.

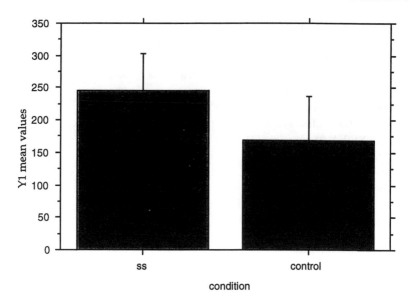

FIG. 21.2. Cell means for *Y1* variable for both short-shift (ss) and control conditions (*n* = 8). Error bars = standard error.

The *Y1* variable produced significantly higher scores during the short-shift condition, $F(1, 7) = 37.4$, $p < .01$. The mean values for the *Y1* variable for both conditions is shown in Fig. 21.2.

In addition, both the *S* function, $F(1, 7) = 7.36$, $p < .01$, and the standard deviation of steering, $F(1, 7) = 22.28$, $p < .01$, was significantly higher during the short-shift condition.

There was also a marginal effect on the mean speed of the vehicle due to the influence of the short-shift condition. Mean speed was lower in this condition than in the control condition, $F(1, 7) = 2.94$, $p < .1$.

Effects due to the influence of journey SECTION were apparent for the standard deviation of the accelerator pedal, $F(5, 40) = 3.71$, $p < .01$, and the standard deviation of speed, $F(5, 40) = 2.93$, $p < .05$. Both increased during the latter phase of the test route. As this effect was not affected by CONDITION, it was hypothesized that the effect was due to idiosyncrasies of traffic flow and/or road geometry.

Psychophysiology and Subjective Ratings

EEG data was quantified by power spectrum analysis and averaged into 100-sec "real-time" windows. Each bandwidth was quantified and converted into a ratio score of alpha + theta/beta. The 100-sec windows were averaged to produce six journey sections. The MANOVA analysis revealed no significant differences due to either experimental CONDITION or SECTION.

Subjective self-ratings were collected on at least seven occasions, at the beginning and end of the experimental journey, and every 30 min in-between. Therefore, each subject delivered a minimum of seven ratings during each journey. Subjects were asked to provide self-ratings on a number of bipolar scales, each containing nine points.

A MANOVA analysis was performed on the data, testing the effect of CONDITION and the number of rating scale SAMPLES (which effectively functioned as time-on-task). Significant effects were found due to the effect of CONDITION (Wilks' Lambda = .88, $p < .05$) and effect of SAMPLES (Wilks' Lambda = .49, $p < .01$).

Subjects self-rated higher levels of boredom during both the short-shift condition, $F(1, 7) = 8.96$, $p < .01$, and as the journey progressed, $F(6, 42) = 4.11$, $p < 01$. Self-ratings of stress showed a significant increase during the short-shift condition, $F(1, 7) = 5.9$, $p < .01$. Subjects rated lethargy as higher during the short-shift condition, $F(1, 7) = 7.9$, $p < .01$. Self-rated effort was significantly higher in the short-shift condition, $F(1, 7) = 3.94$, $p < .05$, and increased as the journey progressed, $F(6, 42) = 2.33$, $p < .05$.

Subjects rated sleepiness higher in the short-shift condition, $F(1, 7) = 10.98$, $p < .01$, and as the journey progressed, $F(6, 42) = 4.5$, $p < .01$ (see Fig. 21.3). During the short-shift condition the subjects rated their levels of

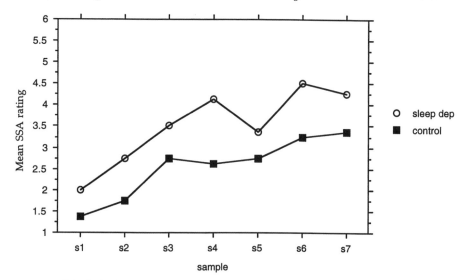

FIG. 21.3. Cell means of self-rated sleepiness for both short-shift (sleep dep.) and control conditions and by each section of the experimental journey ($n = 8$).

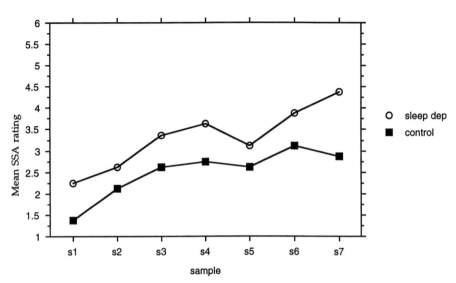

FIG. 21.4. Cell means of self-rated inattention for both short-shift (sleep dep.) and control conditions and by each section of the experimental journey ($n = 8$).

inattention as higher than the control condition, $F(1, 7) = 7.58$, $p < .01$. Additionally, self-rated inattention increased as the journey progressed, $F(6, 42) = 4.51$, $p < .01$ (see Fig. 21.4).

Correlation Analyses

The previous analyses illustrated how each individual data group was affected by the experimental manipulation. This section deals with the interrelations between the vehicle data, the EEG, and the subjective self-ratings.

Several vehicle variables demonstrated sensitivity to the effects of the experimental manipulation. However, these analyses do not provide detail on the reliability and sensitivity of any variable to changes in EEG activity. Therefore, correlation coefficients were calculated to assess the degree of association between the steering variables, subjective self-assessment, and EEG activity. As before, the data was averaged to yield six analysis windows. Correlation coefficients were calculated across the subject group within each analysis window. This analysis was conducted only on data from the "short-shift" condition. Pearson's correlation coefficient was used in all analyses. Each figure includes an approximation of the critical value of r required to achieve significance. The critical r was estimated on the basis of a one-tailed hypothesis at the 1% significance level.

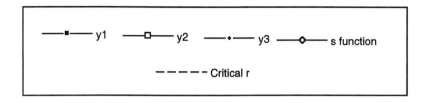

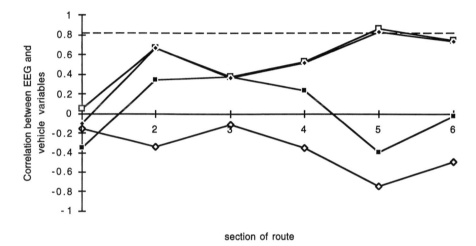

section of route

FIG. 21.5. Correlation coefficients between dedicated steering measures and EEG activity ($n = 8$).

The first analysis was performed on the four measures of steering that purport to indicate driver fatigue. The results of this analysis are shown in Fig. 21.5.

The *Y1* and *S* function variables show a poor correspondence with EEG activity, however, the *Y2* and *Y3* measures demonstrate a consistent positive correlation, which achieves significance during the latter stage of the journey.

A correlation analysis of the relation between subjective data and specialized steering variables was performed. This aim of this analysis was to test the correspondence between self-rated sleepiness and performance-based indicators of driver fatigue. The results are shown in Fig. 21.6.

The results of these analyses shows that the *Y2* and *Y3* variables have the higher degree of association with the subjective sleepiness than the other steering measures. The same analysis was applied to psychophysiology and subjective sleepiness. An average positive correlation of .52 was found, which failed to reach significance.

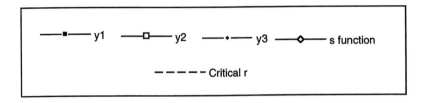

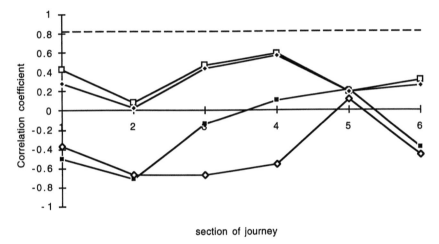

section of journey

FIG. 21.6. Correlation coefficients between four steering measures and subjective SLEEPINESS ($n = 8$).

DISCUSSION

The experiment revealed three measures of steering behavior that were sensitive to the main manipulation of shiftwork (Figs. 21.2–21.4). However, only a marginal effect was found for speed control where drivers in the short-shift condition decreased their average speed throughout the experimental session. This finding was in broad agreement with Fuller's (1978) discovery that convoy drivers strategically increased headway to a lead vehicle due to the influence of time-on-task; that is, drivers adopt more cautious driving practices when under the influence of fatigue. The failure of the other speed control measures (i.e., accelerator pedal standard deviation, speed standard deviation, and acceleration/deceleration standard deviation) to change systematically due to the influence of shiftwork may indicate an overwhelming influence of transient (i.e., vehicle flow) and static (i.e., road layout) features of traffic environment. The MANOVA analysis revealed that both accelerator pedal standard deviation and speed standard deviation were influenced by the effect of journey section, and presumably those features of the environment associated with each.

The subjective self-assessment (SSA) data depicted the expected trend of increased impairment during the short-shift journey. Prior to each experimental journey, subjects were instructed that they were eligible for one "rest" break of approximately 10 min. The locations of service stations en-route was included on the route navigation sheet, which was visible to subjects throughout each journey. The majority of subjects (with one exception) elected to break during the short-shift condition. This contrasted with the control condition where few subjects elected to break at all. For the short-shift condition, most subjects took a break after driving for approximately 2 hr. This coincided with declines of self-assessed sleepiness and inattentiveness at section five, as illustrated in Figs. 21.3 and 21.4. It was noticeable that this alleviation of subjective impairment had both dissipated and slightly accrued at the next elicitation of subjective data collected 30 min later. Therefore, the effect of a 10-min break, which usually involved getting out of the vehicle and taking a walk, provided only short-lived respite from the subjective experience of impairment.

The correlation analyses were performed in order to assess the intrasession estimates of association between performance-based indicators of driver fatigue and their psychophysiological and subjective correlates. These analyses revealed a number of surprising results. It appears that those steering measures, which were sensitive to the influence of fatigue (i.e., Y1, S function), were not reliably associated with either psychophysiological or subjective measures of fatigue (Figs. 21.5 and 21.6) over the course of the trial. However, the Y2 and Y3 functions, which failed to exhibit statistically significant differences between the two conditions, showed a higher degree of association than the other steering measures. The positive association between Y2/Y3 and EEG indicators of fatigue peaks during the last two sections of the journey (Fig. 21.5). This finding may indicate differences of relative sensitivity between the steering variables. Those variables, which measured the magnitude of steering change (i.e., Y1, steering SD, and the S function), illustrated an expected increase during the short-shift condition but were not associated with EEG or subjective measures of fatigue. The Y2 and Y3 variables measured the frequency of steering adjustments (i.e., frequency of steering update), and as such were more sensitive to both EEG and subjective measures.

This finding indicates a degree of independence between sensitivity and reliability with important implications for the selection and development of vehicle-based fatigue indicators. It is apparent from these findings that an indicator may be sensitive but not reliable, and vice versa. Unfortunately, a feasible DIM system must fulfil both criteria to achieve an acceptable degree of accuracy.

CONCLUSIONS

The acceptability of DIM technology will depend, to a large extent, on the degree of accuracy associated with any given system. Perceptions of system accuracy will critically influence the degree of cooperation and trust imbued by the system. In order for a DIM system to be effective, it must be accurate. The accuracy of vehicle-based indicators of fatigue may be expressed in terms of the strength of association between these indicators and psychological fatigue. A sensitive indicator is capable of accurately detecting a fatigue "signal" against the background "noise" of driving behavior. A reliable indicator is capable of consistent association with measures of psychological fatigue within a designated time frame. The field study demonstrated the independence of both qualities within a test group of steering-based indicators of fatigue. It is concluded that candidate indicators of fatigue detection must fulfil both criteria in order to function within an acceptable DIM system.

ACKNOWLEDGMENTS

This research was funded by the CEC under the DRIVE Program in the DETER project (V2009). The author would like to acknowledge the contributions of Stephen Hirst and John Richardson during the planning, running, and analysis stages of the study; those members of the Leicestershire Constabulary who acted as subjects and their superior officer, Chief Inspector G. Compton for his help; and also the assistance of David Rogers, who acted as co-driver for the experiment.

REFERENCES

Allen, R. W., Parseghian, Z., Kelly, S., & Rosenthal, T. J. (1994). *An experimental study of driver alertness monitoring* (Paper No. 508). Hawthorne, CA: Systems Technology, Inc.

Aoki, M., Katahara, S., & Nakai, K. (1994). Motion detection for eyes and eyelids from face image sequences for driver monitoring. *Proceedings of the 27th ISATA, Aachen, Germany, 1994* (pp. 215–229). Automotive Automation Ltd.

Artaud, P., Planque, S., Lavergne, C., Cara, H., de Lepine, P., Tarriere, C., & Gueguen, B. (1994, May). *An on-board system for detecting lapses of alertness in car driving.* Paper presented at 14th International Conference on Enhanced Safety of Vehicles, Munich, Germany.

Brekke, M., & Sherman, P. (1994). A critical evaluation of factors associated with steering wheel data when used for identifying driver drowsiness. *Proceedings of the 27th ISATA, Aachen, Germany, 1994* (pp. 223–229). Automotive Automation Ltd.

Brookhuis, K. A., & De Waard, D. (1993). The use of psychophysiology to assess driver status. *Ergonomics, 39*(9), 1099–1110.

Brown, I. D. (1994). Driver fatigue. *Human Factors, 36*(2), 298–314.

Drory, A. (1985). Effects of rest and secondary task on simulated truck driving task performance. *Human Factors, 27*(2), 201–207.

Elling, M., & Sherman, P. (1994). Evaluation of steering wheel measures for drowsy drivers. *Proceedings of the 27th ISATA, Aachen, Germany, 1994* (pp. 207–214). Automotive Automation Ltd.

Fairclough, S. H. (Ed). (1994). Driver state monitor. *Deliverable 5 (330A).* Project V2009 (DETER).

Fuller, R. G. C. (1978). Effects of prolonged driving on heavy goods vehicle driving performance. *Driver fatigue in road traffic accidents: Contributions to workshops on physiological, psychological and sociological aspects of the problem* (pp. 151–160). Commission of the European Communities Rep. No. EUR 6065 EN.

Gabrielsen, K., & Sherman, P. (1994). Drowsy drivers, steering data and random processes. *Proceedings of the 27th ISATA, Aachen, Germany, 1994* (pp. 231–240). Automotive Automation Ltd.

Jaspers, H. H. (1958). The ten-twenty electrode system of the International Foundation. *EEG Clinical Neurophysiology, 10,* 371–375.

Mackie, R. R., & Wylie, C. D. (1991). Countermeasures to loss of alertness in truck drivers: Theoretical and practical considerations. In M. Vallet (Eds.), *Le maintien de la vigilance dans les transports, actes des journees d'etude de l'Institut National de Recherche sur les Transports et leur Securite, Bron 18–19 October 1990* (pp. 113–141). Caen: Paradigme.

Maycock, G. (1995). *Driver sleepiness as a factor in car and HGV accidents* (TRL Rep. 169). Crowthorne: TRL.

McDonald, N. (1989). Fatigue and driving. *Alcohol, Drugs and Driving, 5,* 185–192.

Riemersma, J. B. J., Sanders, A. F., Wildervanck, C., & Gaillard, A. W. (1977). Performance decrement during prolonged night driving. In R. R. Mackie (Ed.), *Vigilance: Theory, operational performance and physiological correlates* (pp. 41–58). New York: Plenum.

Skipper, J. H., & Wierwille, W. W. (1986). An investigation of low-level stimulus-induced measures of driver drowsiness. In A. G. Gale, M. H. Freeman, C. M. Haslegrave, P. Smith, & S. P. Taylor (Eds.), *Vision in vehicles* (pp. 139–148). Amsterdam: Elsevier Science.

Tilley, D. H., Erwin, C. W., & Gianturco, D. T. (1973, January). *Drowsiness and driving: Preliminary report of a population survey.* Paper presented at the SAE International Automotive Engineering Congress, Detroit, MI.

Tock, D., & Craw, I. (1992, September). *Blink rate monitoring system for a driver awareness system.* Paper presented at British Machine Vision Conference, Leeds, UK.

Treat, J. R. (1980). *A study of precrash factors involved in traffic accidents.* Ann Arbor, MI: Highway Safety Research Institute.

Ueno, H., Kaneda, M., & Tsukino, M. (1994). Development of drowsiness detection system. *Proceedings of the 1994 conference on Vehicle Navigation and Information Systems (VNIS)* (pp. 15–20). Yokohama, Japan. (IEE Catalog No. 94CH35703).

Wierwille, W. W. (1994, May). *Overview of research on driver drowsiness definition and driver drowsiness detection.* Paper presented at 14th International Conference on Enhanced Safety of Vehicles, Munich, Germany.

Wierwille, W. W., & Ellsworth, L. A. (1994). Evaluation of driver drowsiness by trained observers. *Accident Analysis and Prevention, 26*(5), 571–581.

Wierwille, W. W., Wreggit, S. S., & Knipling, R. R. (1994). Development of improved algorithms for on-line detection of driver drowsiness. *Proceedings of the 1994 International Conference on Transportation Electronics* (pp. 331–340). Warrendale, PA: Society of Automotive Engineers.

Chapter 22

Attention Switching Time:
A Comparison Between Young
and Experienced Drivers

Simon A. Moss
Thomas J. Triggs
Monash University, Australia

Intelligent vehicle highway systems (IVHS) are designed to assist the driver by providing task-relevant information (National Highway Traffic Safety Administration, 1993). The introduction of these systems, however, may augment the mental workload imposed by driving. Specifically, IVHS can be regarded as an additional task, requiring drivers to alternate their attention between the system and the driving task. This requirement has the potential to disrupt, rather than enhance, overall performance, particularly in those individuals who cannot rapidly switch their focus of attention.

The present study examined the effect of age, mental workload, and driving experience on attention switching time. Attention switching can be regarded as a critical aspect of the driving task. In particular, the driver must continually shift their attention from one spatial location to another. Moreover, driving entails numerous cognitive processes, such as response selection, memory, and planning. These processes each demand attention. Unfortunately, humans cannot attend to all of these processes simultaneously. Hence, drivers must rapidly shift their attention from one set of processes to another.

Thus, driving performance will vary inversely with switching time. This claim was verified by Kahneman, Ben-Ishai, and Lotan (1973), who found a significant positive correlation between switching time and accident rate.

SWITCHING TIME AND AGE

The relation between switching time and age, outside the driving domain, is unclear. Braune and Wickens (1985) found that interaural switching time, that is, the time required to shift attention from one ear to the other, is a positive function of age. Analogous results were obtained by Barrett, Mihal, Panek, Sterns, and Alexander (1977).

In a study conducted by Hartley, Kielry, and McKenzie (1987; cited in McDowd & Birren, 1990), however, attention switching time, within the visual domain, did not vary with age. Hence, the nature of the task may dictate the relation between switching time and age. Alternatively, methodological differences may be responsible for this inconsistency.

SWITCHING TIME AND MENTAL WORKLOAD

Several studies have shown that a rise in mental workload prolongs switching time. For instance, in a study conducted by Weber, Burt, and Noll (1986), subjects were instructed to alternate their attention between items held in memory and items displayed on a screen. Switching time, using this paradigm, was shown to be a positive function of memory load.

Analogous results were obtained by Meiselman (1974). Subjects were required to switch their attention between the visual and auditory modalities. Meiselman observed that a reduction in mental effort prolonged switching time. Moreover, as proposed by Kahneman (1973), a rise in mental workload reduces the availability of mental effort. These claims, taken together, suggest that a rise in mental workload will prolong switching time.

Mental workload, of course, may not influence all forms of attention switching. Switching from one spatial location to another, for example, may be independent of mental workload.

SWITCHING TIME AND DRIVING EXPERIENCE

Several studies have shown that the mental workload imposed by driving attenuates with experience. In a study conducted by Brown and Poulton (1961), for instance, subjects performed an auditory detection task while driving a car. Experienced drivers outperformed their novice counterparts on the auditory detection task. This finding, according to Brown and Poulton, indicates that the mental workload imposed by the driving task varies inversely with experience. Moreover, as stated earlier, switching time has been shown to vary with mental workload. These findings, taken together, suggest that switching time, while operating a vehicle, may be slower in novice drivers vis-à-vis experienced drivers.

Driving experience, however, is typically confounded with age; in general, novice drivers are younger than their experienced counterparts. Moreover, as discussed, switching time may be a positive function of age. Accordingly, switching time, while driving, may be faster in novice motorists.

In the present study, switching time was examined in isolation, hereafter referred to as *baseline switching time*, and while operating a vehicle. This technique separates the effects of age and experience. For instance, suppose that switching time is a positive function of age. Baseline switching time will thus be faster in younger drivers, relative to their more experienced counterparts. Moreover, suppose that switching time is an inverse function of experience. Any retarding effect of driving on switching time will thus be more pronounced in the younger drivers.

To examine the relation between switching time and driving experience, the present study extended the paradigm devised by Sperling and Reeves (1980). This paradigm was readily adaptable to the driving domain. Subjects operated a driving simulator. A stream of continuously changing numerals was located in the visual scene peripheral to the roadway. Subjects, upon hearing the appropriate auditory cue, switched their attention from the roadway to the numerals and reported the first perceived digit. The subject's response was used to estimate the time required to switch attention.

The process of switching attention from one spatial location to another entails several components (LaBerge, 1973). In particular, switching time, as measured in the present study, presumedly encompasses the time associated with recognizing the auditory cue; the time associated with selecting, generating, and initiating a response; the time associated with disengaging attention from the driving task; the time associated with shifting attention from the driving task to the number stream; the time associated with entering the latter channel; and the minimum duration that each digit must be exposed before identification can proceed. For convenience, however, these components will be referred to collectively as *switching time*.

METHOD

Subjects

Two groups of subjects were used: young, inexperienced drivers and somewhat older, experienced drivers. The young drivers possessed, at most, 2 years driving experience, and comprised 8 males and 4 females, between ages 18 and 20. The experienced drivers possessed, at least 6 years driving experience, and comprised 7 males and 5 females, between ages 25 and 40. None of the 24 subjects had previously operated a driving simulator.

Apparatus

The experiment utilized a fixed-based car simulator, developed by Systems Technology Incorporated (for a detailed description, see Stein, 1990). The simulator consisted of four components. The first component, a VGA monitor, displayed the visual scene. This monitor was positioned 1.3 m from the subject. The displayed image was 40 cm in width and 30 cm in length. The second component, a stereo auditory system, produced the acoustic signals. The third component, a computer system, controlled the visual display, acoustic display, and the additional auditory cues. The final component, the cabin of a Falcon XEA sedan, was connected to the computer via several cables and force transducers.

Lateral position was controlled via the steering wheel. Rotations of the steering wheel were converted to electrical signals and then transformed into software language. Using a complex algorithm—the vehicle dynamics mathematical model—this information was used to determine lateral position.

The auditory cue was 500 Hz and lasted 100 msec. The volume of this cue was considered loud, but not uncomfortable.

Tracks

Each trial comprised a single driving track and lasted approximately 2.5 min. Six different tracks were used throughout the experiment. Each track comprised the same roadway elements; however, the order in which the various elements were encountered varied across the tracks. Subjects encountered each track on two occasions and thus performed 12 trials. The tracks were presented in a random order. The same track, however, was never used on consecutive trials.

The various roadway elements can be subdivided into four categories: *right curves*; *left curves*; *overtaking maneuvers*; and vehicle-free linear road segments, hereafter referred to as *vehicle-free straights.*

The length of each curve varied from 270 m to 300 m. Moreover, the minimum radius of each curve varied from 167 m to 250 m. Vehicle speed was preset at 90 kph (25 m/sec); hence, each curve required between 10.8 sec and 12 sec to negotiate. The overtaking maneuver entailed shifting into the right lane, passing a stationary vehicle, and returning to the left lane. These maneuvers were performed exclusively on linear road segments. Finally, the vehicle-free straights ranged from 60 m to 100 m in length and thus lasted between 2.4 sec and 4 sec.

Each road-element type appeared three times within a single track. During two of the three occasions, the auditory cue was presented. These cues corresponded approximately with the minimum radius of the curves, the passing phase of the overtaking maneuver, and the halfway point of the vehicle-free straights. The precise timing of the cues, however, was varied to preclude anticipation.

Number Stream

The number stream comprised a single-digit display. Each digit consisted of up to seven line segments. The numbers appeared at a rate of five per second. The number sequence was randomly generated; however, the same digit was not presented more than once within a 1-sec time interval. The numbers were coordinated with the auditory cues; the offset of the cue coincided with the transition between digits.

From the subjects viewpoint, the digits were 3 cm (1.4 degrees) in height. Moreover, the digits were located 5 cm from the top of the screen, 18 cm from the bottom of the screen, 28 cm from the left of the screen, and 10 cm from the right of the screen. The horizontal distance between the digits and the roadway varied throughout the trial. In particular, this distance was contingent on the curvature of the roadway, varying between 5 cm (2.3 degrees) and 11 cm (5.0 degrees), with a median of 8 cm (3.6 degrees).

Switching time, as defined here, was computed using the following formula:

$$\text{Switching time} = 200 \ (n + 1) \pm 100 \text{ msec}, \tag{1}$$

where n denotes the number of digits displayed between the offset of the cue and the presentation of the reported digit. This formula takes into account the rate of presentation and the duration of the auditory cue.

Procedure

The driving and digit-naming tasks were performed both concurrently and alone, yielding a total of three conditions. In the driving-only condition, subjects were instructed to disregard the cues. The task essentially entailed negotiating curves and overtaking vehicles. Subjects were further instructed to drive in the center of the left-hand lane and to keep the accelerator fully depressed. The maximum speed was preset at 90 kph.

In the numbers-only condition, subjects, on hearing the cue, switched their attention from the roadway to the number stream, reported the first perceived digit, and then switched their attention back to the roadway. Moreover, subjects were instructed to ignore the numbers until the appropriate cue was presented. This instruction was reinforced before each trial. Finally, in this condition, subjects did not perform the driving task. Instead, the driving scene was programmed in advance.

In the dual-task condition, subjects performed both the driving and digit-naming tasks simultaneously. Subjects were again instructed to ignore the numbers until the appropriate cue was presented.

The practice session comprised three trials. During these trials, subjects performed both tasks concurrently. The experimental session comprised three trials for each condition, yielding nine trials in total. The order of presentation was counterbalanced across subjects.

Subjects were granted a 20-sec rest period between each trial. Each experimental session lasted approximately 35 min.

Data Treatment

The subject's lateral position was recorded every 200 msec. These recordings were grouped according to road-element type. The standard deviation of lateral position was then calculated for right curves, left curves, and vehicle-free straights. A low standard deviation represents efficient performance, whereas a high standard deviation represents poor performance.

The data pertaining to driver performance were subjected to a three-way analysis of variance (Age × Condition × Road-element type), where both condition and road-element type constituted repeated measures. The significance criterion was set at 5%. The switching time data were also analysed using a three-way analysis of variance (Age × Condition × Road-element type).

RESULTS

Driving Data

Figure 22.1 shows the standard deviation of lateral position pertaining to each road-element type, for both young and experienced drivers. Panel A corresponds to the driving-only condition, whereas panel B corresponds to the dual-task condition.

The standard deviation of lateral position was found to be independent of age, $F(1, 22) = 2.46$, $p > .05$. A significant main effect, however, was obtained for condition, $F(1, 22) = 7.81$, $p < .05$, and road-element type, $F(2, 44) = 4.86$, $p < .05$. The interaction between condition and road-element type, however, was also significant, $F(2, 44) = 4.10$, $p < .05$.

To ascertain the source of this interaction, the simple main effects of condition were analyzed for each road-element type individually. The standard deviation of lateral position was greater in the dual-task condition, relative to the driving-only condition for both left curves, $F(1, 44) = 4.90$, $p < .05$, and right curves, $F(1, 44) = 7.24$, $p < .05$. For the vehicle-free

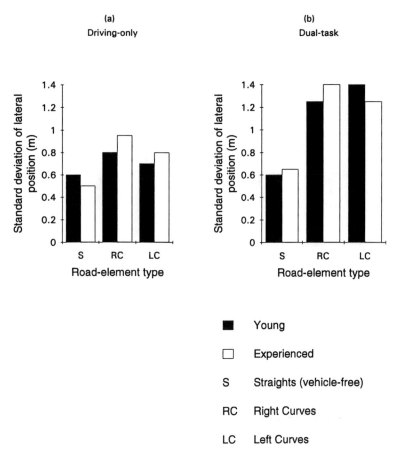

FIG. 22.1. The standard deviation of lateral position as a function of road-element type and driving experience in the (a) driving-only condition, and (b) dual-task condition.

straights, however, the standard deviation of lateral position was constant across conditions, $F(1, 44) = 2.30$, $p > .05$.

None of the remaining interactions were statistically significant.

Switching Time Data

The relation between switching and road-element type, for both young and experienced drivers, is shown in Fig. 22.2. Panel A corresponds to the numbers-only condition, whereas panel B corresponds to the dual-task condition.

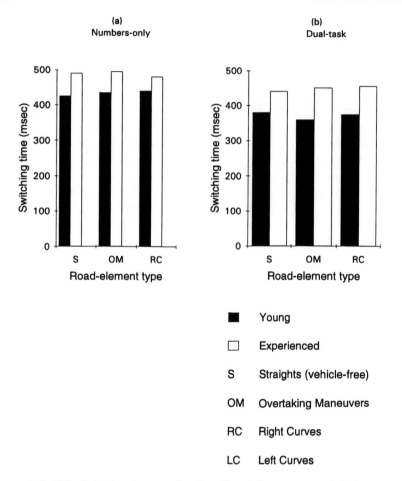

FIG. 22.2. Switching time as a function of road-element type and driving experience in the (a) numbers-only condition, and (b) dual-task condition.

Switching times were faster in younger drivers when compared with their more experienced counterparts, $F(1, 22) = 14.32$, $p < .05$. Contrary to expectations, however, age did not interact significantly with condition, $F(1, 22) = 1.88$, $p > .05$, revealing that the relation between age and switching time was independent of condition.

Interestingly, switching time was faster in the dual-task condition, relative to the numbers-only condition, $F(1, 22) = 11.69$, $p < .01$. Switching time, however, was independent of road-element type, $F(3, 66) = 2.01$, $p > .05$. Moreover, none of the remaining interactions achieved statistical significance; $F < 1$ for Age × Road-element, Condition × Road-element, and Age × Condition × Road-element, respectively.

DISCUSSION

Switching Time and Age

Switching time, in both the numbers-only and dual-task conditions, was faster in the young drivers. This result demonstrates that switching time, as defined in the present study, is a positive function of age.

Notwithstanding this finding, the time required to shift attention may not vary with age. Instead, the observed relation between switching time and age can be ascribed to a variety of other factors. Several studies, for example, have shown that vigilance, the ability to maintain a state of readiness, may deteriorate with age (e.g., Kirchner, 1958). Moreover, as already stated, switching time encompasses several components in addition to shifting attention, including inter alia, recognizing the auditory cue, selecting and initiating the appropriate response, and identifying a digit. Any one of these components may vary with age (see, e.g., Newman & Spitzer, 1983; Stelmach, Goggin, & Garcia-Colera, 1987).

The Impact of Driving on Switching Time

Interestingly, switching time was faster in the dual-task condition, relative to the numbers-only condition. Hence, a rise in mental workload expedited, rather than impaired, performance on the digit-naming task.

The notion of arousal can be invoked to accommodate this finding. In particular, the level of arousal was presumedly heightened during the dual-task condition. This rise in arousal may have reduced switching time.

Several possible explanations can account for the proposed relation between arousal and switching time. First, arousal may directly facilitate one or more of the processing stages that underlie attention switching, including inter alia, auditory recognition, response selection and initiation, digit identification, or attention switching per se.

The second explanation concerns the relation between arousal and visual dominance. Shapiro, Egerman, and Klein (1984), for instance, showed that visual dominance—the tendency to process visual information in preference to acoustic information—is reduced at high levels of arousal. Hence, a rise in arousal may attenuate the reaction time to auditory cues and thus curtail switching time.

The third explanation concerns the relation between arousal and central dominance. Several studies have shown that, during heightened arousal, central dominance—the tendency to process foveal information in preference to parafoveal information—is reduced (e.g., Shapiro & Johnson, 1987). In the present study, the digit stream was located peripheral to the roadway, and hence, the numbers can be regarded as parafoveal information. Accordingly, a reduction in central dominance could account for the purported relation between arousal and switching time.

Switching Time and Driving Experience

Contrary to expectations, the effect of driving on switching time did not vary with experience. This finding suggests that switching time, while operating a vehicle, does not vary with driving experience per se.

Two explanations can be invoked to accommodate this result. First, the mental workload imposed by the driving task used in this experiment may not have varied with experience. The driving task, for instance, may not have been sufficiently challenging to differentiate between novice and experienced drivers.

Second, switching time, as defined in the present study, may not vary with the mental workload imposed by driving. This notion was supported by the finding that switching time was independent of road-element type. Conceivably, switching between spatial locations may not demand attentional resources. Alternatively, driving and switching may consume attention from separate resource pools (see, e.g., Navon & Gopher, 1979). Accordingly, the amount of attention available for switching will be independent of the mental workload imposed by driving.

Methodological Issues Arising From the Present Study

Several important methodological issues emerge from the present study. First, subjects were instructed to disregard the number stream until the onset of the appropriate cue. Although unlikely, subjects may not have followed this instruction either deliberately or inadvertently. In particular, subjects may have focused on the number stream before hearing the cue. This strategy, of course, can only be adopted by those subjects who can simultaneously distribute their attention across both tasks. Hence, switching time, as defined in the present study, may be contingent on a variety of extraneous factors.

Second, the simulated driving task was somewhat simple, and thus, may not have differentiated between novice and experienced drivers. Perhaps, under these conditions, the mental workload imposed by driving was not a function of experience. Of course, under different conditions, mental workload, and thus switching time, may vary with driving experience.

A further study was conducted to examine these issues. Overall, 24 subjects participated in this experiment: 12 young drivers and 12 experienced drivers. In this study subjects were presented with two, rather than one, number streams. These number streams were located to the left and right of the roadway, respectively. On hearing the appropriate cue, subjects: (a) switched their attention from the driving task to one of the two number streams, hereafter referred to as the *primary switch*; then (b) switched their attention from the first number stream to the second, hereafter referred to as the *secondary switch*; and finally, (c) reported the first perceived digit in each of the two number streams, respectively. Conceivably, primary switching time, as suggested earlier, could be attenuated by switching attention before the appropriate cue is presented. Secondary switching time, however, should not be influenced by this strategy and thus is not contingent on such extraneous factors.

Moreover, in this study the driving task was more challenging and perhaps more realistic. First, subjects were required to modulate their speed via the accelerator and brake pedals. Second, all overtaking maneuvers were performed on curves, rather than straights.

Nonetheless, these amendments did not modify the pattern of results. In particular, switching time was faster in the dual-task condition vis-à-vis the single-task condition, $F(1, 22) = 6.95$, $p < .05$. Moreover, switching time did not vary across road-element type, $F(3, 66) = 2.01$, $p > .05$. Finally, the effect of driving on switching time did not vary with experience; that is, the interaction between age and condition was not significant, $F(1, 22) = 2.35$, $p > .05$. Hence, the results of the present study cannot be ascribed to premature switching. In addition, the original findings generalize to more demanding conditions.

CONCLUSIONS

In summary, the present study showed that switching time while operating a vehicle is a positive function of age, but independent of driving experience per se. Moreover, the driving task expedited, rather than retarded, switching time. These results discredit the notion that switching time is a positive function of mental workload, and thus, are encouraging for IVHS being used by younger drivers. In particular, any additional tasks required by a high technology system should not, according to the present results, retard switching time. Although these systems may increase the required rate of switching, and thus influence complex driving performance, they should not differentially disadvantage young drivers.

The present study, however, entails several limitations. First, these findings may not generalize to real-world driving. Second, the present study examined only one aspect of switching, namely, switching between spatial locations. Other forms of switching, such as switching from one cognitive process to another, may yield conflicting results.

REFERENCES

Barrett, G. V., Mihal, W. L., Panek, P. E., Sterns, H. L., & Alexander, R. A. (1977). Information processing skills predictive of accident involvement for younger and older commercial drivers. *Industrial Gerontology, 4,* 173–182.

Braune, R., & Wickens, C. D. (1985). The functional age profile: An objective decision criterion for the assessment of pilot performance capacities and capabilities. *Human Factors, 27,* 681–693.

Brown, I. D., & Poulton, E. C. (1961). Measuring the spare "mental capacity" of car drivers by a subsidiary task. *Ergonomics, 4,* 35–40.

Kahneman, D. (1973). *Attention and effort.* Englewood Cliffs, NJ: Prentice-Hall.

Kahneman, D., Ben-Ishai, R., & Lotan, M. (1973). Relation of a test of attention to road accidents. *Journal of Applied Psychology, 58,* 113–115.

Kirchner, W. K. (1958). Age differences in short-term retention of rapidly changing information. *Journal of Experimental Psychology, 55,* 352–358.

LaBerge, D. (1973). Identification of two components of the time to switch attention: A test of a serial and parallel model of attention. In S. Kornblum (Ed.), *Attention and performance* (Vol. 4, pp. 71–85). New York: Academic Press.

McDowd, J. M., & Birren, J. E. (1990). Aging and attentional processes. In J. E. Birren & K. W. Schaie (Eds.), *Handbook of the psychology of aging* (pp. 222–233). San Diego: Academic Press.

Meiselman, K. C. (1974). Time taken to switch attention between modalities as a function of mental effort. *Perceptual and Motor Skills, 39,* 1043–1046.

National Highway Traffic Safety Administration (NHTSA). (1993). *Addressing the safety issues related to younger and older drivers.* Report to Congress.

Navon, D., & Gopher, D. (1979). On the economy of the human-processing system. *Psychological Review, 86,* 214–255.

Newman, C. W., & Spitzer, J. B. (1983). Prolonged auditory processing time in the elderly: Evidence from a backward recognition-masking paradigm. *Audiology, 22,* 241–252.

Shapiro, K. L., Egerman, B., & Klein, R. M. (1984). Effects of arousal on human visual dominance. *Perception and Psychophysics, 35,* 547–552.

Shapiro, K. L., & Johnson, T. L. (1987). Effects of arousal on attention to central and peripheral visual stimuli. *Acta Psychologica, 66,* 157–172.

Sperling, G., & Reeves, A. (1980). Measuring the reaction time of a shift of visual attention. In R. S. Nickerson (Ed.), *Attention and performance* (Vol. 8, pp. 347–360). Hillsdale, NJ: Lawrence Erlbaum Associates.

Stein, A. C. (1990). *A part task driving simulator based on low-cost Microcomputer Technology: Proceedings of the 1990 Image V conference.* Pheonix, AZ: IMAGE Society.

Stelmach, G. E., Goggin, N. L., & Garcia-Colera, A. (1987). Movement specification time with age. *Experimental Aging Research, 13,* 39–46.

Weber, R. J., Burt, D. B., & Noll, N. C. (1986). Attention switching between perception and memory. *Memory and Cognition, 14,* 238–245.

Chapter 23

The Role of Standards
for In-Vehicle MMI

Andrew Parkes
University of Leeds, UK

National and international standards are being raised for in-vehicle man–machine interaction (MMI). That is, not just criteria for the physical characteristics of the interface displays and controls, but also the form and content of the "dialogue" between the driver and the vehicle. The scope of such standards will grow ever wider as the potential of display and control devices, and the range of applications and services within the vehicle, increases. Such developments are controversial, and two questions need to be answered: Can useful standards be developed? Should such standards be developed?

Much of the material presented here was stimulated by a panel discussion at the International Ergonomics Association Symposium on IVHS/RTI (1994). This discussion drew together academics and industrial speakers from Europe, the United States, and Japan, and covered the history of in-vehicle standards, current initiatives, and hopes for the future. Although stimulated by a particular event in time, it is hoped that this discussion of the role of standards in in-vehicle MMI raises general issues for the contribution of human factors and ergonomics in the systems design process.

WHAT SORT OF STANDARDS ARE THERE?

There are two terms that occur frequently in the discussion of in-vehicle MMI: transport information and control systems (TICS) and road transport informatics (RTI). Neither term is defined precisely. Both are seen as technical terms covering some of the areas of interest of the large collaborative programs in Europe (e.g., advanced transport telematics) and the United States (e.g., intelligent vehicle highway systems). These terms, and the names of the research programs, are continually evolving, with new ones entering the arena at frequent intervals. However, it appears that a loose consensus has emerged amongst professionals in the area, with the term intelligent transportation systems (ITS) acting as an umbrella for all vehicle-based and infrastructure-supported developments, and TICS MMI referring to those interactions between driver and vehicle of interest to this discussion.

There are three types of standard relevant to MMI in the vehicle: *product, performance,* and *procedural* (Howarth, 1987).

Product Standards

Product standards give the designer specific recommendations for product characteristics, such as size, shape, color, and position. As a result, these types of standard are easy to understand, and enable easy evaluation of the finished product. However, by their nature, they apply to specific parts of a system, and conformance does not in itself provide any guarantee that the whole system will be usable, efficient, or safe for the driver. Product standards are also technology dependent. For example, recommendations for character height for text information presentation are not absolute; they depend on the color combinations used, the contrast ratios of foreground and background, the resolution of the display device, and the stability of the image. As such, a standard that is reasonable with one type of display technology may soon become outdated as technological improvements are made. Product standards may also suffer from being too prescriptive, and reduce the degree of innovation in design that might be regarded as healthy.

Performance Standards

Performance standards specify the user performance levels that should be achieved with a particular system. For example, some commentators feel that drivers should be able to obtain the information they require from a display, without taking their eyes from the forward view of the traffic situation for more than a certain length of time. Although fraught with controversy over the exact value to be prescribed and the measurement technique to be employed (Parkes, 1995), the approach has an intuitive appeal. It is not concerned with commenting directly on the format of the display or on the supporting technology, but rather the focus is solely on the abilities of the driver population (however defined) to assimilate the information provided.

Such standards have great potential, but are proving difficult to produce. Although they are technology independent, and encourage innovative design, they give the designer no direct guidance, either on what to produce or on how to perform an appropriate evaluation of that product. Extending the example used previously, a hypothetical performance standard might say something like the following: Drivers should be able to determine their speed from the speedometer display with an average glance duration of under 2 sec. But the designer might immediately ask a series of difficult questions, such as:

How do I measure a glance duration?

How do I choose representative subjects?

What driving tasks should the driver engage in at the same time?

Can I do this test in a simulator rather than on the real road?

What type of simulator is appropriate?

What evidence supports this 2-sec criteria?

Are two glances of 1 sec each, in quick succession, the same as a single glance of 2 sec?

How bad is it if I get figures of 2½ sec?

Does the same criterion stand for glances to a dynamic map display?

How much is this kind of test going to cost?

How long will it take?

Several things become immediately apparent. First, the ergonomics standards community have very few authoritative answers to these questions. And, second, it is unrealistic to expect the designers of systems to perform these evaluations for themselves. To maximize the validity of such performance testing, it would be necessary for it to be conducted by experts who are familiar with the relevant methods and metrics. It is important to remember that performance testing focuses on the user of the system. Thus, there are human subjects who must be dealt with according to strict codes of ethical conduct. Such trials can be expensive in terms of time and resources. Certain types of performance testing might require specialist equipment and data analysis software (e.g., if psychophysiological monitoring is to be involved). Formal evaluations will also entail substantial time spent running trials themselves (screening, briefing, familiarization, trial, debriefing). It is obviously more appropriate for design and development teams to spend their time in activities in their own areas of expertise, rather than organizing, and running human factors evaluations. This would require manufacturers to establish their own specialist facilities, or to utilize accredited test centers.

The *validity* of performance testing, in an area as new as TICS MMI, is contentious. Many readers will be well versed in established notions of content validity, construct validity, and concurrent validity; each needing to be satisfied before a test can be deemed valid in the accepted scientific sense. *Content validity* refers to the constituent materials of the test, each item should be deemed appropriate to the test, and it should form part of a cohesive whole. An often-cited example is of mathematical testing; here, items should not be presented in a way such that advanced verbal abilities of the subject are critical to the person taking the test to understand what is required. Nor should the test purport to measure general mathematical ability if only algebra is addressed. *Construct validity* is demonstrated when the test items capture the hypothetical quality or trait it was designed to measure. *Concurrent validity* refers to the correlation between performance of a prescribed group on the test, and known skill levels and performance of that group in real task situations.

In practice, performance standards (and indeed other types of standard) may be described as having *consensual validity*. That is, the more persons who concur with a proposition, the greater the likelihood that the proposition is true. Although the history of science is full of wonderful examples of paradigm shift (i.e., a point in time at which previously held truths are rejected in favor of a new proposition), international standards are generated in a more mundane world where consensus by a relatively small number of interested experts is sufficient, and indeed the only practical possibility, for standard formulation within a time frame that is useful to the commercial community.

Procedural Standards

Procedural standards can describe the program of analysis and testing, which must be followed during the design process, to provide assurance that best practice has been followed in an attempt to maximize usability and safety. Such standards can prescribe the knowledge gathering, testing, and documentation processes to support the design. They are common in areas where there is a strong single purchaser, who can use the standard as a means of ensuring the supplier complies with contract objectives (e.g., military purchases of complex systems). Such standards are obviously much less common where there are strong suppliers and a relatively disparate purchasing population, as occurs with the private vehicle market. In this case, reliance would be placed on a strong certification body acting as the champion of the end user. Although such standards have not been developed to date for road vehicle MMI, there are examples of movements in that direction. Most notably, the sponsors of the large research programs in Europe (e.g., transport telematics), where an emphasis is placed on fully documented test and evaluation procedures, and who are producing their own guideline documentation for potential projects to follow. Two of the procedural standards best known in the human factors community are not directly focused on automotive systems, but nevertheless give a useful lead on how such standards might develop.

ISO 9000 is widely regarded as a quality standard. Many organizations aim for accreditation to the standard, and indeed, use it as a marketing tool to promote the perception of high quality work. This is a good example of a common misperception of the power and authority of standards. Adherence to ISO 9000 does not, and cannot, guarantee a good quality product—be that an item for manufacture, or a research report. The standard is primarily concerned with identifying the decision-making process in an activity, and ensuring that all decisions are made by an appropriate person in the organization (based on a documented task analysis), and that the person has had the required training. Conducting activities in accord with the standard should ensure that all decisions are traceable, and evidence of the reasoning for all decisions is available. The standard in itself cannot prevent poor decisions being taken, and although appearing daunting to the uninitiated, is simply an accumulation of best practice put together by experts experienced in complex project management. Depending on the size of the project work, adherence to the standard may or may not be cost effective. However, if a complex project is undertaken, particularly one involving safety critical systems, the procedures documented are an excellent source of guidance to the project team, and can be viewed as insurance by the customer, who knows that a thoroughly professional approach has been adopted.

A second procedural standard of note is ISO 9241, which deals with ergonomic requirements for office work with visual display terminals. In particular Part 11, guidance on specifying and measuring usability, provides a high-level framework that could form the basis for a similar standard tailored specifically to the usability of TICS MMI. *Usability* is defined in ISO 9241 as a function of efficiency, effectiveness, and satisfaction. This maps fairly well onto the declared program aims of DRIVE and PROMETHEUS, which sought improvements in efficiency, safety, and acceptability of transport systems. The standard seeks to explain how to identify the information that should be taken into account when considering the usability of products. A heavy emphasis is placed on describing the context of use of the product or system, and identifying user requirements in terms of precise and meaningful measures. Declaration of the formal description of the context of use of the system should allow such a context to be reproduced in subsequent evaluation trials, and user requirements as targets with actual numerical values, allows objective verification.

Finally, in the transport telematics domain, the DRIVE 2 program supported work to demonstrate the feasibility and effectiveness of the application of system safety techniques to ITS. It produced a framework for "prospective systems safety analysis" and for "retrospective system safety evaluation." The focus was on system software build and documentation, but the approach to the identification and the rigorous examination of potential failure modes for systems, provides useful analogies for the domain of in-vehicle MMI in the future.

DO MANY RELEVANT STANDARDS EXIST AT PRESENT?

Currently, the existing standards that are directly relevant to systems design (see Parkes & Ross, 1991, for review) fall into six main categories: driver hand control reach, drivers' eye ellipses, location of hand controls, location of indicators and tell-tales, symbols for controls, and symbols for indicators and tell-tales.

The standards relating to the physical characteristics of the driver will still be relevant when advanced in-vehicle telematics systems are introduced, but other categories will require much more work. Yet, other areas will have to be addressed for the first time. For example, the introduction of new applications such as route guidance (RG) or adaptive cruise control (ACC) may require a whole new set of controls and symbols. Also, standards may need to be developed for tactile displays and auditory information, or for virtual image visual displays, or for controls activated by voice command. These are areas where it is difficult to find useful information that could be adapted from other task domains, such as office systems. This is because either the nature of the displays and controls is different, or that the context of usage is so distinct. Consequently, there is much original research to be conducted before a sufficient body of knowledge exists for scientific consensus to be reached on these interesting, but demanding, topics.

In addition to the need to collect empirical data to support the generation of criteria for in-vehicle MMI standards, thought must be given to new ways of formulating standards. They will need to encompass the complexity of systems in future vehicles that may be adaptive, multimodal, integrated with other systems, or even variable in terms of the locus of control between driver and vehicle for such actions as vehicle speed or heading.

Scientific consensus needs to be multidisciplinary, and include input from manufacturers, designers, psychologists, specialists in driver behavior, and accident researchers. However, given the lack of directly comparable scientific studies, many divergent opinions exist, and consensus will only emerge slowly, if at all.

WHAT PRACTICAL STEPS CAN BE TAKEN?

The immense difficulties in producing valid and supportive standards should not be taken as a reason for holding back from attempting to move the area forward. Standards need to be developed, not as inhibitors to the introduction of new technology to the vehicle, but as support materials for those involved in design and development.

Compliance with standards can only become mandatory by national or international directive. For example, European directives are issued by the Council of the European Communities, and each member state must bring into force the laws, regulations, and administrative provisions necessary to comply with the directive. It is interesting to note that the European directive with the closest area of coverage to that of RTI is that of "minimum safety and health requirements for work with display screen equipment" (Commission of European Communities [CEC], 1990). This directive explicitly excludes "drivers cabs" and "computer systems on board a means of transport" from the terms of reference, precisely because it opens up so many difficult questions.

However, progress is being made. Technological advances in programs (such as PROMETHEUS, DRIVE, ITS America, and VERTIS) have also stimulated human factors activities, and led to collaborative efforts in areas common to several applications. For example, in PROMETHEUS, a specific working group (WG4) attempted to define a "Driver Model" against which various MMI designs could be tested. The results, however, proved discouraging, and even attempts at producing simplified models for parts of the driving task proved to be so complicated that they could be of little practical value to the design process. A change in approach became necessary as specific design rules for MMI were not available.

The role of human factors specialists in PROMETHEUS became more of an awareness raising lobby group. However, the attempts by WG4 to bring human factors considerations into the development projects were often regarded as mere interference by parts of the engineering community. A problem being faced not only in PROMETHEUS but also other current large programs was that in many cases, the design and development community expected from human factors specialists detailed specifications on the MMI for systems that were barely specified in engineering terms (Hallén, 1994). There are of course certain principles that can be derived from the huge banks of knowledge under the various labels of psychology, cognitive engineering, human factors, cybernetics, and physiology, and there is no shortage of general texts on display and control design. However, the level of detail necessary for taking a product to successful implementation in the market is not, and cannot, be made available. Product-based standards or guidelines can only ever make sure a product achieves some minimum level of efficiency.

The strongest message that always emerges from the human factors community is to consider the requirements and limitations of the end users of the system from the very earliest stages of that system design. At present, the greatest contribution comes in areas where a structured approach is taken to identifying the key stakeholders in a system design, then user requirements are identified in an objective form that can later be used as the basis for evaluation, and sufficient resources are devoted to user testing and redesign before market launch. Each of these stages would be aided by the availability of validated performance and process standards.

INTERNATIONAL STANDARDS BODIES

In 1991, discussions were started in Europe on how to bring PROME-THEUS and DRIVE projects into commercially viable products, and what support in terms of standardization was needed to achieve successful implementation (Hallén, 1994). This resulted in a new technical committee (TC278) being established within CEN (Comité Européen de Normalisation). In all 13, working groups were created in TC278, one of which (WG10) was devoted to MMI. Hallén (1994) observed that the area of MMI seemed to be controversial from the outset, with debate over the desirable level of standardization due to competitive concerns—that is, how much of the MMI should be standardized, and how much should be left to the discretion of the individual manufacturer to establish competitive advantage in the marketplace.

A strong driving force behind this working group's standardization efforts was the concern that uncontrolled introduction of TICS could result in potentially unsafe systems on the road. It was clear that the performance of a technically efficient system is reduced if usability issues are not sufficiently attended to; and whereas it would be expected that good design would evolve naturally through competition in the marketplace, the time lag and the consequences of systems that are difficult to use, may be unacceptable. The first meeting of this group was December 1992, and from the beginning the intention was to produce performance standards, wherever possible, and to avoid overreliance on product standards.

Previous work on standardization for road vehicles had been the task of International Standards Organization (ISO, 1989). The standards produced within this group have been based on long-term experience of the systems being standardized, such as interior packaging procedures, symbols, control location, and so on. Hallén (1994) characterized the efforts for standardization of TICS as quite different, in that very few systems exist in the market, and it may even be difficult to predict what TICS functions and support systems will be developed over the coming years. Hallén also raised an important point:

> TICS must not be regarded in isolation when standards for these new systems are considered, but one has to bear in mind that TICS will be operated by the driver in parallel with conventional and traditional systems in the vehicle. It is also conceivable that standards developed for TICS may be applicable to other, already existing systems that are not related to TICS. (p. 2)

At present, there are two technical committees of the ISO that look at MMI issues of new telematics systems in vehicles. In broad terms, one has the primary responsibility for in-vehicle systems relevant to road transport informatics (RTI), and the other has the primary responsibility for overall system aspects, as well as infrastructure aspects for RTI. However, at the individual subcommittee and working group level, this distinction is a little blurred.

The terminology of standards bodies, the heavy use of acronyms, and the mazelike structure of committees, subcommittees, and working groups can be confusing even to those experts initiated into the process and contributing at the working level. Although the names and reporting structures of the various bodies are of little direct importance to the wider research and design communities, a review of the work plans and time scales for the development of standards is useful.

ISO TC22 Road Vehicles SC13
"Ergonomics Applicable to Road Vehicles"

This subcommittee is divided into a number of working groups.

WG8 Traffic Informatics Control Systems on Board—MMI. WG8 is the extension of the original CEN group that emerged from PROMETHEUS. The original composition of the ISO working group and the workplan reflected the needs and interests of the predominantly European membership. Since its inception, the WG has increased in breadth of membership to include a fully international membership, and the workplan has been rationalized. The following priorities have been established:

First Priority

- Requirements for visual information presentation

- Requirements for auditory information, including tonal and vocal coding
- Dialogue management/user-system interaction
- Visual demand measurement methods

Second Priority

- Principles for choice of information presentation mode
- Message presentation
- Symbols (in conjunction with WG5)
- Evaluation methods

In addition, some supporting activities are being completed, with internal technical reports dealing with issues of terminology, definition of parts of the driving task, and descriptions of types of information that will be available to the driver.

It is worth noting that work items of the first priority are intended to be enabling standards that are application independent. This means that the standards are intended to be equally relevant to the development of any future system to be introduced to the vehicle. To explain by way of the example of visual demand used previously, the standard will make no attempt to provide explicit pass/fail criteria in terms of maximum average glance duration, or frequency for particular displays, such as route guidance. Instead, the standard will (after achieving consensus among the experts involved in forming the draft) provide definitions of many of the concepts used when visual demand is considered, and explain data capture techniques and metrics used for reporting demand in a way that will lead to reliable comparison with reference values. As such, the standard needs to consider recommendations for subject selection and screening, context of evaluation, and data reduction and analysis.

In short, the standard will not state that "drivers should be able to determine their speed from the speedometer display with an average glance duration of under two seconds." Rather, it will provide a source document for those who wish to consider visual demand as part of an evaluation. The standard will describe best practice in demand measurement technique, and thus act as an enabling document, ensuring that results from evaluation adhering to the standard can be readily compared.

Other first priority items lend themselves to more prescriptive design guidance. For example, the standard on requirements for visual information presentation will give guidance for designing and specifying the image quality of information to drivers from in-vehicle TICS. The recommendations are independent of display technology (with the caveat that head-up displays are not covered) but are not intended to be applicable to the display of pictorial images (or, therefore, maps).

The standard for auditory information presentation will have two parts. The first will be a set of guidelines for in-vehicle messages from TICS systems. It will aim to outline message features and functional features that have to be taken into consideration to give drivers comprehensible and useful information and to prevent auditory or mental overload. The second part of the standard will present a procedure for auditory information assessment by a driver panel.

It can be argued that this group is taking a top–down approach to standard development. That is, it attempts to draw together broad human factors principles that could be applied to the design and evaluation of any TICS system.

ISO TC204 Transport Informatics and Control Systems

WG13 Human Factors and MMI. This working group originally focused on off-board information presentation (e.g., variable message signs) but has been restructured by the membership to meet the pressing needs of industry and program management. The current workplan has four main work items: human factors and safety TICS bibliography, taxonomy of driver–vehicle transactions associated with advanced navigation and route guidance, operational standards for driver–vehicle control systems, and integration of driver information and control systems. It can immediately be seen that the approach of this group is different from ISO TC22 SC13 WG8. It is worth looking briefly in more detail at the four substantive work items.

The human factors and TICS bibliography will encompass research and development publications worldwide. It is recognized that human factors and MMI standards must be based on well-founded research material. Therefore, it is essential to create an up-to-date index of research in this field. Hopefully, a result of this work will be a fully annotated bibliography available as a key word accessible PC database.

The second major activity relates specifically to the application of route navigation and guidance. To create a basis for common understanding, a glossary of terms will be produced to define various driver–vehicle interactions associated with navigation and guidance. Such interactions can then be organized into a hierarchical taxonomy of functions, tasks, and activities. This is seen as the first step toward a coherent set of guidelines for the design of these systems.

The functional performance of driver–vehicle control systems (DVCS) should be standardized with respect of the information provided to the driver, the control input by the driver, and the response of the system under various driving scenarios. These elements need standardization to reduce the difficulties that may be experienced if drivers move from one vehicle to another. There should be consistency in the way DVCS work so that expectations established in one vehicle successfully transfer to other vehicles. At present a successful standard approach to the positioning of control devices such as the steering wheel and pedals exists, but drivers often make errors with minor switchgear when changing vehicles. However, the most serious consequence may be inadvertent activation of the windshield wipers when the desired response was to turn on the headlights. When we move to systems that operate support or partial control over functions, such as lateral and longitudinal control of the vehicle itself, then the consequences of driver confusion are potentially more serious, and it is reassuring to note that these issues are being given such a high priority.

The final work item of this group is perhaps the most ambitious. As new TICS systems are deployed, drivers will have an increasing number of displays and controls, and the potential for a corresponding increase in workload. It is important that these systems are integrated in a way that will minimize driver workload, be easy to learn, and facilitate safe and convenient operation. The activities in this final work item are intended to lead to a list of TICS functions, their definitions, the relations between them, and a framework for a functional interface integration specification. This work takes an innovative approach to utilizing information theory to determine precise information content values for a range of conceivable messages, and by weighting the values in terms of "criticality" to determine message priorities and maximum values. A starting point for this work, therefore, is a complete description of information sources and display and control mediums.

If the approach of the ISO TC22 group could be characterized as top–down, then the approach of this ISO TC204 Group can be characterized (or perhaps oversimplified) as bottom–up. It is certainly true that the focus is on the requirements for the development of particular applications, or combination of applications. The contrast in approach between the two groups is interesting, and raises general issues for the provision of human factors knowledge to the developers of TICS. Should information be packaged relevant to a particular application, or should it be generic? Certainly, in practice, a designer is likely to be focused on a particular product. If all relevant standards and guidelines for design and evaluation (of a collision avoidance system, for example) are contained in one compendium document, then one aspect of the job at least is simplified. If, however, the designer is faced with the task of integrating a collision avoidance system, a traffic information system, and a route guidance system, then perhaps generic information relating to visual or auditory displays could be easier to assimilate. The two approaches are not mutually exclusive, and of course much of the information contained within the standards from the two different approaches should be common, or at least corroborative.

CONCLUSIONS

With all standards, it is important to recognize that they have to be realistic in scope, and they are not intended to act as some legislative harness constraining innovative design. The standards themselves are not mandatory; compliance only becomes a formal requirement if the particular nation–state puts legislation in place to make it so. Serving on standards committees is a largely voluntary activity, and active participation comes from individuals with genuine interests in the particular fields, but they also have to represent their parent national and commercial interests. Such interests tend to be conservative, with a reluctance to become too specific, or be seen in any way to be restrictive. As such there is the danger that standards, though well intentioned, can become so general that they have little impact on the design community.

There are several long-term requirements of TICS MMI standards. They must aid the design of TICS and make the difficult compromise between being specific enough to give clear guidance, and general enough to be widely applicable; and they must be able to encompass differing end user populations with their differing cultural attributes (e.g., in relation to control stereotypes) and performance abilities (e.g., perceptual abilities of young and elderly). They must also guide system developers through the maze of difficulties inherent in the attempts to introduce new technologies, without imposing demands that will slow the progress down, nor constrain free competition of technological excellence in the marketplace. Standards need to be put in place to help designers resolve problems associated with the allocation of function between the driver and the vehicle, and also for the integration of systems with each other around the central primary driving task. Finally, standards need to show how valid evaluations of design options can be performed.

If performance standards for in-vehicle MMI can be developed, they will encourage and facilitate good design. However, we are still some way away from fully documented and validated performance standards. In the meantime, product and process standards will continue to evolve, and they certainly have the potential to minimize the occurrences of poor design of displays and controls; but, in the near future, there is little they can contribute to the design of complex interactive systems for use in the dynamic traffic environment.

Although the ultimate goal of fully validated and comprehensive performance-based standards may still be several years from being achieved (if indeed possible at all), it is important for experts working in the relevant committees to provide usable statements as early as possible. Full validation of guidelines and recommendations can only be achieved through the continuing accrual of empirical evidence. However, there is a case to be made for taking state-of-the-art human factors knowledge as the basis for informed expert opinion on these issues, and disseminating as widely as possible to the potential user audience.

An example of both the advantages and disadvantages of making early statements is given by the UK Code of Practice on Driver Information Systems (Department of Transport [DOT], 1994). The UK Department of Transport had become increasingly concerned about the potential impact of TICS on driver behavior and road safety, and was also aware of the increasing availability of business equipment available as portable units and adaptable for in-vehicle use. The department decided to commission a reference document to cover the design, installation, and use of such equipment, especially that incorporating displays. This document has been produced in two major sections, the first being a "Code of Practice," which in essence is a series of high-level statements about pertinent issues for consideration if a product is to achieve high levels of efficiency and safety. The second section is of "Design Guidelines," and serves as an introduction to best ergonomics practice.

The UK DOT recognized this as an interim document, because

> in a number of areas there is insufficient underpinning research to be able to specify clearly and unambiguously what constitutes a safe in-vehicle system. . . . However in some areas the need for advice is pressing. As there is much material and a large body of expert knowledge to contribute to descriptions of what makes for safer systems, the Department believes it is sensible to make this Guide available now. (DOT, 1994, unnumbered)

This code of practice underwent several revisions before being widely promulgated. There was much debate about how detailed the code could or should be, with the result that consensus was only reached when a number of more stringent recommendations were removed. The result is a noncontroversial document containing little new material for a human factors specialist. However, the value in the document lies in the fact that it serves as a ready single source to non-human factors specialists, has resulted from a long consultative process involving a very diverse range of expert opinion, and has been the catalyst for discussion and awareness raising in the manufacturing community. The need for such a document has also been emphasized by the way it has been taken on board by the European Conference of Ministers of Transport (ECMT) and further promulgated as a "Statement of Principles" to the wider European community.

Examples of the high-level statements contained in the documents include the following:

> The system must be designed so that it does not unduly distract the driver, nor give rise to potentially dangerous driving behavior by the driver or other road users. (p. ii)
>
> Any display should not aim to visually entertain the driver. (p. iii)
>
> The system should not produce patterns or sounds liable to startle the driver. (p. iii)
>
> The system should not obstruct or interfere with existing vehicle controls or instrumentation, especially those required for safe driving. (DOT, 1994, p. v)

Such statements or principles as these are probably self-evident to the human factors community, and may appear to the design community to be little more than expressions of common sense (and how often has that charge been leveled at ergonomics in general?). However, it is clear that if a system is produced that does indeed comply with the full set of principles, then it is likely to be an example of best practice as it is commonly understood. The principles can also be used as a simple checklist to help identify likely problem areas with a design.

Two examples of the design guidelines are as follows:

> Devices producing tone output can be used to give feedback, for example where there is no tactile feedback in a control, a tone output can provide confirmation that input has been successful. They can also be used to alert the driver to the presence of new information such as a warning that requires immediate action, or to direct attention to a visual output or the start of a speech message. They are best used to convey simple messages where speech may be lost in background noise or the drivers ability to hear speech may be impaired and in such cases a visual display should also be provided.
>
> Frequencies should range from 500 to 3000 Hz where hearing is at its most sensitive. If the system may be used by older people, as is the case with passenger cars, the upper limit should be reduced to 2000 Hz to allow for possible loss of upper frequency hearing ability. (DOT, 1994, p. 20)

Debate will probably continue about the wording of such guidelines, and the numerical values should be the subject of regular review in the light of empirical evidence. However, they exist in the public domain, and as such are a reasonably accessible source for designers and ergonomists alike. As a minimum, the existence of such a document could act to raise awareness of important issues to be considered when designing TICS.

The UK Code of Practice can be regarded as the domain of ISO TICS MMI standards in miniature, and enables an answer to be given to the two questions posed at the start of the review: Can standards be developed? Should standards be developed? First, it is not possible to provide concrete product or performance standards that will guarantee safe systems are developed. However, it is possible to produce standards that incorporate best practice and state-of-the-art knowledge, which can lead to improved design and identify clearly unsafe systems.

It is a mistake to hope to place standards at the same position on some scale of verisimilitude as natural laws of physics. Standards can be useful tools, even if they simply represent an informed consensus view. In many cases, standards are designed not to tell designers what to do, but how to do it. They should be presented in terms of the best possible advice that can be given at that point in time, and formulated to stimulate techno-logical development rather than impede it. As such, the distinction between guidelines and traditional notions of standards is blurred, and will probably continue to be blurred in the domain of much of human factors and ergonomics. If a hard view of requiements for standards is taken, in which all methods, tools, and metrics could be demonstrated to have stood the test of time and meet all the criteria of scientific validity, then, in practice, little progress can be made.

TICS MMI is a rapidly moving, highly pragmatic domain. It has the added characteristic of involving humans in complex and potentially haz-ardous environments, and a history of poor designs emerging into the public arena. Because the consequences of poor design are potentially serious and, for the individual even catastrophic, there is an important role for standardization to encourage good design.

ACKNOWLEDGMENTS

The author is grateful to Tom Dingus, Iowa University; Tsuneomi Yano, Zexel; and Peter Hancock, Minnesota University, for their contributions to the panel discussion that served as the stimulus for this chapter. Special acknowledgment is due to Anders Hallén of Volvo Car Corporation for his presentation and written material that has been drawn on in this review, and to Francois Hartemann of Renault and Gene Farber of Ford US who convene ISO TC22 SC13 WG8 and ISO TC204 WG13, respectively, and who are the driving forces behind the achievements of these groups.

REFERENCES

Commission of European Communities (CEC). (1990). *Council directive of 29 May 1990 on the minimum safety and health requirements for work with display screen equipment* (Fifth individual directive within the meaning of Article 16(1) of Directive 87/391/EEC).

Department of Transport (DOT). (1994). *Driver information systems: Code of practice and design guidelines (Revision D)*. United Kingdom: Department of Transport.

Hallén, A. (1994, August). *In-vehicle MMI standardisation in Europe: Past, present and future.* Paper presented at the International Ergonomics Association 12th Triennial Congress, Volvo Car Corporation, Sweden.

Howarth, C. I. (1987). Psychology and information technology, in information technology and people? In F. Blackler & D. Oborne (Eds.), *Designing for the future* (pp. 1–19). Cheltenham: British Psychological Society.

International Organization for Standardization (ISO). (1989). Procedures for the technical work of ISO. *IEC/ISO Directives—Part 1*. Geneva: Author.

Parkes, A. M. (1995). The contribution of human factors guidelines and standards to usable and safe in-vehicle systems. In J. Pauwelussen (Ed.), *Smart vehicles* (pp. 393–402). Lisse, Holland: Swets & Zeitlinger.

Parkes, A. M., & Ross, T. (1991). The need for performance based standards in future vehicle man machine interfaces. *Advanced telematics in road transport* (Vol. 2, pp. 1312–1321). Amsterdam: Elsevier Science.

Author Index

411

Subject Index

A

Abstract visual display, in collision avoidance systems study, 206, 207
Acceptability, of driver impairment monitoring systems, 365
Accuracy, of traffic information, 1, 8
Adaptable and intelligent systems, 170–172
Advanced Traveler Information Systems (ATIS)
  decision aids and guidelines for, 23–24
  decision tree approach to, 27
  design process for, 25, 26, 27
  displays, 58–59
  navigational preference/driver acceptance of, 345–346, 360–361
    study methodology, 347–352
    study results, 352–360
  private driver functions
    centrality measures, 79
    cliques, 81
    cluster analysis, 81
  questions for designers of, 24
  route guidance research and, 1–5
  trade study analysis approach to, 27–29, 30, 31
    conceptual design development, 43
    display format allocation analysis, 37, 39, 40, 41–43
    display format trade study analysis results, 53–58
    display modality selection results, 44–48
    functional information grouping, 32, 33

information criticality assessment while driving analysis, 36
information items for given conceptual system, 31–32
information-type categorization, 35
sensory modality allocation, 33–35
sensory modality/trip status trade study analysis results, 52–53
total display information-processing overload assessment, 44
trip status allocation, 37, 38, 39
trip status analysis results, 48–51
Advanced travel information systems/commercial vehicle operations (ATIS/CVO), 63–64
  functional description of, 64–65
  definition and framework for, 66
  development and application of, 66–69, 70
  human factors issues in, 69
    commercial vehicle operations, 75–76
    in-vehicle motorist services/information systems, 71–72
    in-vehicle routing/navigation systems, 70–71
    in-vehicle safety/warning systems, 73–74
    in-vehicle signing/information systems, 72–73
  information flow between immediate hazard warning and, 76
  network analysis of functional relations in, 76–77, 82–83
    centrality analysis, 79–80

417

relation between chromaticity difference
and, 158

M

Maneuvers/procedures, in entering Automated
Highway Systems study, 313–314
Man–machine interface (MMI), 185
design of, driver age and, 186–187
evaluation of, 189, 199–200
apparatus, 190
assimilation of message content,
197–199
discussion, 195–199
eye glance behavior, 192–195
memory recall performance, 192–193
memory retention of messages, 196–197
message display design, 190–191
methodology, 189–192
procedure, 191–192
results, 192–195
subjects, 189
MANOVA, *see* Multivariate analysis of variance
Map orientation
for easy judgment of turn direction, 159–160
on navigation displays, 160
Map sketch method, 278, 280
instructions for, 276–277
Maplike knowledge, cognitive map and, 276
Maps, in navigational preference/driver acceptance of ATIS, 349
Maps/notes, in route guidance systems field
study, 118
Mean penalty costs, 9, 17
Mean purchased information costs, 9, 11
Mean rated interlink trust, 19
Mean rated self-confidence, 12
Mean rated trust, 9, 10, 11–12
minus self-confidence, 12–13, 14
Memory
changes in, with age, 187
retention of messages in, 196–197
Memory recall performance, 192–193, 194
Mental cost of navigation, 294–295
Mental load, influences on, 178–180
Mental workload, switching time and, 382
Message(s)
in ATIS, 47
content,
assimilation of, 197–199
displays, 47
display design, 190–191
in-vehicle
length of, 187–188
timing of, 188–189
Methodological issues

in attention switching time study, 390–391
comparisons, 21
in route guidance systems, 117
Minimum braking, in collision warnings study,
209
MM display, discussion on, 125–126
MMI, *see* In-vehicle man–machine interaction;
Man–machine interface
Model of users of intelligent interface, 174
Monitoring systems, *see* Driver impairment
monitoring systems
Motorist services
in Advanced Traveler Information Systems,
47–48
in display format trade study analysis, 57–58
Moving map-based system, 118
Multivariate analysis of variance (MANOVA),
in automatic versus interactive navi-
gation systems study, 301

N

Navigation
definition of, 287
ease of, 299–300
looking behavior during, 296–297
mental cost of, 294–295
Navigation aids, 349–350, 351
performance with, 302–304, 305
ratings of, 356–357, 358, 359–360
Navigation assistance, quality of driving with,
295
Navigation questionnaire, 347, 348, 349
Navigation systems
attention diversion time for, 156
automatic versus interactive, 287–289,
305–306
overview of experiments, 289–290
performance with alternative navigation
aids, 302–305
quality driving with route guidance assistance study, 290–297
screen versus voice guidance, 297–302
at different levels of aggregation and abstraction, 66
urban area driving with, 85–86, 87, 94–95
car navigation systems used, 88
data acquisition, 87, 88
discussion, 92–94
driver use of information, 88–90
landmarks used, 90, 91, 92
methodology, 86–88
results, 88–92
subjects, 88
Navigational errors
in ATIS, 355, 356